Texts and
Monographs
in Physics

F. J. Ynduráin

Quantum Chromodynamics

An Introduction to the Theory of Quarks and Gluons

Springer-Verlag
New York Berlin Heidelberg Tokyo

F. J. Ynduráin
Departamento de Física Teórica
Universidad Autónoma de Madrid
Canto Blanco, Madrid - 34
Spain

Library of Congress Cataloging in Publication Data
Ynduráin, F. J.
Quantum chromodynamics.
(Texts and monographs in physics)
Bibliography: p.
Includes index.
1. Quantum chromodynamics. 2. Gluons. 3. Quarks.
I. Title. II. Series
QC793.3.Q35Y59 1983 530.1'42 83-6700

Typeset by Computype, Inc., St. Paul, MN.
Printed and bound by R. R. Donnelley & Sons, Harrisonburg, VA.
Printed in the United States of America.

9 8 7 6 5 4 3 2 1

ISBN 0-387-11752-0 Springer-Verlag New York Berlin Heidelberg Tokyo
ISBN 3-540-11752-0 Springer-Verlag Berlin Heidelberg New York Tokyo

To Elsa, Marcos, and Elena

Preface

It has been almost thirty years since Yang and Mills (1954) performed their pioneering work on gauge theories, and it is probably safe to say that we have in our hands a good candidate for a theory of the strong interactions, based precisely on a non-Abelian gauge theory. While our understanding of quantum chromodynamics (QCD) is still incomplete, there have been sufficient theoretical developments, many of them enjoying a degree of support from experimental evidence, to justify a reasonably systematic treatise on the subject.

Of course, no presentation of QCD can claim to be complete, since the theory is still in the process of elaboration. The selection of topics reflects this: I have tried to discuss those parts of the theory that are more likely to endure, and particularly those developments that can, with a minimum of rigor, be derived from "first principles." To be sure, prejudice has also influenced my choice: one necessarily tends to give more attention to subjects with which one is familiar, and to eschew unfamiliar ones. I will not pause here to point out topics which perhaps should have been included* (see, however, Section 46); the list of references should fill in the gaps.

*The one I regret most is lattice QCD. At the time I wrote the first draft of this book, lattice QCD had not undergone the spectacular development we have recently witnessed. Unfortunately, a detailed introduction to a treatment of lattice QCD included at a later stage would have caused excessive delay in publication of this book.

This work grew out of graduate courses I have been teaching for the last few years: the book is intended to reflect the pedagogical and introductory nature of those lectures. With this aim in mind, I have tried to write a self-contained text which avoids as far as possible the maddening circumventions of sentences like "it can be shown" or "as is well known." However, I have assumed the reader to have a basic knowledge of field theory and particle phenomenology, and have no doubt that occasional recourse to the literature will be necessary.

What this book owes to the standard reviews and articles on the subject should be apparent and is recorded in the references. I have directly benefited from collaboration with my colleagues: A. González-Arroyo, C. Becchi, S. Narison, J. Bernabeu, E. de Rafael, R. Tarrach, and particularly, C. López and P. Pascual (who also spotted several mistakes in a preliminary version of this work), to name only a few. I also wish to acknowledge the invaluable secretarial help of Antoinette Malene.

Contents

CHAPTER I

Generalities

Anaximander, student of Thales of Miletus, maintained that the primeval substance of things should be called the unbound: he was the first to use this word for the substratum. He pointed out, however, that this primeval substance, of which all elements in heaven and the worlds are made, is not water or any of the so-called "elements" we find around, but a different, limitless substance which fills all of them.

ANAXIMANDER, 546 B.C.E.

1 The Rationale for QCD

Historically, quantum chromodynamics (QCD) originated as a development of the quark model. In the early sixties it was established that hadrons could be classified according to the representations of what today we would call flavor $SU_F(3)$ (Gell-Mann, 1961; Ne'eman, 1961). This classification presented a number of features that are worth noting. First of all, only a few, very specific representations occurred; they were such that they built representations of a group $SU(6)$ (Gürsey and Radicatti, 1964, Pais, 1964) obtained by adjoining the group of spin rotations $SU(2)$ to the internal symmetry group, $SU_F(3)$. However, neither for $SU_F(3)$ nor $SU(6)$ did the fundamental representations (3 and $\bar{3}$ for $SU_F(3)$) appear to be realized in nature. This led Gell-Mann (1964*a*) and Zweig (1964) to postulate that physical hadrons are composite objects, made up of three *quarks* (baryons) or of a quark-antiquark pair (mesons). These three quarks are now widely known as the three *flavors*, u (up), d (down), and s (strange); the first two carry the quantum numbers of isospin, the third strangeness. It has been found that precisely those representations of $SU_F(3)$ occur that may be obtained by reducing the products $3 \times 3 \times 3$ (baryons) or $3 \times \bar{3}$ (mesons); when the spin 1/2 of the quarks is taken into account, the $SU(6)$ scheme is obtained. In addition, the mass differences of the hadrons may be understood by assuming

$$m_d - m_u \approx 4 \text{ MeV}, \qquad m_s - m_d \approx 150 \text{ MeV}, \tag{1.1}$$

together with eventual electromagnetic radiative corrections. The electric charges of the quarks, in units of the proton charge, are:

$$Q_u = \tfrac{2}{3}, \qquad Q_d = Q_s = -\tfrac{1}{3}. \tag{1.2}$$

That hadrons are composite objects was a welcome hypothesis on other grounds, too. For example, it is known that the magnetic moment of the proton is $\mu_p = 2.79 \times e\hbar/2m_p$, instead of the value $\mu_p = e\hbar/2m_p$ expected if it were elementary. The values of the magnetic moments calculated with the quark model are, on the other hand, in reasonable agreement with experimental results.

These successes began a massive search for quarks that still goes on. None of the candidates found to this date has been confirmed, but at least we have a *lower* bound (of the order of a dozen GeV) for the mass of free quarks, which seems to imply that hadrons are very tightly bound states of quarks. This picture can be challenged, however, on at least two grounds. First, the fundamental state of a composite system is one in which all relative angular momenta vanish. Thus, the Δ^{++} resonance had to be interpreted as made up of

$$u\uparrow, u\uparrow, u\uparrow, \tag{1.3}$$

(where the arrows stand for spin components) at relative rest. However, this is preposterous: being spin one-half objects, quarks should obey Fermi–Dirac statistics and their states should be antisymmetric, which is certainly not the case in (1.3). Second, one can use current algebraic techniques (Gell-Mann, Oakes, and Renner, 1968; Glashow and Weinberg, 1968; Leutwyler, 1974) to calculate m_s/m_d with the result

$$m_s/m_d \simeq 20, \tag{1.4}$$

which is a flat contradiction of (1.1) for quarks of a few GeV mass.

With respect to the first objection, a possible solution was proposed by Greenberg (1964), who assumed that quarks obey parastatistics of rank three. It is known that such parastatistics can be disposed of by taking ordinary Fermi-Dirac statistics and introducing a new internal quantum number,[1] which Gell-Mann and his collaborators[2] called "*color*," so that each species of quark may come in any of the three colors $i = r, y, v$ (red, yellow, violet). Then, one can reinterpret the Δ^{++} as

$$\sum \epsilon^{ikl}(u^i\uparrow, u^k\uparrow, u^l\uparrow),$$

or perfectly anti-symmetric. In addition, the absence of states with, say, two or four quarks (so-called "exotics") could be explained by postulating that all physical hadrons are colorless, i.e., they are singlets under rotations in color space:

$$U_c : q^i \to \sum_k U_c^{ik} q^k, \qquad U_c^+ U_c = 1. \tag{1.5}$$

[1]In fact, a color quantum number was first introduced by Han and Nambu (1965).

[2]Fritzsch and Gell-Mann (1972); Fritzsch, Gell-Mann, and Leutwyler (1973).

If we take these transformations of determinant 1 so as to eliminate a trivial overall phase, they build a new group, namely, color $SU_c(3)$. Now, the singlet representation only appears in $3_c \times 3_c \times 3_c$ (baryons) or $3_c \times \bar{3}_c$ (mesons), and this explains why we have these particles, and no exotics, which we do not find in nature.

We will not yet discuss a solution to the second difficulty, but rather make it worse by digressing to current algebra. If quarks are elementary, one must build *currents* out of quarks. Thus, the electromagnetic (em) current is:

$$J^{\mu}_{\text{em}} = \tfrac{2}{3}\bar{u}\gamma^{\mu}u - \tfrac{1}{3}\bar{d}\gamma^{\mu}d - \tfrac{1}{3}\bar{s}\gamma^{\mu}s + \tfrac{2}{3}\bar{c}\gamma^{\mu}c; \tag{1.6a}$$

and the charged weak current (θ_C is the Cabibbo angle),

$$\begin{gathered} J^{\mu}_{2} = \bar{u}\gamma^{\mu}\frac{1-\gamma_5}{2}d_{\theta} + \bar{c}\gamma^{\mu}\frac{1-\gamma_5}{2}s_{\theta}, \\ d_{\theta} = d\cos\theta_C + s\sin\theta_C; \qquad s_{\theta} = -d\sin\theta_C + s\cos\theta_C, \end{gathered} \tag{1.6b}$$

Summing over omitted color indices is understood, and we have also taken into account the contribution of the c (charmed) quark. Gell-Mann (1962, 1964*b*) then postulated that, at short distances, the commutation relations of these currents appear as if the quark fields entering into them were free:

$$\mathscr{L}_{\text{quarks}} \approx \mathscr{L}_0(x) = \sum_{q=u,d,\ldots}\sum_j \bar{q}^j(x)(i\not\partial - m_q)q^j(x). \tag{1.7}$$

It was difficult to understand how this could be so, but the hypothesis met with spectacular success in the Adler–Weissberger sum rule, the Cabibbo–Radicati sum rule, and the calculations by Sirlin and others of radiative corrections to β decay in nuclei.

Another view of the quark model came from deep inelastic scattering experiments. Here a virtual photon, or W, with large invariant mass, Q^2, and high energy, ν, was scattered off some target (a proton, for example). One found the surprising result (which had been anticipated by Bjorken [1969]) that the cross section was of the form

$$\frac{d\sigma}{d\Omega dk_0'} = \frac{\alpha}{4m_p k_0^2 \sin^4\theta/2}\left\{W_2\cos^2\frac{\theta}{2} + 2W_1\sin\frac{\theta}{2}\right\}, \tag{1.8a}$$

where, if we write

$$\begin{gathered} f_1(x, Q^2) = 2xW_1, \\ f_2(x, Q^2) = \frac{\nu}{m_p^2}W_2, \\ x = Q^2/\nu, \\ \int d^4z \langle p|[J^{\mu}(z), J^{\nu}(0)]|p\rangle e^{iq\cdot z} \approx -g^{\mu\nu}W_1 + \frac{1}{m_p^2}p^{\mu}p^{\nu}W_2, \end{gathered} \tag{1.8b}$$

the f_i are approximately independent of Q^2 for $Q^2 \to \infty$, when x has a fixed value (Bjorken scaling). Feynman showed how this could be interpreted if we consider that as Q^2, $\nu \to \infty$ (which, in view of (1.8*b*), means short distances), we consider the proton to be made up of parts, "partons", that do not interact among themselves. It took only one step to identify these partons with quarks, which again appear to be free at short distances, thus creating another puzzle.

Clearly, all these difficulties are dynamical and can therefore only be solved by building a theory of strong interactions, so we come to the crux of the matter: which are the interactions among hadrons? A remarkable fact of hadron physics is that in spite of the variety of hadrons (compare for example the π, K masses), interactions among them (coupling constants and high energy cross-sections, where one can neglect mass differences) are flavor-independent. This means that whatever agency causes quarks to interact, it must act equally on u or d, s or c.

In the meantime, renormalizable, unified theories of weak and electromagnetic interactions had been constructed by Glashow, Weinberg, Salam and Ward, and others. Weinberg (1973*a*) and Nanopoulos (1973) have shown that, to avoid catastrophic violations of parity to order α, one needs strong interactions to act on quantum numbers other than flavor. These were among the reasons that led physicists to consider the possibility that whatever glued the quarks (the *gluons*) interacted precisely with color to which weak and electromagnetic interactions are blind (cf., [1.6]). One takes eight vector gluons, B_a^μ, $a = 1$ to 8 in the adjoint representation of $SU_c(3)$ interacting universally with all quark flavors:

$$\mathscr{L}_1 = \mathscr{L}_0 + g\sum_q \sum_{ika} \bar{q}^i(x)\gamma_\mu t_{ik}^a q^k(x) B_a^\mu(x), \tag{1.9}$$

where $\mathscr{L}_0$ is still given by (1.7) and the t^a matrices are $t^a = (1/2)\,\lambda^a$, with λ the Gell-Mann matrices; they generate the fundamental representation of $SU_c(3)$, and satisfy the commutation relations:[3]

$$[t^a, t^b] = i\sum_c f^{abc} t^c. \tag{1.10}$$

The color and vector character of gluons has the extra virtue of explaining the split between the masses of the Δ_{33} and the nucleons, N (De Rújula, Georgi, and Glashow, 1975).

A further step is taken if it is realized that, for *massless* vector fields, a non-Abelian gauge theory (first introduced by Yang and Mills, 1954) possesses hideous infrared singularities that could prevent the liberation of individual quarks and gluons. Thus, we can at last be reconciled to (1.1) and (1.4): one cannot see individual quarks because they cannot escape due

[3]For group theoretic relations, see Appendix C. We will write color indices as subscripts or superscripts arbitrarily: $f^{abc} = f_{abc}$, $t_a^{ik} = t_{ik}^a$, etc.

to their interactions, not because they are heavy. This is the hypothesis of confinement. Now, we modify (1.9) to:

$$\mathscr{L}_{\text{QCD}} = \mathscr{L}_1 - \tfrac{1}{4}\sum_a G_a^{\mu\nu}(x) G_{a\mu\nu}(x),$$
$$G_a^{\mu\nu} = \partial^\mu B_a^\nu - \partial^\nu B_a^\mu + g \sum f_{abc} B_b^\mu B_c^\nu \tag{1.11}$$

We get an extra bonus: as occurs in all non-Abelian gauge theories, the coupling constant g is automatically universal. Equation (1.11) is the standard QCD Lagrangian, which will be our starting point in the following sections.

Until now the entire construction has been rather fragile. It consists of a set of assumptions, culminating in (1.11), in which each hypothesis takes us further away from the real world (pions, protons, etc.) into a fictitious realm (quarks and gluons) with a set of predictions that hardly outnumber the assumptions. However, the situation changed radically in the early seventies. At that time 't Hooft (unpublished), Politzer (1973), and, independently, Gross and Wilczek (1973*a*, 1973*b*, 1974), proved that in a theory such as (1.11), the effective coupling constant vanishes at short distances (*asymptotic freedom*) and increases at long distances. Thus in one stroke they explained the success of current algebra and the parton model, and made confinement probable. What is more, the corrections to the free-field behavior of quarks are calculable; when calculated, they are found in systematic agreement with experiments—as much, indeed, as the accuracy of the calculations (and of the experimental data!) allows. By and large, there is a good chance that QCD is the theory of strong interactions.[4]

Another important property of QCD, which is perhaps not sufficiently emphasized in most presentations, is its character as a *local field theory*, which leads (at least if the ideas about confinement really work) to local observables. To be precise, the expected pattern is the following. The fields in the Lagrangian (1.1) are defined in a Hilbert space, $\mathfrak{H}_{\text{QCD}}$, made up of quarks and gluon states, and built from a perturbation treatment of (1.1), for example. The quarks and gluons are given there by local fields, $q(x)$, $B(x)$. If confinement ideas are correct, however, it is only a subspace $\mathfrak{H}_{\text{Ph}}$ that contains physical states. That is, if we solved the theory exactly, only color singlet operators would survive. These include *currents*, like

$$\sum \bar{q}^i \gamma^\mu (1 \pm \gamma_5) q'^i,$$

or other composite operators: the operator for a π or for a proton,

$$\sum \bar{u}^i \gamma_5 d^i, \qquad \sum \epsilon^{ijk} u^i u^j d^k,$$

etc. The point is that these operators, though composite, are still local; so if

[4] Skeptical viewpoints may be found in Preparata (1979).

the picture is right, observable operators in the physical Hilbert space, $\mathfrak{H}_{\mathrm{Ph}}$, are local. This is sufficient[5] to derive all the standard results of old-fashioned hadron physics—fixed t dispersion relations, Froissart-like bounds, etc.—whose success, when tested experimentally, is very impressive.

One more property of QCD which, although at a more speculative level than the ones mentioned is still worth noting, is that of allowing naturally grand unifications. Because $SU(3)$ of color is a larger group than the electroweak standard $SU(2) \times U(1)$, it follows that an energy scale may exist where all couplings are equal. While the size of the scale (10^{14} GeV) lies well beyond our reach in the foreseeable future, the predictions of grand unified schemes that are testable are not in disagreement with experimental results.

2 Perturbative Expansions; *S*-Matrix and Green's Functions; Wick's Theorem

In this section we review very briefly a few basic topics of relativistic field theory. Of course this is by no means intended to cover the subject. The subjects of this section are presented merely to establish the notation and to outline the prerequisites for understanding what will follow; the details will have to be sought elsewhere.[6]

A field theory may be specified by giving the relevant Lagrangian. If $\Phi_i(x)$ are the fields in the theory, the Lagrangian is a function of $\Phi_i(x)$ and their space-time derivatives, $\partial\Phi_i(x)$. It is customary to split the *Lagrangian*, $\mathscr{L}$ ($\mathscr{L}$ is actually the Lagrangian density) into $\mathscr{L}_0$ and $\mathscr{L}_{\mathrm{int}}$, where $\mathscr{L}_0$ is obtained from $\mathscr{L}$ by setting all interactions equal to zero, and $\mathscr{L}_{\mathrm{int}}$ is defined as $\mathscr{L}_{\mathrm{int}} \equiv \mathscr{L} - \mathscr{L}_0$. For example, in QCD, the Lagrangian is (1.11) and

$$\mathscr{L}_0 = \sum_q \bar{q}(x)(i\partial\!\!\!/ - m_q)q(x) - \frac{1}{4}\sum_a (\partial^\mu B_a^\nu(x) - \partial^\nu B_a^\mu(x)) \times (\partial_\mu B_{a\nu}(x) - \partial_\nu B_{a\mu}(x)).$$

Besides the basic, or elementary, fields that enter into the theory $\Phi_i(x)$ (the q, B for QCD), we often require *composite operators*, usually local combinations of the Φ_i, i.e., combinations involving *finite* products of the $\Phi_i(x)$ and its derivatives at the same point. For example, in QCD we will use the *currents*, $q(x)\gamma^\mu q'(x)$. Of course, $\mathscr{L}(x)$ itself is a composite local operator.

[5] See Epstein, Glaser, and Martin (1969) and Bogolubov, Logunov and Todorov (1975) from which one can trace the relevant literature.

[6] For example, in standard textbooks like Bogoliubov and Shirkov (1959); Bjorken and Drell (1965) or Itzykson and Zuber (1980).

With local fields or, more generally, local operators (whether elementary of composite), we can form new local operators. The simplest method is by ordinary multiplication; but there are two other types of product that we will consider repeatedly. These are the *Wick product* and the *time-ordered product*. The *Wick*, or *normal product*, is defined primarily for free basic fields, as follows. Expand the Φ_i in creation-destruction operators:

$$\Phi_i(x) = \sum_n C_i^{(n)}(x)a_n + \sum_n \bar{C}_i^{(n)}(x)\bar{a}_n^+ ;$$

where a, $\bar{a}$ may or may not coincide. For example, if Φ is q,

$$q(x) = \frac{1}{(2\pi)^{3/2}} \sum_\sigma \int \frac{d\vec{p}}{2p^o} \left\{ e^{-ip\cdot x}u(p,\sigma)a(p,\sigma) + e^{ip\cdot x}v(p,\sigma)\bar{a}^+(p,\sigma) \right\},$$

where u, v are the standard Dirac spinors, and a^+ ($\bar{a}^+$) creates particles (antiparticles). Then, the Wick product, $:\Phi_1(x_1)\Phi_2(x_2):$ is obtained by placing all creators to the left of all annihilators as if they commuted/anti-commuted if the fields Φ_i corresponded to bosons/fermions:

$$:\Phi_1(x_1)\Phi_2(x_2): \equiv \sum_{n,n'} \left\{ C_1^{(n)}(x_1)C_2^{(n')}(x_2)a_n a_{n'} + \bar{C}_1^{(n)}(x_1)\bar{C}_2^{(n')}(x_2)\bar{a}_n^+ a_{n'}^+ \right.$$
$$\left. + \bar{C}_1^{(n)}(x_1)C_2^{(n')}(x_2)\bar{a}_n^+ a_{n'} + (-1)^\delta C_1^{(n)}(x_1)\bar{C}_2^{(n')}(x_2)\bar{a}_{n'}^+ a_n \right\},$$

$\delta = 1$ for fermions, $\delta = 0$ for bosons.

The extension to Wick products of more factors, $:\Phi_1(x_1) \ldots \Phi_n(x_n):$, or to Wick products of Wick products like $:(:\Phi_1(x_1)\Phi_2(x_2):)(:\Phi_3(x_3)\Phi_4(x_4):):$ is straightforward—one always expands in creators/annihilators and writes creators at the left of annihilators as if they commuted/anticommuted.

The proof is not totally trivial, yet it is not too difficult to verify that the Wick product of local operators at the same point is itself local[7], i.e., if $O_1, \ldots, O_n$ are local, so is $:O_1(x) \ldots O_n(x):$.

Another important property of the Wick product is that it is regular, that is to say, for any states a, b, the matrix elements of a Wick product $\langle a|:O_1(x_1) \ldots O_n(x_n)|b\rangle$ are regular functions of $x_1, \ldots, x_n$.

The *time-ordered product*, or *T-product* of local (elementary or composite) operators $O_1(x_1) \ldots O_n(x_n)$ is defined as follows:

$$TO_1(x_1) \ldots O_n(x_n) \equiv T\{O_1(x_1) \ldots O_n(x_n)\} = (-1)^\delta O_{i_1}(x_{i_1}) \ldots O_{i_n}(x_{i_n}).$$

Here the permutation is such that, at the right-hand side, the times are ordered, i.e., $x_{i_1}^0 \geqslant x_{i_2}^0 \geqslant \cdots \geqslant x_{i_n}^0$, and δ is the number of transpositions of the indices corresponding to fermion operators which were necessary to bring $1, \ldots, n$ to $i_1, \ldots, i_n$. Otherwise stated, the time-ordered product

[7] For our purposes a local operator $O_\alpha(x)$, is one transforming locally under Poincaré transformations: $U(a,\Lambda)O_\alpha(x)U^{-1}(a,\Lambda) = \sum P_{\alpha\alpha'}(\Lambda)O_{\alpha'}(\Lambda x + a)$, and commuting with itself at spatial separation.

$TO_1(x_1)\ldots O_n(x_n)$ is obtained by rearranging the operators in the natural sequence of times as if they commuted (for boson operators) or anticommuted (for fermion operators). For example, for two factors,

$$Tq_1(x)q_2(y) = \theta(x^0 - y^0)q_1(x)q_2(y) - \theta(y^0 - x^0)q_2(y)q_1(x)$$

or

$$Tq_1(x)B_2(y) = \theta(x^0 - y^0)q_1(x)B_2(y) + \theta(y^0 - x^0)B_2(y)q_1(x).$$

Note that boson-fermion operators are always taken to be commuting. As is known, the time-ordered product is relativistically invariant.

The S matrix is the operator that transforms free states at time $-\infty$ into free states at time $+\infty$. S may be obtained in terms of the interaction Lagrangian using Matthews's formula:

$$S = T\exp i\int d^4x\, \mathscr{L}^0_{\text{int}}(x). \tag{2.1a}$$

Here $\mathscr{L}^0_{\text{int}}(x)$ is the interaction Lagrangian with all fields in it taken as if they were *free*, and in normal order. The time-ordered exponential is a formal device; it actually is defined by its series expansion,

$$\begin{aligned} S &= T\exp i\int d^4x\, \mathscr{L}^0_{\text{int}}(x) \\ &\equiv 1 + i\int d^4x\, \mathscr{L}^0_{\text{int}}(x) + \cdots \\ &\quad + \frac{i^n}{n!}\int d^4x_1\ldots d^4x_n\, T\mathscr{L}^0_{\text{int}}(x_1)\ldots\mathscr{L}^0_{\text{int}}(x_n) + \cdots. \end{aligned} \tag{2.1b}$$

Many times, instead of matrix elements of S, we will require the matrix elements of currents (or products of currents) or more general composite operators. These may be obtained by adding a fictitious extra term to the interaction. For example, suppose we require

$$\langle a|TJ_1^\mu(x)J_2^\nu(y)|b\rangle, \tag{2.2}$$

where the J are weak or electromagnetic currents [see (1.6)]. We then change $\mathscr{L}_{\text{int}}$ to

$$\mathscr{L}^\phi_{\text{int}} = \mathscr{L}_{\text{int}} + J_{1\mu}(x)\phi_1^\mu(x) + J_{2\mu}(x)\phi_2^\mu(x), \tag{2.3}$$

where the ϕ are c-number auxiliary fields. We expand again,

$$\begin{aligned} &\langle a|T\exp\int d^4x\, \mathscr{L}^\phi_{\text{int}}(x)|b\rangle \\ &\quad = \langle a|b\rangle + i\langle a|\int d^4x\left\{\mathscr{L}^0_{\text{int}}(x) + \sum_i J^0_{i\mu}(x)\phi_i^\mu(x)\right\}|b\rangle \\ &\quad + \cdots + \frac{i^n}{n!}\langle a|\int d^4x_1\ldots d^4x_n\, T \\ &\quad \times\left\{\mathscr{L}^0_{\text{int}}(x_1) + \sum_i J^0_{i\mu}(x_1)\phi_i^\mu(x_1)\right\}\times\cdots \\ &\quad \times\left\{\mathscr{L}^0_{\text{int}}(x_n) + \sum_i J^0_{i\mu}(x_n)\phi_i^\mu(x_n)\right\}|b\rangle + \cdots. \end{aligned}$$

We let the ϕ be infinitesimal and keep only terms $O(\phi)$, $O(\phi^2)$. The last are of the form

$$\frac{i^n}{n!}\langle a|\int d^4x_1 \ldots d^4x_n \sum_{ij} T\mathscr{L}^0_{\text{int}}(x_1)\ldots\left[\mathscr{L}^0_{\text{int}}(x_i)\right]\ldots$$
$$\times\left[\mathscr{L}^0_{\text{int}}(x_j)\right]\ldots\mathscr{L}^0_{\text{int}}(x_n)J^0_{1\mu}(x_i)J^0_{2\nu}(x_j)|b\rangle\phi^\mu_1(x_i)\phi^\nu_2(x_j),$$

where $[\mathscr{L}]$ means that we have dropped the bracketed term. Letting $\phi_{i\mu}(x)=\epsilon_{i\mu}\delta(x-y_i)$, differentiating with respect to ϵ_1, ϵ_2, and setting $\epsilon_1=\epsilon_2=0$, we get the Gell-Mann–Low equation,

$$\langle a|TJ^\mu_1(x)J^\nu_2(y)|b\rangle=\frac{\delta^2}{\delta\phi_{1\mu}(x)\delta\phi_{2\nu}(y)}$$
$$\times\langle a|T\exp i\int d^4z\left\{\mathscr{L}^0_{\text{int}}(z)+\sum_i J^0_{i\lambda}(z)\phi^\lambda_i(z)\right\}|b\rangle$$
$$=\sum_{n=0}^\infty\frac{i^n}{n!}\langle a|\int d^4x_1\ldots d^4x_n\,T\mathscr{L}^0_{\text{int}}(x_1)\ldots$$
$$\times\mathscr{L}^0_{\text{int}}(x_n)J^{0\mu}_1(x)J^{0\nu}_2(y)|b\rangle. \tag{2.4}$$

To identify the right-hand side with (2.2) we have used the formula (proved in Bogoliubov and Shirkov, 1959; see also Sections 39, 42 below; functional derivatives are defined in Appendix H)

$$\left.\frac{\delta^2 S_\phi}{\delta\phi_{1\mu}(x)\delta\phi_{2\nu}(y)}\right|_{\phi=0}=TJ^\mu_1(x)J^\nu_2(y). \tag{2.5}$$

We next turn to the two requirements of relativistic invariance and unitarity. If (a,Λ) is a transformation in the Poincaré group then

$$U(a,\Lambda)SU^{-1}(a,\Lambda)=S; \tag{2.6}$$

and S is relativistically invariant. It is also unitary,

$$S^+S=SS^+=1. \tag{2.7}$$

If we write

$$S=1+i\mathscr{T},$$

where $\langle a|\mathscr{T}|b\rangle$ is the so-called transition amplitude, then (2.7) may be rewritten in terms of $\mathscr{T}$

$$\operatorname{Im}\langle a|\mathscr{T}|b\rangle=\frac{1}{2}\sum_{\text{all }c}\langle c|\mathscr{T}|b\rangle\langle c|\mathscr{T}|a\rangle^*. \tag{2.8}$$

[We have assumed time reversal invariance to derive Equation (2.8).] When expanding in powers of g, Equations (2.6) and (2.8) imply relations order by order in perturbation theory. As (2.6) is linear, it must hold for each order; but, because of its nonlinearity, (2.8) mixes different orders. For example, if we let

$$\mathscr{T}=g\sum_{n=0}^\infty g^n\mathscr{T}_n,$$

then the second order constraint is

$$\operatorname{Im}\langle a|\mathscr{T}_2|b\rangle = \frac{i}{2}\sum_{\text{all } c}\{\langle c|\mathscr{T}_0|b\rangle\langle c|\mathscr{T}_2|a\rangle^* + \langle c|\mathscr{T}_2|b\rangle\langle c|\mathscr{T}_0|a\rangle^* + \langle c|\mathscr{T}_1|b\rangle\langle c|\mathscr{T}_1|a\rangle^*\}. \tag{2.9}$$

We will finish by briefly introducing reduction formulas. Consider a scattering amplitude, say, $a + b \to a' + b'$ with a, a' bosons with fields ϕ_a, $\phi_{a'}$. We may write the scattering amplitude as

$$\langle a', b'|S|a, b\rangle = \lim_{\substack{t' \to +\infty \\ t \to -\infty}} \langle a', b', t' \,|\, a, b, t\rangle.$$

Now, if p_i is the momentum of particle i, we may use the formula (see, e.g., Bjorken and Drell [1965] for more details on the following discussion)

$$a^+(p_a) = \lim_{t \to -\infty} \frac{i}{2(2\pi)^{3/2}} \int d\vec{x}\, e^{-ip_a \cdot x} \overleftrightarrow{\partial}_0 \phi^+(x),$$

to write, after some manipulations, *reduction formulas*, for example,

$$\langle a', b'|S|a, b\rangle = \frac{i}{(2\pi)^{3/2}} \int d^4x\, e^{-ip_a \cdot x} \times (\partial^2 + m_a^2)\langle a', b'|\phi_a^+(x)|b\rangle.$$

We will not prove this, or give a full set of reduction formulas, which may be found in Bjorken and Drell [1965]; but shall at least present a few typical cases. If we also "reduce" a', we would obtain

$$\langle a', b'|S|a, b\rangle = \frac{i}{(2\pi)^{3/2}} \times \frac{-i}{(2\pi)^{3/2}} \int d^4x \int d^4y\, e^{-ip_a \cdot x} e^{ip_{a'} \cdot y} \times (\partial_x^2 + m_a^2)(\partial_y^2 + m_{a'}^2)\langle b'|T\phi_{a'}(y)\phi_a^+(x)|b\rangle.$$

If we continue reducing, we ultimately obtain the Fourier transform of the vacuum expectation value (VEV) of the T-product of four fields:

$$\langle 0|T\phi_{a'}(y)\phi_{b'}(z)\phi_a^+(x)\phi_b^+(w)|0\rangle.$$

The extension to spinor or vector fields is easy. For example, if we replace the scalar a by a fermion with momentum p_a and spin σ, and denote its field by ψ, we obtain

$$\langle a', b'|S|(p_a, \sigma), b\rangle = \frac{i}{(2\pi)^{3/2}} \int d^4x \langle a', b'|\bar{\psi}(x)|b\rangle(\overleftarrow{i\partial\!\!\!/} + m_a)u(p, \sigma)e^{-ip_a \cdot x},$$

etc.

Finally, we turn to Wick's theorem. An expression like (2.1b) allows us to calculate, order by order in perturbation theory, the S matrix elements (or current matrix elements, or Green's functions). The tool that allows us to do so is Wick's theorem. Consider the time-ordered product of two free fields, $T\Phi_1^0(x_1)\Phi_2^0(x_2)$. We may expand the Φ_i in creation-destruction operators,

$$\Phi_i(x) = \frac{1}{(2\pi)^{3/2}} \int \frac{d\vec{k}}{2k^0}$$
$$\times \sum_\sigma \left\{ e^{-ik\cdot x}\xi_+(k,\sigma)a_+(k,\sigma) + e^{ik\cdot x}\xi_-(k,\sigma)a_-^+(k,\sigma) \right\};$$

here σ is the spin; the $\xi_\pm$ are the corresponding wave functions, and $a_\pm$, $a_\pm^+$ are the destruction-creation operators for particles $(+)$ and antiparticles $(-)$. The commutation relations (the symbol [,] is to be interpreted as the anticommutator for fermions)

$$\left[a_\pm(k,\sigma), a_\pm^+(k',\sigma') \right] = 2\delta_{\sigma\sigma'} k^0 \delta(\vec{k} - \vec{k}'),$$
$$\left[a_+, a_-^+ \right] = 0$$

can be applied to check that the difference

$$T\Phi_1^0(x_1)\Phi_2^0(x_2) - :\Phi_1^0(x_1)\Phi_2^0(x_2): \equiv \underbrace{\Phi_1^0(x_1)\Phi_2^0(x_2)}$$

is a *c*-number, called the *contraction*. Thus it coincides with its vacuum expectation value (the *propagator*):

$$\underbrace{\Phi_1^0(x_1)\Phi_2^0(x_2)} = \langle 0 | T\Phi_1^0(x_1)\Phi_2^0(x_2) | 0 \rangle \equiv \langle T\Phi_1^0(x_1)\Phi_2^0(x_2) \rangle_0 .$$

Applying this repeatedly to, say, (2.1) we find that $T\mathscr{L}_{\text{int}}^0 \ldots \mathscr{L}_{\text{int}}^0$ may be written as a combination of contractions times fully normal ordered products of operators. As the matrix elements of these may be easily calculated, the full result for each term in the perturbative expansion can be found. The Feynman rules are such that they summarize the manipulations, allowing us to write the result directly. For QCD, they are as shown in Appendix D (see also Section 42, where some of the Feyman rules are explicitly deduced).

CHAPTER II

QCD as a Field Theory

"A slow sort of country!" said the Queen. "Here, you see, it takes all the running you can do to stay in the same place."

LEWIS CARROLL, 1896

3 Gauge Invariance

Let us consider the set of fields that we have postulated for QCD, viz., three $q^i(x)$ for each flavor of quark, and eight $B_a(x)$. The first set builds up the fundamental representation of $SU(3)$, viz., if U is a 3×3 unitary matrix of determinant unity, then the q^j transform simply as

$$U : q^j(x) \to \sum_k U_{jk} q^k(x).$$

It is possible to write any matrix in $SU(3)$, U, in terms of the eight generators of its Lie algebra, t^a. These 3×3 matrices are given in Appendix C; we have

$$U = \exp\left\{ -ig \sum_a \theta_a t^a \right\}.$$

The θ_a are the parameters of the group, and the factor g is introduced for future convenience. Thus, representing q^j by a vertical three-matrix, we have

$$q(x) \to e^{-ig\Sigma\theta_a t^a} q(x).$$

For B, we have to consider the *adjoint* (dimension 8) representation of $SU(3)$. Let C^a be the corresponding matrices, with elements $C^a_{bc} = -if_{abc}$

(see Appendix C for the explicit values). We then have,

$$B^{\mu}(x) \to e^{-ig\Sigma\theta_a C^a} B^{\mu}(x).$$

If the θ_a are constant independent of the space-time point, the analysis is complete; we have a *global* $SU(3)$ invariance. However, as we know from quantum electrodynamics (QED), we have an interest in extending the transformation to parameters $\theta_a(x)$ which depend upon the space-time point. We thus define the (local) *gauge transformations*,

$$q(x) \to e^{-ig\Sigma\theta_a(x)t^a} q(x). \tag{3.1a}$$

Similarly, we generalize the usual QED transformations and define

$$B^{\mu}(x) \to e^{-ig\Sigma\theta_a(x)C^a} B^{\mu}(x) - \partial^{\mu}\theta(x), \tag{3.1b}$$

or, for infinitesimal θ,

$$\begin{aligned} q^j(x) &\to q^j(x) - ig\sum_{a,k} \theta_a(x) t^a_{jk} q^k(x), \\ B^{\mu}_a(x) &\to B^{\mu}_a(x) + g\sum_{b,c} f_{abc}\theta_b(x) B^{\mu}_c(x) - \partial^{\mu}\theta_a(x). \end{aligned} \tag{3.1c}$$

We will assume invariance under the transformations (3.1) (in fact, the Lagrangian (1.11) has this property built in). As we shall see, this invariance forces the fields to appear in very precise combinations, and it will be clear at the end of this section that (1.11) is indeed the most general Lagrangian invariant under (3.1) *and* involves no constants with dimensions of a *negative* power of the mass (cf., however, sections 38 and following).

Let us consider the transformation properties of the derivative of a field, say $\partial^{\mu} q(x)$. From (3.1c),

$$\begin{aligned} \partial^{\mu} q^j(x) \to \partial^{\mu} q^j(x) &- ig\sum t^a_{jk}\theta_a(x)\partial^{\mu} q^k(x) \\ &- ig\sum t^a_{jk}(\partial^{\mu}\theta_a(x)) q^k(x). \end{aligned}$$

We see that it transforms differently from the field itself. To obtain an invariant Lagrangian, all derivatives must appear in covariant combinations:

$$D^{\mu} q^j(x) \equiv \sum_k \left\{ \delta_{jk}\partial^{\mu} - ig\sum_a B^{\mu}_a(x) t^a_{jk} \right\} q^k(x); \tag{3.2}$$

where D^{μ} is called the (gauge) *covariant derivative*. The proof that D^{μ} is covariant is simple. Using matrix notation,

$$\begin{aligned} D^{\mu} q(x) \to \partial^{\mu} q(x) &- ig\sum t^a\theta_a(x)\partial^{\mu} q(x) - ig\sum t^a(\partial^{\mu}\theta_a(x)) q(x) \\ &- g^2\sum B^{\mu}_a(x) t^a t^b \theta_b(x) q(x) - ig\sum B^{\mu}_a(x) t^a q(x) \\ &- ig^2\sum f_{abc} t^a \theta_b(x) B^{\mu}_c(x) q(x) + ig\sum(\partial^{\mu}\theta_a(x)) t^a q(x). \end{aligned} \tag{3.3a}$$

Because

$$t^a t^b = t^b t^a + [t^a, t^b]$$

and

$$[t^a, t^b] = i\sum f^{abc} t^c,$$

the right-hand side of (3.3) is

$$D^\mu q(x) - ig\sum t^a \theta_a(x) D^\mu q(x), \tag{3.3b}$$

as we wished to prove. Similarly, the *covariant curl* of the field B is[8]:

$$(D^\mu \times B^\nu)_a \equiv G_a^{\mu\nu} = \partial^\mu B_a^\nu - \partial^\nu B_a^\mu + g\sum f_{abc} B_b^\mu B_c^\nu . \tag{3.4}$$

In terms of these, we can write (1.11) in a manner that is manifestly gauge invariant. Dropping the QCD index, we have

$$\mathscr{L} = \sum_q \{ i\bar{q}(x) \not{D} q(x) - m_q \bar{q}(x) q(x) \} - \tfrac{1}{4}(D \times B)^2. \tag{3.5}$$

Here $(D \times B)^2$ is short for the *pure Yang–Mills* component,

$$(D \times B)^2 \equiv G^2 = \sum_a G_a^{\mu\nu} G_{a\mu\nu}; \qquad \mathscr{L}_{\text{YM}} \equiv -\tfrac{1}{4}(D \times B)^2.$$

The importance of gauge invariance is threefold. First, as is clear from the proof of (3.3), it requires universality of the coupling; i.e. one single constant g couples all quarks to gluons and self-couples the last. Second, 't Hooft (1971) has proved that a non-Abelian theory is renormalizeable, but only if it is gauge invariant. Third, it has been shown by Coleman and Gross (1973) that *only* a non-Abelian theory can be asymptotically free.

At first sight, it looks as if Equation (3.5) could be carried over to the quantum stage directly simply by reinterpreting the fields as quantum fields. However, as we know from QED, this is not so. It is clear from gauge invariance that the fields B are undefined, since we may effect transformations like (3.1) that will alter the commutation relations. Of course, this is related to the fact that the particles corresponding to the fields B, being massless, have only two degrees of freedom; whereas the fields B^μ have four independent components. To effect the quantization, we will be forced to select definite representatives of each gauge class (*gauge fixing*), which breaks manifest gauge invariance. Because of the gluon self-interactions, we expect this to cause more trouble than in the Abelian case and, indeed, we will see that Lorentz covariant gauges require the introduction of extra, nonphysical fields[9] (*ghosts*) to restore gauge invariance and unitarity. Alternatively, we may choose ghost-free gauges (axial gauges) which, however, break manifest Lorentz invariance.

To complete this, and before considering the quantized theory, we write the equations of motion that follow from (3.5), at the classical level.

[8] Again, the analogy of $G_a^{\mu\nu}$ with the electromagnetic field strength tensor, $F^{\mu\nu} = \partial^\mu A^\nu - \partial^\nu A^\mu$ is apparent.

[9] Peculiar gauges with ghosts may also be constructed for Abelian theories.

Euler–Lagrange equations for a general field Φ are gotten by requiring stationary *action*, $\mathscr{A} = \int d^4 \times \mathscr{L}$:

$$\partial_\mu \frac{\partial \mathscr{L}}{\partial(\partial_\mu \Phi)} = \frac{\partial \mathscr{L}}{\partial \Phi};$$

hence, we obtain

$$\begin{gathered} \bar{q}(x)(\overleftarrow{i\not{D} + m}) = 0, \qquad (i\not{D} - m)q(x) = 0, \\ D_\mu G_a^{\mu\nu}(x) \equiv \partial_\mu G_a^{\mu\nu}(x) + g\sum f_{abc} B_{b\mu}(x) G_c^{\mu\nu}(x) = 0. \end{gathered} \tag{3.6}$$

4 Canonical Quantization; Gauge Fixing; Covariant Gauges

Let us start by trying to quantize the *free* gluon fields. The free-gluon (Yang–Mills) Lagrangian is:

$$\begin{gathered} \mathscr{L}^0_{\mathrm{YM}} = -\frac{1}{4}\sum G_a^{0\mu\nu} G^{0a}_{\mu\nu}, \\ G_a^{0\mu\nu} = \partial^\mu B_a^{0\nu} - \partial^\nu B_a^{0\mu} \end{gathered} \tag{4.1}$$

and the index 0 denotes free fields. Equation (4.1) is similar to the Lagrangian for eight uncoupled electromagnetic fields; as such it is invariant under the *free* gauge transformations:

$$B_a^{0\mu} \to B_a^{0\mu} - \partial^\mu \theta_a. \tag{4.2}$$

We expect all the problems and benefits associated with gauge invariance. In particular, since B is undefined, it will be impossible to quantize (4.1) directly. In fact, suppose we want to implement the standard canonical quantization procedure. We define momenta conjugate to the B_a^0; dropping the indices 0 that denote free fields,

$$\pi_a^\mu(x) = \frac{\partial \mathscr{L}_{\mathrm{YM}}}{\partial(\partial_0 B_{a\mu})} = G_a^{\mu 0}: \tag{4.3}$$

we see that $\pi_a^0(x)$ vanishes identically. The *canonical commutation relations* are

$$[\pi_a^\mu(x), B_b^\nu(y)]\delta(x^0 - y^0) = -i\delta_{ab} g^{\mu\nu}\delta(x-y): \tag{4.4}$$

the $B_a^0(x)$ commute with all operators and are thus c-numbers.

At this point, two paths lie open to us. We may choose a gauge in which the nonphysical degrees of freedom are absent. It is quite clear that this violates manifest Lorentz invariance. Or we may treat all the B^μ in the same manner. Since this introduces nonphysical degrees of freedom, we will be forced to work in a space with indefinite metric. We shall discuss physical gauges later on and for the moment consider covariant ones.

As is known from the case of the electromagnetic field (and on the level at which we are now working, there is no difference), we cannot have the Lorentz condition, $\partial_\mu B_a^\mu = 0$, and at the same time keep covariant commutation relations. Therefore, we have to give up $\partial B = 0$ as an operator statement. We then introduce a *Gupta–Bleuler* space $\mathfrak{H}_{GB}$ where Equation (4.4) is realized in its previously stated form. We shall see that this implies an indefinite metric for $\mathfrak{H}_{GB}$. Physical vectors are those for which

$$\langle \Phi_{ph}|\partial_\mu B^\mu_a(x)|\Phi_{ph}\rangle = 0. \tag{4.5}$$

If we identify vectors that differ by a vector of zero norm, i.e.,

$$|\Phi_{ph}\rangle \sim |\Phi'_{ph}\rangle = |\Phi_{ph}\rangle + |\Phi^{(0)}\rangle, \tag{4.6}$$

when $\langle \Phi^{(0)} \mid \Phi^{(0)}\rangle = 0$, we finally obtain the space of physical vectors $\mathfrak{L}$.

To maintain Equation (4.4), we have to modify the Lagrangian (4.1). We may do this by adding a term $-(\lambda/2)\sum_a(\partial_\mu B^\mu_a)^2$ (*gauge fixing*):

$$\mathscr{L}_{\lambda YM} = -\frac{1}{4}\sum_a G^{\mu\nu}_a G_{a\mu\nu} - \frac{\lambda}{2}\sum_a(\partial_\mu B^\mu_a)^2. \tag{4.7}$$

This should have no physical consequences, at least in the free-field case, because the term added vanishes between physical vectors, as in equation (4.5). The momenta conjugate to the B are now

$$\pi^\mu_{\lambda a}(x) = G^{\mu 0}_a(x) - \lambda g^{\mu 0}\partial_\nu B^\nu_a(x), \tag{4.8}$$

which is not zero and hence we can keep (4.4). However, there arises an indefinite metric. For example, consider (4.4) with $\mu = 0$:

$$\lambda\left[\partial_\mu B^\mu_a(x), B^\nu_b(y)\right]\delta(x_0 - y_0) = i\delta_{ab}\delta_{0\nu}\delta_4(x-y). \tag{4.9}$$

Its sign is undefined. To see this more clearly, we consider momentum space. Let us take the case $\lambda = 1$ and introduce a canonical $\epsilon^{(\rho)}(k)$ tetrad associated to any lightlike vector k:

$$\epsilon^{(0)}_\mu = \delta_{\mu 0};$$

$$\epsilon^{(i)}_0 = 0, \qquad \vec{\epsilon}^{(i)}\cdot\vec{k} = 0, \qquad i = 1,2, \qquad \epsilon^{(3)}_\mu = \frac{1}{k^0}k_\mu - \delta_{\mu 0}; \tag{4.10}$$

$$\epsilon^{(i)}_\mu \epsilon^{(j)\mu} = -\delta_{ij}, \qquad i,j = 1,2,3.$$

Of these, only $\epsilon^{(i)}$, $i = 1,2$ correspond to physical zero-mass particles; $\epsilon^{(3)}$ is longitudinal, and $\epsilon^{(0)}$ corresponds to a zero-spin object. We may expand B into creation and annihilation operators,

$$B^\mu_b(x) = \frac{1}{(2\pi)^{3/2}}\int\frac{d\vec{k}}{2k^0}\sum_\rho\left\{e^{-ik\cdot x}\epsilon^{(\rho)\mu}(k)a_\rho(b,k) + e^{ik\cdot x}\epsilon^{(\rho)\mu}(k)^* a^+_\rho(b,k)\right\}. \tag{4.11}$$

From (4.4), we find

$$\left[a_\mu(b,k), a^+_\nu(b',k')\right] = -g_{\mu\nu}\delta_{bb'}2k^0\delta(\vec{k}-\vec{k}'). \tag{4.12}$$

Thus, $\langle 0|a_0(k)a^+_0(k)|0\rangle$ is negative in the gauge we are considering.

From (4.12), we may calculate the propagator. If

$$\langle TB^\mu_a(x)B^\nu_b(0)\rangle_0 = D^{\mu\nu}_{ab}(x),$$

then writing the answer for an arbitrary value of λ,

$$D_{ab}^{0\mu\nu}(x)=\delta_{ab}\frac{i}{(2\pi)^4}\int d^4k\,e^{-ik\cdot x}\,\frac{-g^{\mu\nu}+(1-\lambda^{-1})k^\mu k^\nu/(k^2+i0)}{k^2+i0}\,. \tag{4.13a}$$

We have introduced the notation

$$\langle fg\ldots h\rangle_0\equiv\langle 0|fg\ldots h|0\rangle,$$

which will be used consistently in the following. It is convenient to write $1-1/\lambda=\xi$; this simplifies the expression for the propagator. In momentum space,

$$D_{ab}^{\mu\nu}(k)=i\delta_{ab}\frac{-g^{\mu\nu}+\xi k^\mu k^\nu/(k^2+i0)}{k^2+i0}\,. \tag{4.13b}$$

An especially simple case is the *Fermi–Feynman* gauge, $\xi=0$; also useful is the *Landau* or *transverse* gauge, $\xi=1$.

Actually, and for $\lambda\neq 1$, Eqs. (4.13) have to be obtained somewhat indirectly, because for physical, massless gluons, the term $k^\mu k^\nu/k^2$ is infinite. The solution is obtained by introducing a fictitious mass, M. With it, we obtain, in momentum space,

$$D_{ab}^{\mu\nu}(k,M)=\frac{-g^{\mu\nu}+(1-\lambda^{-1})k^\mu k^\nu/(k^2-\lambda^{-1}M^2+i0)}{k^2-M^2+i0}\,i\delta_{ab}\,;$$

taking the limit $M\to 0$, Eqs. (4.13) follow.

In QED, because the photons do not possess self-interactions, one can work with covariant gauges without additional considerations. In QCD, self-interactions cause further complications. This will be seen in the next section.

5 Unitarity; Lorentz Gauges, Ghosts; Physical Gauges

i Covariant Gauges

The fact that not all states in the space where fields are defined correspond to physical vectors means that we have to be careful with unitarity. The unitarity condition (2.7) or (2.8) is valid only for physical states. To extend it to our case, we introduce the projector into physical states, P:

$$P\mathfrak{H}_{\rm GB}=\mathfrak{L},\qquad P^2=P^+=P. \tag{5.1}$$

The unitarity condition then reads

$$(PSP)(PSP)^+=P; \tag{5.2}$$

if the Lagrangian is Hermitian, the S matrix is unitary in $\mathfrak{H}_{\rm GB}$, so we find that (5.2) will be satisfied if and only if S commutes with P. In QED, this is automatic for the gauges defined previously. In QCD, such is not the case because, except for the case $g = 0$, gauge transformations involve interactions. This means that the Lagrangian

$$\mathscr{L}^{\xi} = \sum_q \{ i\bar{q}\not{D}q - m_q\bar{q}q \} - \frac{1}{4}(D \times B)^2 - \frac{\lambda}{2}(\partial B)^2, \qquad \xi = 1 - 1/\lambda, \tag{5.3}$$

obtained by adjoining the gauge-fixing term to (3.5) is not complete as it stands and will have to be modified.

To see how this modification comes about, we will check, in the particular case of the Fermi–Feynman gauge, how (5.2) is violated. Consider second-order quark-antiquark scattering.

The Feynman diagrams that contribute are those of Figure 1. It is not difficult to see that diagrams b and c are not problematic and only diagram

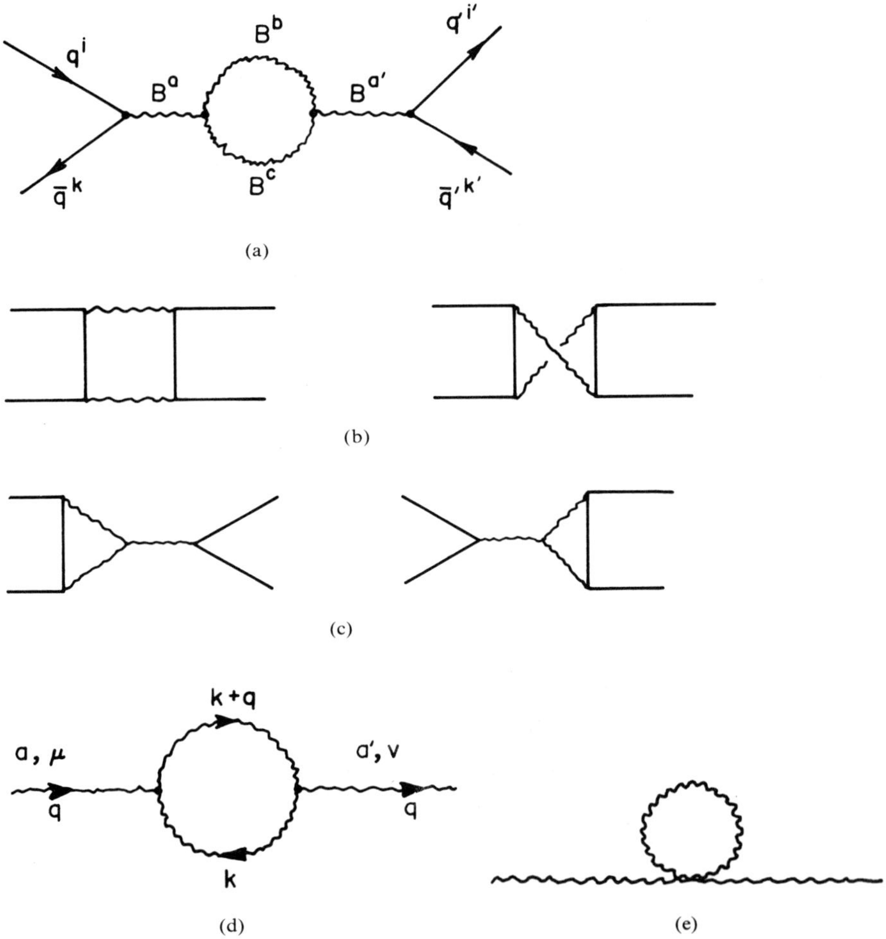

Figure 1. (*a–c*), Diagrams for *qq* scattering; (*d*), gluon loop; (*e*) tadpole.

a will cause trouble. We will calculate a in dimension D (see below, Section 7), and at the end we will take the physical limit, $D \to 4$. The corresponding amplitude (see the routing of momenta in Figure 1[10]) is

$$\mathcal{T}_4 = \frac{-g^2}{(2\pi)^2} \sum \bar{v}_k \gamma_{\mu'} u_i t^a_{\text{tr}} \frac{-ig^{\mu'\mu}}{q^2} \Pi_{aa'\mu\nu} \frac{-ig^{\nu\nu'}}{q^2} \bar{u}'_{k'} \gamma_{\nu'} v'_{i'} t^{a'}_{\text{tr}} \delta(P_i - P_f), \quad (5.4a)$$

where

$$\Pi^{\mu\nu}_{aa'}(q) = \frac{-ig^2}{2} \sum f^{abc} f^{a'bc} \int \frac{d^D k}{(2\pi)^D} \cdot \frac{1}{k^2(k+q)^2}$$
$$\times \left\{ \left[-(2k+q)^\mu g_{\alpha\beta} + (k-q)_\beta \, g^\mu_\alpha + (2q+k)_\alpha \, g^\mu_\beta \right] \right.$$
$$\left. \times \left[-(2k+q)^\nu g^{\alpha\beta} + (k-q)^\beta g^{\nu\alpha} + (2q+k)^\alpha g^{\nu\beta} \right] \right\}. \quad (5.4b)$$

Using the relation $\Sigma ff = \delta^{aa'} C_A$ (cf. Appendix C), and carrying out the usual manipulations we obtain

$$\Pi^{\mu\nu}_{aa'} = \delta_{aa'} C_A \frac{g^2}{32\pi^2}$$
$$\times \left\{ \left[\frac{19}{6} N_\epsilon - \frac{1}{2} - \int_0^1 dx \, (11x^2 - 11x + 5) \log\left(-x(1-x)q^2 \right) \right] q^2 g^{\mu\nu} \right.$$
$$- \left[\frac{11}{3} N_\epsilon + \frac{2}{3} - \int_0^1 dx \, (-10x^2 + 10x + 2) \right.$$
$$\left. \left. \times \log\left(-x(1-x)q^2 \right) \right] q^\mu q^\nu \right\}; \quad (5.5)$$

$$N_\epsilon \equiv \frac{2}{\epsilon} - \gamma_E + \log 4\pi, \qquad \epsilon = 4 - D \to 0.$$

This is divergent, but that is not the difficulty worrying us at present. Unitarity tells us that $\operatorname{Im} \mathcal{T} = (1/2) \mathcal{T} \mathcal{T}^+$. Now, $\operatorname{Im} \mathcal{T}$ is obtained from (5.4), replacing Π by $\operatorname{Im} \Pi$ which, according to (5.5), yields

$$\operatorname{Im} \Pi^{\mu\nu}_{aa'}(q) = \delta_{aa'} C_A \frac{g^2}{32\pi} \theta(q^2) \left\{ -\frac{19}{6} q^2 g^{\mu\nu} + \frac{22}{6} q^\mu q^\nu \right\}, \quad (5.6)$$

which is finite even for $D = 4$. This should be equal to $(1/2)\Sigma_{c,\text{phys}} \mathcal{T} |c, \text{phys}\rangle\langle c, \text{phys}| \mathcal{T}^+$, i.e., to the square of the amplitude for $q\bar{q} \to BB$ with *physical* gluons BB (see Figure 2). If we make use of the Feynman rules, we see that the expression for this is similar to $\operatorname{Im} \mathcal{T}$ with the replacement of $\operatorname{Im} \Pi^{aa'}_{\mu\nu}(q)$ by

$$\delta_{aa'} C_A \sum_{\substack{\eta_1, \eta_2 \\ k_1 + k_2 = q}} \mathcal{A}_\mu(k_1, k_2; \eta_1, \eta_2) \mathcal{A}^*_\nu(k_1, k_2; \eta_1, \eta_2), \quad (5.7a)$$

[10] A potential "tadpole" contribution (Figure 1e) has been omitted because *in dimensional regularization*, $\int d^D k \, (k^2 + i0)^{-1} \equiv 0$. Cf. Section 7.

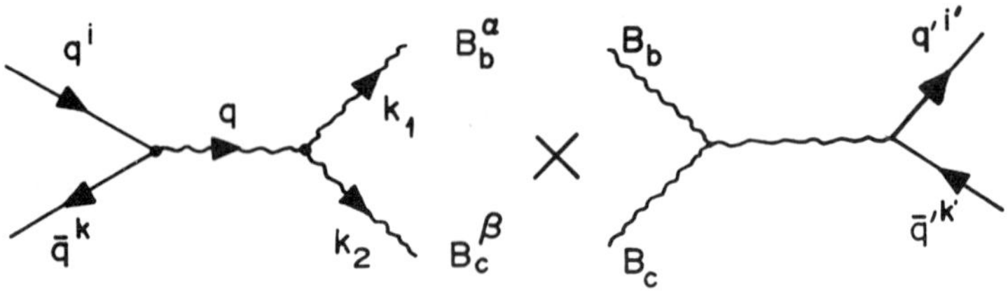

Figure 2. Imaginary part of $\mathcal{T}$.

where $\eta = \pm 1$, are the physical helicities of the gluons, and

$$\mathscr{A}^{\mu} = \left[(k_1 + q)_{\beta}\, g_{\alpha}^{\mu} - (q + k_2)_{\alpha}\, g_{\beta}^{\mu} + (k_2 - k_1)^{\mu} g_{\alpha\beta} \right] \epsilon_p^{\alpha}(k_1, \eta_1) \epsilon_p^{\beta}(k_2, \eta_2). \tag{5.7b}$$

Here the ϵ_p are the polarization vectors for physical gluons given by, e.g.,

$$\epsilon_p^{\alpha}(k, \eta) = \frac{1}{\sqrt{2}} \left\{ \epsilon^{(1)\alpha}(k) + i\eta\epsilon^{(2)\alpha}(k) \right\},$$

with the $\epsilon^{(i)}$ of (4.10). Because the gluons are physical, the ϵ_p verify

$$k_{\alpha} \epsilon_p^{\alpha}(k, \eta) = 0; \qquad k^2 = 0,$$

so (5.7b) may be written as (note that $q = k_1 + k_2$)

$$\mathscr{A}^{\mu} = \left[2k_{1\beta} g_{\alpha}^{\mu} - 2k_{2\alpha} g_{\beta}^{\mu} + (k_2 - k_1)^{\mu} g_{\alpha\beta} \right] \epsilon_p^{\alpha}(k_1, \eta_1) \epsilon_p^{\beta}(k_2, \eta_2),$$

and it is then easy to check that

$$q_{\mu} \mathscr{A}^{\mu} = 0.$$

Unitarity cannot be satisfied in the space of physical gluons. Indeed, from (5.6),

$$q_{\mu} \Pi_{aa'}^{\mu\nu}(q) \neq 0.$$

Of course, what happens is that $\mathscr{L}_{\text{int}}^{\xi}$ sends physical states into nonphysical ones. This fact was first noted by DeWit (1964) and by Feynman; the solution was given by Feynman (1963) in particular cases, and by Fadeyev and Popov (1967) in general. The idea is to introduce extra nonphysical particles (*ghosts*) that will cancel exactly the nonphysical states produced by $\mathscr{L}_{\text{int}}^{\xi}$. We thus modify $\mathscr{L}^{\xi}$ by adding the ghost term

$$\mathscr{L}_{\text{all}}^{\xi} = \mathscr{L}^{\xi} + \sum \left(\partial_{\mu} \bar{\omega}_a(x) \right) \left(\delta_{ab} \partial^{\mu} - g f_{abc} B_c^{\mu}(x) \right) \omega_b(x), \tag{5.8}$$

with $\mathscr{L}^{\xi}$ given by (5.3). The fields $\omega, \bar{\omega}$ are of spin zero, but they satisfy Fermi–Dirac statistics.[11] Since they will never appear in initial or final states (they are, by hypothesis, nonphysical!) this need not worry us.

Let us proceed with our analysis, introducing the ghost contribution. Since ghosts only couple to gluons, they will only modify Figure 1*a*, which

[11] It is convenient at times, although not necessary, to think of $\bar{\omega}, \omega$ as mutually adjoint. We will discuss ghosts further in Sections 41 and 42.

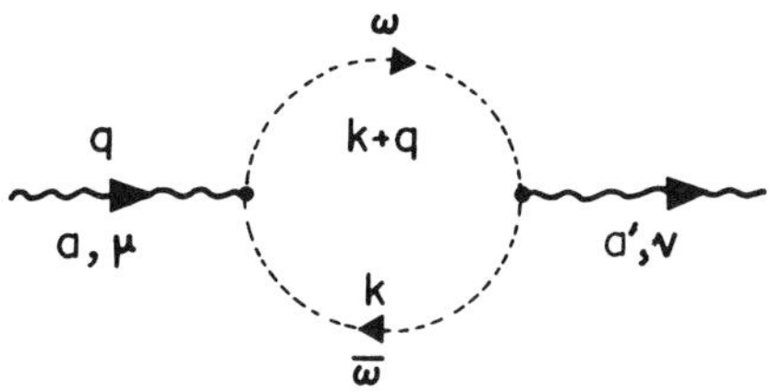

Figure 3. Ghost loop.

suits us. Their contribution to Π is easily evaluated to be, with the conventions of Figure 3,

$$\Pi^{\mu\nu}_{(\mathrm{Ghost})aa'} = \delta_{aa'} C_A i g^2 \int \frac{d^D k}{(2\pi)^D} \cdot \frac{k^\mu (k+q)^\nu}{k^2 (k+q)^2}$$

$$= \frac{\delta_{aa'} g^2}{32\pi^2} C_A \left\{ \left[\tfrac{1}{6} N_\epsilon + \tfrac{1}{6} - \int_0^1 dx \cdot x(1-x) \log(-x(1-x)q^2) \right] q^2 g^{\mu\nu} \right.$$

$$\left. - \left[-\tfrac{1}{3} N_\epsilon + 2 \int_0^1 dx \cdot x(1-x) \log(-x(1-x)q^2) \right] q^\mu q^\nu \right\}.$$

After some simple manipulations, and using the integration formulas of Appendix B, we find:

$$\Pi^{\mu\nu}_{(\mathrm{all})aa'} = \delta_{aa'} \frac{g^2 C_A}{32\pi^2} (-g^{\mu\nu} q^2 + q^\mu q^\nu) \left\{ -\tfrac{10}{3} N_\epsilon - \tfrac{62}{9} + \tfrac{10}{3} \log(q^2) \right\}, \tag{5.9}$$

which certainly verifies the transversality condition,

$$q_\mu \Pi^{\mu\nu}_{(\mathrm{all})aa'} = q_\nu \Pi^{\mu\nu}_{(\mathrm{all})aa'} = 0. \tag{5.10}$$

It is left to the reader to check that $\mathrm{Im}\,\Pi \sim \sum \mathscr{A} \mathscr{A}^*$. We will henceforth drop the index "all" and consider the QCD Lagrangian in a covariant (*Lorentz*) gauge to be (5.8), i.e.,

$$\mathscr{L}^\xi_{\mathrm{QCD}} = \sum_q \{ i\bar{q} \not{D} q - m_q \bar{q} q \} - \tfrac{1}{4} (D \times B)^2 - \tfrac{\lambda}{2} (\partial B)^2$$

$$+ \sum (\partial_\mu \bar{\omega}_a)(\delta_{ab} \partial^\mu - g f_{abc} B^\mu_c) \omega_b, \qquad \xi = 1 - 1/\lambda. \tag{5.11}$$

From now on, we will also drop the index QCD from $\mathscr{L}^\xi_{\mathrm{QCD}}$ as given by (5.11).

ii Physical Gauges

Since the appearance of ghosts was caused by the fact that the projection over physical states P does not commute with the QCD Lagrangian in a Lorentz gauge, it may appear that the problem does not arise if we chose a gauge with only physical gluons so that the whole field Hilbert space is physical. As we already know at the level of QED, we cannot simultaneously have positivity, locality and manifest Lorentz invariance; so out of necessity, we have to work with a non-covariant gauge. A Coulomb gauge[12] still has ghosts, but ghost-free gauges exist if we require

$$n \cdot B = 0, \qquad n^2 \leqslant 0. \tag{5.12}$$

For $n^2 < 0$, one considers *axial gauges*;[13] $n^2 = 0$ is a *lightlike gauge*.[14] Since n is an external vector, manifest Lorentz invariance is lost; of course, gauge invariance guarantees that the result for *physical* quantities is independent of n, hence Lorentz invariant.

Let us begin with an axial gauge. The Lagrangian is

$$\mathscr{L}_n = \sum_q \{ i\bar{q}\not{D}q - m_q\bar{q}q \} - \frac{1}{4}(D \times B)^2 - \frac{1}{2\beta}(n \cdot B)^2, \tag{5.13}$$

and the limit $\beta \to 0$ is to be taken, so the condition (5.12) holds as an operator statement over the entire Hilbert space. The propagator that corresponds to (5.13) is

$$i\frac{-g^{\mu\nu} - k^\mu k^\nu(n^2 + \beta k^2)/(k \cdot n)^2 + (n^\mu k^\nu + n^\nu k^\mu)(n \cdot k)^{-1}}{k^2 + i0}, \tag{5.14}$$

which in the limit as $\beta \to 0$, becomes

$$i\frac{-g^{\mu\nu} - n^2\big(k^\mu k^\nu/(k \cdot n)^2\big) + (n^\mu k^\nu + n^\nu k^\mu)/k \cdot n}{k^2 + i0}. \tag{5.15}$$

The extension of the theory to axial gauges is nontrivial; for details we refer the reader to Kummer (1975). Here we will only make one-loop calculations for which the problems do not exist.

For lightlike gauges, it is convenient to introduce the so-called "null" coordinates: for any vector, v,

$$v^{\pm} = \frac{1}{\sqrt{2}}(v^0 \pm v^3), \qquad \underline{v} = \begin{pmatrix} v^1 \\ v^2 \end{pmatrix}; \qquad v^\alpha = v^{\pm} \text{ or } v^i \quad (i = 1,2).$$

We also define the metric

$$g_{+-} = g_{-+} = 1, \qquad g_{++} = g_{--} = 0, \qquad g_{ij} = -\delta_{ij}, \quad \text{and} \quad i,j = 1,2.$$

Note that

$$v \cdot w = v_+ w_- + v_- w_+ - \underline{v}\,\underline{w} = v_\alpha v^\alpha.$$

For a lightlike vector, we may choose u with $\underline{u} = 0$, $u_- = 0$, $u_+ = 1$. Then the supplementary condition $u \cdot B = 0$ may be written as

$$B^a_-(x) = 0. \tag{5.16}$$

[12] Moreover, it presents extra complications. The formulation of QCD in a Coulomb gauge may be found in Christ and Lee (1980).

[13] Axial gauges are discussed in Kummer (1975) and work quoted there.

[14] See, for example, Tomboulis (1973) and references therein.

The propagator is

$$i\frac{P^{\mu\nu}(k,u)}{k^2+i0}=i\frac{-g^{\mu\nu}+(u^\mu k^\nu+u^\nu k^\mu)/(u\cdot k)}{k^2+i0}, \tag{5.17}$$

which will be recognized as the specialization of (5.15) to $n=u$, $u^2=0$. In terms of the null coordinates, (5.17) may be rewritten as

$$\frac{P^{\alpha\beta}}{k^2}=\frac{-g^{\alpha\beta}+(\delta^\alpha_- k^\beta+\delta^\beta_- k^\alpha)/k_-}{k_\alpha k^\alpha+i0}.$$

As an example of the use of a lightlike gauge, we consider the second-order gluon propagator. In this gauge

$$\begin{aligned}\Pi^{\mu\nu}_{l,ab}=&\frac{-ig^2C_A\delta_{ab}}{2}\int\frac{d^Dk}{(2\pi)^D}\cdot\frac{1}{k^2(k+q)^2}\\&\times\left[-(2k+q)^\mu g^{\alpha\beta}+(k-q)^\beta g^{\mu\alpha}+(2q+k)^\alpha g^{\mu\beta}\right]P_{\alpha\rho}(k,u)\\&\times\left[-(2k+q)^\nu g^{\rho\sigma}+(k-q)^\sigma g^{\nu\rho}+(2q+k)^\rho g^{\nu\sigma}\right]P_{\sigma\beta}(k+q,u).\end{aligned}$$

We will consider only the divergent and logarithmic part. This enormously simplifies the calculation, and we find

$$\Pi^{\mu\nu}_{l,ab}(q)=\frac{11C_Ag^2\delta_{ab}}{3\times16\pi^2}(-q^2g^{\mu\nu}+q^\mu q^\nu)\{N_\epsilon-\log(-q^2)+\text{constant terms}\}. \tag{5.18}$$

We note that it is transverse; no ghosts are required. It should also be noted that the propagator is "self-reproducing" under (5.18) in the sense that

$$\frac{P^{\mu\alpha}(q,u)}{q^2}\{-q^2g_{\alpha\beta}+q_\alpha q_\beta\}\frac{P^{\beta\nu}(q,u)}{q^2}=\frac{P^{\mu\nu}(q,u)}{q^2}. \tag{5.19}$$

6 The Becchi–Rouet–Stora Transformations

In the previous section, we showed that the Lagrangian without ghosts violated unitarity in the physical space of states. Since gauge invariance guarantees that this should not occur, it is clear that the phenomenon must be due to the introduction of the gauge-fixing term which by its very nature is not gauge invariant. One may then wonder whether the ghosts may not be interpreted as a term that restores gauge invariance. This is indeed so, as will be discussed in the present section.

Let us begin with QED.[15] The Lagrangian in a covariant gauge,

$$\mathcal{L}^\xi=\bar\psi(i\not{D}-m)\psi-\frac{1}{4}F_{\mu\nu}F^{\mu\nu}-\frac{\lambda}{2}(\partial_\mu A^\mu)^2, \tag{6.1}$$

[15] We follow the discussion of de Rafael (1977, 1979).

where

$$F^{\mu\nu} = \partial^\mu A^\nu - \partial^\nu A^\mu, \qquad D^\mu = \partial^\mu ieA^\mu,$$

is not gauge invariant because of the gauge-fixing term $-(\lambda/2)(\partial A)^2$. Gauge invariance may, however, be restored by means of the following trick. Add a term

$$\mathscr{L}_\omega = -\tfrac{1}{2}(\partial_\mu \omega)\partial^\mu \omega \tag{6.2}$$

to $\mathscr{L}^\xi$ with ω a massless field without interactions. If we generalize the gauge transformations to include the field ω, writing for infinitesimal $\theta(x) = \epsilon\omega(x)$,

$$\psi(x) \to \psi(x) + ie\epsilon\omega(x)\psi(x), \qquad A^\mu(x) \to A^\mu(x) - \epsilon\partial^\mu\omega(x),$$
$$\omega(x) \to \omega(x) - \epsilon\lambda\partial_\mu A^\mu(x), \tag{6.3}$$

then, up to a four-divergence,

$$\mathscr{L}^\xi_{\mathrm{QED}} = \mathscr{L}^\xi + \mathscr{L}_\omega \tag{6.4}$$

is invariant under (6.3). The restoration of gauge invariance was comparatively easy here; because A has no charge, and thus no self-interactions, we could take ω to be real and free. However, the simplicity of $\mathscr{L}_\omega$ does *not* mean that it has no deep consequences. In fact, one may show that the transformations (6.3) generate all the Ward identities of QED—which, in particular, ensure that the interaction does not lead from physical to nonphysical states. For example, we will show how the transversality condition for the photon propagator can be deduced from (6.3) and (6.4). (Of course, it may also be verified by direct computation of the vacuum polarization.)

Consider

$$\langle TA_\mu(x)\omega(0)\rangle_0$$

Effecting a generalized gauge transformation, we find, to first order in ϵ,

$$\lambda\langle TA_\mu(x)(\partial_\nu A^\nu(0))\rangle_0 = \langle T(\partial_\mu\omega(x))\omega(0)\rangle_0 .$$

A Fourier transformation gives

$$\begin{aligned}
&\int d^4x\, e^{iq\cdot x}\langle TA^\mu(x)\partial_\nu A^\nu(0)\rangle_0 \\
&\quad = iq_\nu\int d^4x\, e^{iq\cdot x}\langle TA^\mu(x)A^\nu(0)\rangle_0 = iq_\nu D^{\mu\nu}(q) \\
&\quad = \frac{-1}{\alpha}\int d^4x\, e^{iq\cdot x}\langle T(\partial^\mu\omega(x))\omega(0)\rangle_0 \\
&\quad = \frac{i}{\lambda}q^\mu\int d^4x\, e^{iq\cdot x}\langle T\omega(x)\omega(0)\rangle_0 \\
&\quad = \frac{1}{\lambda}\cdot\frac{q^\mu}{q^2+i0} .
\end{aligned} \tag{6.5}$$

The last equality holds true because, since the field ω is free, its propagator is a free-field propagator. Therefore, we have proved that if we write $D^{\mu\nu}$ as a sum of a transverse and a nontransverse (longitudinal) part,

$$D^{\mu\nu}(q) = (-q^2 g^{\mu\nu} + q^\mu q^\nu) D_{\text{tr}}(q^2) + \frac{q^\mu q^\nu}{q^2} D_L(q^2), \tag{6.6}$$

then

$$D_L(q^2) = \frac{-1}{\lambda} \cdot \frac{i}{q^2 + i0}, \tag{6.7}$$

i.e., the nontransverse part of D remains as in the free-field case. Recall that, for the free-field,

$$D^{0\mu\nu}(q) = i \frac{-g^{\mu\nu} + (1 - \lambda^{-1}) q^\mu q^\nu / (q^2 + i0)}{q^2 + i0}.$$

Otherwise stated, if in perturbation theory we write

$$D^{\mu\nu}(q) = D^{(0)\mu\nu}(q) + \frac{e^2}{4\pi} D^{(2)\mu\nu}(q) + \ldots,$$

all the $D^{(n)\mu\nu}(q)$ values satisfy the transversality condition

$$q_\mu D^{(n)\mu\nu}(q) = 0, \qquad n = 2, 4, \ldots,$$

so the equivalent of (5.10) is automatic here.

For a non-abelian theory the generalization of (6.3) are the so-called Becchi–Rouet–Stora (1974, 75) transformations. They extend gauge invariance to the ghost fields, and leave invariant (up to a four-divergence) the full QCD Lagrangian (5.11). As for QED, they generate the analogue of the Ward identities, the *Slavnov* (1975)–*Taylor* (1971) identities. The BRS transformations for QCD are, for infinitesimal ϵ, assumed to be an anticommuting, x independent c-number[16]:

$$\begin{aligned}
B_a^\mu &\to B_a^\mu - \epsilon \sum \{ \delta_{ab} \partial^\mu - g f_{abc} B_c^\mu \} \omega_b, \\
q &\to q - i\epsilon g \sum t^a \omega_a q, \\
\omega_a &\to \omega_a - \frac{\epsilon}{2} g \sum f_{abc} \omega_b \omega_c, \\
\bar{\omega}_a &\to \bar{\omega}_a + \epsilon \lambda \partial_\mu B_a^\mu .
\end{aligned} \tag{6.8}$$

Using them it is easy to derive, by the same methods as for QED, the result analogue to (6.7). If we write

$$D_{ab}^{\mu\nu}(q) = \delta_{ab}(-g^{\mu\nu} q^2 + q^\mu q^\nu) D_{\text{tr}} + \delta_{ab} \frac{q^\mu q^\nu}{q^2} D_L, \tag{6.9}$$

then

$$D_L = -\frac{1}{\lambda} \cdot \frac{i}{q^2 + i0}. \tag{6.10}$$

[16] Hence $\epsilon^2 = 0$, $\epsilon\omega = -\omega\epsilon$, $\epsilon q = -q\epsilon$, $\epsilon B = B\epsilon$, etc. Remember also that the ω are fermions so we may take $\omega_b \omega_c = -\omega_c \omega_b$.

Therefore, with

$$D_{ab}^{\mu\nu} = \sum_{n=0}^{\infty} \left(\frac{g^2}{4\pi} \right)^n D_{ab}^{(n)\mu\nu},$$

we see that if we do recall that, say,

$$D_{ab}^{(2)\mu\nu} = \sum D_{aa'}^{(0)\mu\mu'} \Pi^{(2)}_{a'b'\mu'\nu'} D_{b'b}^{(0)\nu'\nu},$$

with $\Pi^{(2)}$ the second-order vacuum polarization, we have

$$q_\mu \Pi_{ab}^{(2)\mu\nu} = 0.$$

This, of course, we had checked in Equations (5.9), (5.10).

A last important point is that all of the above derivations are formal; that is, we neglected to consider, when manipulating propagators, for example, that they are singular functions. To actually verify the identities, one has to check that they go through the renormalization program (see below, Sections 7, 8, and 9,). Indeed, some formal identities do break down; an example will be found in Section 33; but even those that do not break down may have to be interpreted. This is true for Equation (6.10) because the gauge parameter becomes renormalized.

7 Regularization (Dimensional)

As we saw in the example of Section 5, some amplitudes are divergent. This is due to the fact that field operators are singular objects: it is easy to trace the divergence of the dk integral in (5.4*b*) at large values of k to the occurrence, in position space, of products of field operators at the same space-time point. Because of this, we must, in order to discuss QCD (or indeed any local relativistic field theory), give a meaning to the integrals that appear when we evaluate Feynman diagrams. This is called *regularization*, and it amounts to altering the Lagrangian $\mathscr{L}$ to $\mathscr{L}_\epsilon$ in such a way that $\mathscr{L}_\epsilon$ produces finite answers and, in some sense, as $\epsilon \to 0$, $\mathscr{L}_\epsilon \to \mathscr{L}$. Due to Bohr and Rosenfeld's (1933, 1950) classical work, we know that field operators are intrinsically singular; therefore, any regularization must destroy some physical feature of the theory. Thus, Pauli–Villars regularization destroys hermiticity and gauge invariance for non-Abelian theories; lattice regularization destroys Poincaré invariance, etc. Of course, in the limit as $\epsilon \to 0$, these properties are recovered (if one was careful enough!). Because gauge and relativistic invariance are essential for QCD, we will use dimensional regularization that only destroys scale invariance. The method is related to so-called analytical regularization (Speer, 1968; Bollini, Giambiagi, and González–Domínguez, 1964) and has been thoroughly developed by 't Hooft and Veltman (1972; see also Bollini and Giambiagi, 1972). It amounts to working in an arbitrary dimension, $D = 4 - \epsilon$; the physical limit is, $\epsilon \to 0$. Divergences appear as poles in $1/\epsilon$. As far as the author knows, there is no mathematically sound notion of dimension for D other than that it is a positive integer; but this should not worry us unduly: all we require are interpolation formulas consistent with gauge and Poincaré invariance, applicable to the evaluation of Feynman integrals. This we will accomplish in steps. First, consider a convergent integral of the form $(2\pi)^D \int d^Dk\, f(k^2)$, where typically $f(k^2) = (k^2)^r(k^2 - a^2)^{-m}$, r, m integers, and $d^Dk = dk^0dk^1 \ldots dk^{D-1}$; $k^2 = (k^0)^2 - (k^1)^2 - \cdots - (k^{D-1})^2$. Because f is analytic in the k^0 plane, we can rotate the integration from $(-\infty, +\infty)$ to $(-i\infty, +i\infty)$, a so-called *Wick rotation*. We can recover an integration over $(-\infty, +\infty)$ by then defining the new variable $k^0 \to k^D = ik^0$. Thus we obtain an ordinary Euclidean integral in D dimensions,

$$i\int_{-\infty}^{+\infty} \frac{dk^1}{2\pi} \cdots \int_{-\infty}^{+\infty} \frac{dk^D}{2\pi} f(-k_E^2), \qquad k_E^2 \equiv (k^1)^2 + \cdots + (k^D)^2 \equiv |k_E|^2.$$

If $d^Dk_E = dk^1 \ldots dk^D$, we can introduce polar coordinates, $d^Dk_E = d|k_E| \cdot |k_E|^{D-1}d\Omega_D$. Using the formula $\int d\Omega_D = 2\pi^{D/2}/\Gamma(D/2)$, we finally find

$$\int \frac{d^Dk}{(2\pi)^D} f = \frac{i}{(2\pi)^{D/2}\Gamma(D/2)} \int_0^{\infty} d|k_E| \cdot |k_E|^{D-1} f(-|k_E|^2).$$

The manipulations we have carried out were only valid for $D =$ positive integer; but we can use the last formula to *define* the integral for arbitrary (even complex) D, and arbitrary r, m.

Next, consider the integral of a polynomial in k^μ times $f(k^2)$: we can reduce this to the former situation by symmetric integration, writing, e.g.,

$$\int d^Dk\, f(k^2)k^\mu k^\nu = \frac{g^{\mu\nu}}{D} \int d^Dk\, f(k^2)k^2.$$

Finally, the general case is treated by expanding in a power series in the k^μ. In this way, the integrals of Appendix B, and many more, can be worked out for arbitrary D. For example,

$$\int \frac{d^Dk}{(2\pi)^D} \frac{(k^2)^r}{(k^2 - a^2)^m} = i\frac{(-1)^{r-m}}{(4\pi)^{D/2}} \cdot \frac{\Gamma(r + D/2)\Gamma(m - r - D/2)}{\Gamma(D/2)\Gamma(m)(a^2)^{m-r-D/2}}.$$

If the left-hand side was divergent in, say, the physical case $D = 4$, which occurs when $m - r - D/2 \leqslant 0$, this is reflected in poles in the right-hand side due to poles of the Γ function $\Gamma(m - r - D/2)$, as shown. An arbitrariness in the method is already apparent; we may have multiplied the right-hand side by any function $\varphi(D)$, provided it is analytic in D and $\varphi(4) = 1$. This will be useful later on.

Now we consider spin. It will be convenient to distinguish external and internal lines in Feynman graphs. Later on, we will show that after renormalization, Green's functions, with their external legs amputated, are finite in perturbation theory in the limit as $D \to 4$. Since spin factors in external legs (i.e., factors $u, v, \bar{u}, \bar{v}, \epsilon^\mu \ldots$; cf., Appendix D) are certainly finite for $D = 4$, we may already take them in physical dimension. As for spin effects in internal lines, we have to take $g^{\alpha\beta}$ in dimension D so that, e.g., $g^{\alpha\beta}g_{\alpha\beta} = D$, etc. Likewise, we must consider that we have D γ-

matrices, $\gamma^0, \gamma^1, \ldots, \gamma^{D-1}$. To be totally consistent, we would have to admit that the γ_μ were $2^{D/2} \times 2^{D/2}$matrices (which is the dimension of the corresponding Clifford algebra). But this is not necessary. We are still consistent with gauge invariance if we take the γ_μ to be 4×4 matrices so that, e.g., $\operatorname{Tr} \gamma^\mu \gamma^\nu = 4 g^{\mu\nu}$; and this will be done here. (A method related to dimensional regularization is the so-called dimensional reduction; the interested reader is referred to the original paper of Siegel (1979).)

Thus, the extension of integrals, and Dirac algebra, to arbitrary D is fairly simple; a set of useful formulas are collected in Appendices A and B. Only the introduction of γ_5 is a bit trickier. If, for example, we write $\gamma_5 = i\gamma^0\gamma^1\gamma^2\gamma^3$, it is clear that this will not exist for $D < 4$. The definition $\gamma_5 = i\gamma^0 \ldots \gamma^{D-1}$ may be shown to be inconsistent with gauge invariance (cf., Sec. 33, particularly between Eqs. (33.17) and (33.20), for a discussion). A safe definition seems to be

$$\gamma_5 = \frac{i}{4!} \epsilon^D_{\mu\nu\rho\sigma} \gamma^\mu \gamma^\nu \gamma^\rho \gamma^\sigma,$$

where the symbol ϵ^D coincides with the ordinary antisymmetric one only for $D = 4$. We do not specify it further beyond requiring that it be such that, for arbitrary D,

$$\gamma_5^2 = 1, \quad \text{and} \quad \operatorname{Tr} \gamma_5 \gamma^\alpha \gamma^\beta = 0.$$

(Appendix A). We have thus built fully dimensional regularization; as long as $D \neq$ integer, we see that, with it, all integers appearing in Feynman graphs are finite. It preserves gauge and Poincaré invariance, but it breaks down scale invariance.

The simplest way to see this is to realize that a Feynman integral like the one in (5.4*b*) becomes altered

$$\int \frac{d^4k}{(2\pi)^4} \to \int \frac{d^Dk}{(2\pi)^D}\,.$$

One can rescale all fields and coupling constants accordingly, but it is simpler to utilize the prescription:

$$\int \frac{d^4k}{(2\pi)^4} \to \int d^D\hat{k} \equiv \int \frac{d^Dk\nu_0^{4-D}}{(2\pi)^D}\,, \qquad D = 4 - \epsilon, \tag{7.1a}$$

where

$$\hat{k}^\mu = \nu_0^{4/D-1} k^\mu / 2\pi, \tag{7.1b}$$

thereby explicitly introducing the scale-invariance-breaking arbitrary (but fixed) parameter ν_0 with dimensions of $[\text{Mass}]^1$.

As a first example of these methods, let us calculate the propagator of a quark to second order, in momentum space:

$$S_\xi^{ij}(p) = \int d^4x\, e^{ip\cdot x} \langle T q^i(x) \bar{q}^j(0) \rangle_0\,. \tag{7.2}$$

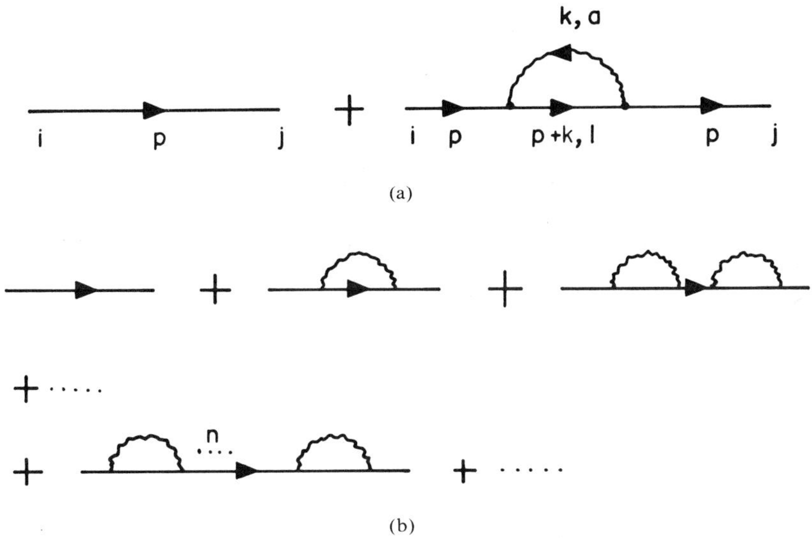

Figure 4. (*a*), Quark propagator; (*b*), iteration.

This is given by the graphs of Figure 4. We have, in an arbitrary gauge, and for dimension $D = 4 - \epsilon$,

$$S^{ij}_{D\xi}(p) = \delta^{ij}\frac{i}{\not p - m + i0} - \frac{1}{\not p - m + i0}\, g^2 \sum_{l,a} t^a_{il} t^a_{lj} \Sigma^{(2)}_{D\xi}(p) \frac{i}{\not p - m + i0}$$

$$+ \text{ higher orders,} \tag{7.3a}$$

where

$$\Sigma^{(2)}_{D\xi}(p) = -i\int d^D\hat{k}\, \frac{\gamma_\mu(\not p + \not k + m)\gamma_\nu}{(p+k)^2 - m^2} \cdot \frac{-g^{\mu\nu} + \xi k^\mu k^\nu / k^2}{k^2}. \tag{7.3b}$$

Writing identically

$$\not k(\not p + \not k + m) = (p+k)^2 - m^2 - (p^2 - m^2) - (\not p - m)\cdot \not k$$

we find

$$\Sigma^{(2)}_{D\xi}(p) = -i\int d^D\hat{k}\left\{\frac{(D-2)(\not p + \not k) - Dm - \xi(\not p - m)}{k^2\left[(p+k)^2 - m^2\right]}\right.$$

$$\left. -\xi(p^2 - m^2)\frac{\not k}{k^4\left[(p+k)^2 - m^2\right]}\right\}.$$

After standard manipulations, this gives (neglecting terms that will vanish as $\epsilon \to 0$)

$$\Sigma^{(2)}_{D\xi}(p) = (\not{p} - m)A_{D\xi}(p^2) + mB_{D\xi}(p^2); \tag{7.4a}$$

$$A_{D\xi} = \frac{1}{16\pi^2}\left\{(1-\xi)N_\epsilon - 1 - \int_0^1 dx\,[2(1-x)-\xi]\log\frac{xm^2 - x(1-x)p^2}{\nu_0^2}\right.$$
$$\left. - \xi(p^2 - m^2)\int_0^1 dx\,\frac{x}{m^2 - xp^2}\right\}; \tag{7.4b}$$

$$B_{D\xi} = \frac{1}{16\pi^2}\left\{-3N_\epsilon + 1 + 2\int_0^1 dx(1+x)\log\frac{xm^2 - x(1-x)p^2}{\nu_0^2}\right.$$
$$\left. - \xi(p^2 - m^2)\int_0^1 dx\,\frac{x}{m^2 - xp^2}\right\}. \tag{7.4c}$$

Here we have defined

$$N_\epsilon = \frac{2}{\epsilon} - \gamma_E + \log 4\pi.$$

In fact, all poles in dimensional regularization appear in this combination. Noting that (see Appendix C) $\sum t^a_{il}t^a_{lj} = C_F\delta_{ij} = \frac{4}{3}\delta_{ij}$, we can insert (7.4) into (7.3) and rearrange it to read:

$$S_{D\xi}(p) = i\{\not{p} - m + g^2 C_F \Sigma^{(2)}\}^{-1}; \tag{7.5a}$$

$$S_{D\xi}(p) = i\,\frac{1 - C_F g^2 A_{D\xi}(p^2)}{\not{p} - m\{1 - C_F g^2 B_{D\xi}(p^2)\}} + \text{higher orders}. \tag{7.5b}$$

Actually, it is easy to see that (7.5a) takes into account exactly the contribution of all the diagrams of Figure 4 and that, replacing $\Sigma^{(2)} \to \Sigma^{\text{exact}}$, is the most general form for S. As we see, there are two divergences:

$$1 - C_F\frac{g^2}{16\pi^2}(1-\xi)N_\epsilon \qquad \text{(from } A\text{)} \tag{7.6}$$

which multiplies the entire S, and

$$1 + 3C_F\frac{g^2}{16\pi^2}N_\epsilon \qquad \text{(from } B\text{)} \tag{7.7}$$

which multiplies m; but both terms are finite, provided we keep $\epsilon \neq 0$.

We end this section with a comment on infrared singularities. In this work, we will be mainly concerned with *ultraviolet singularities*, which appear for $k \to \infty$ and give divergences $\Gamma(\epsilon/2)$; but dimensional regularization also regulates *infrared singularities*, due to the $k \to 0$ region, which give divergences of the type $\Gamma(-\epsilon/2)$. For details, cf., Gastmans and Meuldermans (1973).

8 Renormalization—Generalities

Let us consider the following process. A photon hits a u quark from a proton, and u subsequently decays weakly into $d + e + \nu$ (Figure 5). To the lowest order in weak and electromagnetic interactions, and to zero order in g, we have the diagram of Figure 5a. Gluon corrections may then intervene (see diagrams of Figures 5b, c, and d.) In particular, we see that $S(p)$ will enter into the amplitude with, in obvious notation, $p = p_\gamma + p_u$. Therefore, it looks as if the result for the amplitude is divergent and no sense can be extracted from the theory, at least in a perturbative expansion.

Of course, this is not so. We have been somewhat lax in our formulation. To simplify, let us consider a scalar interaction $\bar{\psi}\phi\phi$ with massless field ϕ. The Lagrangian is

$$\mathcal{L} = \bar{\psi}(i\partial\!\!\!/ - m)\psi + \tfrac{1}{2}\partial_\mu\phi\partial^\mu\phi + g\bar{\psi}\psi\phi. \tag{8.1}$$

As stated previously, the S matrix is given by

$$\begin{aligned} S &= T\exp i\int d^4x\,\mathcal{L}^0_{\text{int}}(x) \\ &= 1 + \sum_{n=1}^{\infty}\frac{i^n}{n!}\int d^4x_1\ldots d^4x_n\,T\mathcal{L}^0_{\text{int}}(x_1)\ldots\mathcal{L}^0_{\text{int}}(x_n), \end{aligned} \tag{8.2}$$

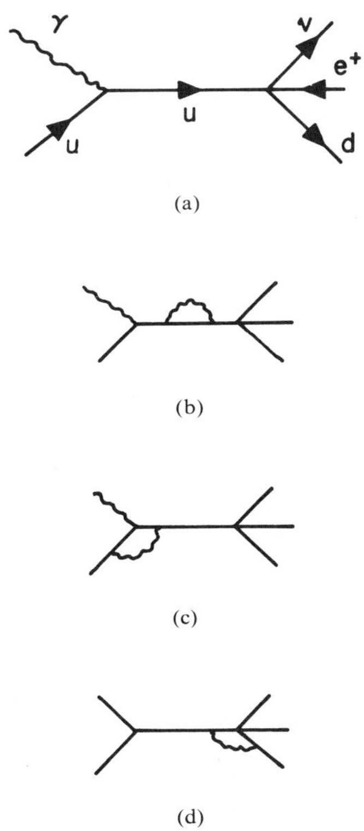

Figure 5. (a–d), The process $\gamma + u \to \nu e^+ d$ and gluon corrections.

where the fields in $\mathscr{L}^0_{\text{int}}$ were to be taken as free, and in normal order, and we identified it with the trilinear term in (8.1) after replacing $\psi \to \psi^0$, $\phi \to \phi^0$:

$$\mathscr{L}^0_{\text{int}} = g :\bar{\psi}^0\psi^0: \phi^0. \tag{8.3}$$

However, this is incorrect. Clearly, the fields in (8.1) are not free; and it is also conceivable that the mass that appears there is not the mass one would have if there were no interactions. This should be apparent from (7.5); the mass has been shifted by the amount

$$m\left\{1 - \tfrac{4}{3} g^2 B_D\right\},$$

and the normalization multiplied by

$$1 - \tfrac{4}{3} g^2 A_D\,.$$

On the grounds of invariance, all possible changes are of two types: of the multiplicative type,

$$\psi \to Z_\psi^{-1/2}\psi_u, \qquad \phi \to Z_\phi^{-1/2}\phi_u, \qquad g \to Z_g g, \qquad m \to Z_m m; \tag{8.4}$$

or the type obtained by adding some invariant extra term to $\mathscr{L}$. In our case, it may be shown that it is necessary to add a term $\lambda(\phi)^4$, but we shall neglect this now. So, taking into account only (8.4), we see that (8.1) becomes the so-called "renormalized" Lagrangian:

$$\begin{aligned}\mathscr{L}^R = {} & Z_\psi^{-1}\bar{\psi}_u i\partial\psi_u - Z_\psi^{-1}Z_m m\bar{\psi}_u\psi_u + Z_\phi^{-1}\partial_\mu\phi_u\partial^\mu\phi_u \\ & + Z_g Z_\psi^{-1} Z_\phi^{-1/2} g\bar{\psi}_u\psi_u\phi_u\,. \end{aligned}\tag{8.5}$$

We find that the interaction, defined as $\mathscr{L}_{\text{int}} \equiv \mathscr{L} - \mathscr{L}_{\text{free}}$, is really

$$\begin{aligned}\mathscr{L}^{R0}_{\text{int}} = {} & :g\bar{\psi}^0_u\psi^0_u\phi^0_u + \left(Z_g^{1/2}Z_\psi^{-1}Z_\phi^{-1/2} - 1\right) g\bar{\psi}^0_u\psi^0_u\phi^0_u \\ & + \left(Z_\psi^{-1} - 1\right)\bar{\psi}^0_u i\partial\psi^0_u - \left(Z_\psi^{-1}Z_m - 1\right) m\bar{\psi}^0_u\psi^0_u \\ & + \left(Z_\phi^{-1} - 1\right)\partial_\mu\phi^0_u\partial^\mu\phi^0_u :, \end{aligned}\tag{8.6}$$

where ψ^0_u, ϕ^0_u are *free* fields, satisfying canonical commutation relations. The terms containing $(Z \ldots - 1)$ factors are called *counterterms*. Clearly, if we expand them in a power series in g, they have to begin at unity, for if g were zero, all Z would equal unity:

$$Z_j = 1 + \sum_{n=1}^{\infty} C_j^{(n)}\left(\frac{g^2}{16\pi^2}\right)^n, \tag{8.7}$$

where the $C_j^{(n)}$ possess a finite Laurent expansion around $\epsilon = 0$ (i.e., are of the form $\sum_{k=0}^{n} a_k^{(n)}\epsilon^{-k} + 0(\epsilon)$). There is another way in which the necessity of counterterms may be seen (Bogoliubov and Shirkov, 1959). If we look at the expansion (8.2), it turns out that because the fields are singular, the product

$$\mathscr{L}^0_{\text{int}}(x_1) \ldots \mathscr{L}^0_{\text{int}}(x_n) \tag{8.8a}$$

is undefined for $x_i = x_j$. Therefore, one can add arbitrary terms

$$p(\partial)\delta(x_1 - x_2)\ldots\delta(x_i - x_j)\ldots\delta(x_{n-1} - x_n), \tag{8.8b}$$

(with p a polynomial in derivatives) to each of (8.8*a*). On analysis, the terms (8.8*b*) are seen to correspond to the counterterms.

How arbitrary are the values of Z? A first condition on them is that $\mathscr{L}^R$ produce finite answers, even in the limit $\epsilon \to 0$. This, however, does not completely fix all the $C_j^{(n)}$ in (8.6*b*). To have a unique theory, we have to give arbitrarily as many independent amplitudes as there are *renormalization constants*, Z.

Let us now return to the QCD Lagrangian. Since QCD is a gauge theory, and we have seen that gauge invariance is essential to keep the theory meaningful, the possible counterterms are strongly restricted: they must respect gauge invariance. A look at the expression for $\mathscr{L}^{\xi}_{\text{QCD}}$, Equation (5.11) shows that the only modifications allowed are the following[17]:

$$\begin{aligned}
q^i(x) &\to Z_F^{-1/2} q_u^i(x),\\
B_a^{\mu}(x) &\to Z_B^{-1/2} B_{ua}^{\mu}(x),\\
\omega_a(x) &\to Z_\omega^{-1/2}\omega_{ua}(s), \qquad \bar{\omega}_a(x) \to Z_\omega^{-1/2}\bar{\omega}_{ua}(x),\\
g &\to Z_g g,\\
m_q &\to Z_{m,q} m_q,\\
\lambda &\to Z_\lambda \lambda.
\end{aligned} \tag{8.9}$$

Gauge invariance forces all the Z values for all the quarks to be equal to one single Z_F, and, likewise, there is one common Z_B for all the gluons. In addition, the potentially different renormalizations of the trilinear $\bar{q}qB$, the trilinear BBB, the quadrilinear $BBBB$ and the ghost $\bar{\omega}\omega B$ couplings must be induced by the *same* Z_g. That this very specific set of Zs is sufficient to render all Green's functions finite is a consequence of the identities (Ward identities for Abelian, Slavnov–Taylor identities for non-Abelian theories) that gauge invariance forces on these Green's functions. As stated earlier, these identities[18] may be generated by the BRS transformations; later on, we will explicitly check a few representative ones.

To end this section, let us introduce a last bit of notation. After the replacements of (8.9), the *renormalized* QCD Lagrangian becomes:

$$\begin{aligned}
\mathscr{L}^{\xi}_{R} = \sum_1 \left\{ i\bar{\tilde{q}}\tilde{\not{D}}\tilde{q} - \tilde{m}_q\bar{\tilde{q}}\tilde{q}\right\} - \frac{1}{4}(\tilde{D}\times\tilde{B})^2 - \frac{\tilde{\lambda}}{2}(\partial\tilde{B})^2\\
+ \sum\left(\partial_\mu\bar{\tilde{\omega}}\right)\tilde{D}^{\mu}\tilde{\omega},
\end{aligned} \tag{8.10a}$$

[17]Note that not all the Z are independent; for example, the Slavnov–Taylor identities give $Z_\lambda = Z_B$; cf., Section 9.

[18]Detailed studies of these identities may be found in Lee (1976) and Fadeyev and Slavnov (1980).

where the tilde means that the corresponding objects embody the appropriate Z factors:

$$\tilde{q} = Z_F^{-1/2} q_u, \quad \tilde{m} = Z_m m, \ldots,$$
$$\tilde{D}\tilde{q} = (\partial - i\tilde{g}t\tilde{B})\tilde{q}, \ldots \text{ etc.} \tag{8.10b}$$

That is to say, $\mathcal{L}_R^\xi$ is formally identical with $\mathcal{L}^\xi$ with the replacement of all objects by renormalized objects. We may also split $\mathcal{L}_R^\xi$, explicitly exhibiting the counterterms

$$\mathcal{L}_R^\xi = \mathcal{L}_{uD}^\xi + \mathcal{L}_{\mathrm{ct}D}^\xi, \tag{8.11a}$$

where $\mathcal{L}_{uD}^\xi$ is the unrenormalized or "bare" part

$$\mathcal{L}_{uD}^\xi = \sum_q \{ i\bar{q}_u \not{D} q_u - m\bar{q}_u q_u \} - \frac{1}{4}(D_u \times B_u)^2 - \frac{\lambda}{2}(\partial B_u)^2$$
$$+ (\partial_\mu \omega_u) D^\mu \omega_u, \tag{8.11b}$$

and

$$\mathcal{L}_{\mathrm{ct}D}^\xi = \mathcal{L}_R^\xi - \mathcal{L}_{uD}^\xi$$
$$= (Z_F^{-1} - 1) i \sum_q \bar{q}_u \not{\partial} q_u$$
$$+ \left(Z_F^{-1} Z_B^{-1/2} Z_g - 1\right) g \sum \bar{q}_u \gamma_\mu t^a q_u B_{ua}^\mu + \ldots \tag{8.11c}$$

We see that, in perturbation theory, the interaction contains not only the terms $g\sum q_u^0 \gamma_\mu t q_u^0 B_u^{0\mu}, \ldots,$ but also $i(Z_F^{-1} - 1)\sum q_u^0 \not{\partial} q_u^0$, etc; the fields $q_u^0, B_u^0, B_u^0, \omega_u^0$ are the ones that verify free-field canonical commutation relations, and thus generate the Feynman rules of Appendix D. It is to be noted that, whereas $\mathcal{L}_{uD}^\xi, \mathcal{L}_{\mathrm{ct}D}^\xi$ each requires regularization, the infinites must compensate in such a way that $\mathcal{L}_R^\xi$ produces finite answers in the limit $D \to 4$. It is far from obvious that there exists a choice of Z values that achieves this, and indeed (at least in perturbation theory), only a limited number of field theories are renormalizable. The proof of the renormalizability of non-Abelian gauge theories, in particular QCD, was first given by 't Hooft (1971).[19] Here we will not go so far, but only check that $\mathcal{L}_R^\xi$ produces finite answers to the lowest orders in perturbation theory.

In our presentation of renormalization theory, based essentially on the discussion of Bogoliubov and Shirkov (1959), finite (*renormalized*) Green functions are obtained for the VEVs

$$\langle 0 | T q_u(x_1) \ldots B_u(g_1) \ldots \omega_u(z_1) \ldots | 0 \rangle$$

in perturbation theory, calculating with the full interaction Lagrangian (including counterterms) of (8.11). The multiplicative character of renormalization, however, allows us to follow a different path. We may neglect

[19] See also Lee and Zinn-Justin (1972). A very up-to-date account may be found in Fadeyev and Slavnov (1980).

the counterterms and simply rescale fields and couplings in the Green's function according to (8.9). We will see this in more detail in the following sections. Also, we should mention that we are renormalizing *perturbatively*. This means that we have to be consistent, and work to the same order, both in the "primitive" interaction and in the counterterms.

9 Renormalization of QCD (One Loop)

i μ-Renormalization

Consider the renormalized QCD Lagrangian. In order to specify it, we have to give the values of Z. To do so, we begin by defining the unrenormalized Green's functions,

$$G_{uD}(x_1, \ldots, x_N),$$

which are calculated with $\mathcal{L}_u^\xi$. If G corresponds to the VEV of a field product:

$$\langle T\Phi_1(x_1) \ldots \Phi_N(x_N)\rangle_0 = G(x_1, \ldots, x_N), \tag{9.1}$$

where the Φ_k are the q_u, ω_u, B_u or, more generally, local operators built from these, then, in perturbation theory,

$$G_{uD}(x_1, \ldots, x_N) = \sum_{n=0}^{\infty} \frac{i^n}{n!} \int d^4z_1 \ldots d^4z_n \times \langle T\Phi_1^0(x_1) \ldots \Phi_N^0(x_N)\mathcal{L}_{uD,\text{int}}^\xi(z_1) \ldots \mathcal{L}_{uD,\text{int}}^\xi(z_n)\rangle_0 . \tag{9.2}$$

The G_{uD} are, generally speaking, divergent as $D \to 4$. The renormalized Green's functions are defined as:

$$G_R(x_1, \ldots, x_N) = \sum_{n=0}^{\infty} \frac{i^n}{n!} \int d^4z_1 \ldots d^4z_n \times \langle T\Phi_1^0(x_1) \ldots \Phi_N^0(x_N)\mathcal{L}_{R,\text{int}}^\xi(z_1) \ldots \mathcal{L}_{R,\text{int}}^\xi(z_n)\rangle_0 . \tag{9.3}$$

What we then require is that the G_R be finite, i.e., that the modifications of the counterterms introduced in (9.3) cancel the singularities of (9.2). In QCD, we have six values of Z; to fix them it will be sufficient to fix *six* independent Green's functions. That the result is independent of our choice of the six functions is a consequence of the Ward–Slavnov–Taylor identities among Green's functions, and this is actually a highly nontrivial part of the renormalization program. For the moment, we shall make a specific choice. We will also work in momentum space and begin with the quark propagator

$$S_{R\xi}(p) = i\{\not p - m + \Sigma(p)\}^{-1}, \qquad \Sigma(p) = (\not p - m)A(p^2) + mB(p^2). \tag{9.4a}$$

Let us choose a spacelike momentum,[20] $\bar p$, $\bar p^2 < 0$. Then we may specify the values of

$$A_{\xi R}(\bar p^2), B_R(\bar p^2). \tag{9.4b}$$

The first will fix Z_F, the second a combination of Z_F, Z_m, Z_λ. Next we turn to the gluon propagator,

$$D^{\mu\nu}_{R\xi}(q) = (-q^2 g^{\mu\nu} + q^\mu q^\nu)D_{R\text{tr}}(q) + g^{\mu\nu}D_{RL}(q). \tag{9.5a}$$

which, also choosing $q = \bar p$ for simplicity and fixing

$$D_{R\text{tr}}(\bar p), \qquad D_{RL}(\bar p), \tag{9.5b}$$

allows us to obtain Z_B and a combination of Z_B, Z_λ. The ghost propagator[21]

$$G_R(p) = \int d^4x\, e^{-ip\cdot x}\langle T\omega(x)\bar\omega(0)\rangle_0, \tag{9.6a}$$

when chosen at $p = \bar p$,

$$G_R(\bar p), \tag{9.6b}$$

yields Z_ω. The missing condition that will allow us to fix Z_g may be taken to be any vertex: $\bar q qB$, BBB, $BBBB$ or $\bar\omega\omega B$. Here we will select the first. If we define the "amputated" vertex V by

$$\int d^4x\, d^4y\, e^{-ip_1\cdot x}e^{ip_2\cdot y}\langle q^k_\beta(y)B^a_\mu(0)\bar q^j_\alpha(x)\rangle_0$$

$$= \sum D^{ab}_{\mu\nu}(p_2 - p_1)S^{ki}_{\beta\alpha'}(p_2)V^{il;b,\nu}_{R\xi,\alpha'\beta'}(p_1, p_2)S^{lj}_{\beta'\alpha}(p_1), \tag{9.7a}$$

$$V^{il,b,\nu}_{R\xi;\alpha'\beta'} = it^b_{il}\gamma^\nu_{\alpha'\beta'} + \cdots,$$

then we can specify, with $\bar p^2 = -\mu^2$, $\mu^2 > 0$:

$$V_{R\xi}\big|_{p_1^2 = p_2^2 = (p_1 - p_2)^2 = -\mu^2}. \tag{9.7b}$$

The implementation of the renormalization program is greatly facilitated by the fact, already noted, that $\mathscr{L}^\xi_R$ may be obtained from $\mathscr{L}^\xi_{uD}$ by simply effecting the replacements of (8.9). To calculate any Green's function, we begin by writing it more explicitly, in momentum space and after so-called "*amputation*", as

$$\Gamma(p_1, \ldots, p_{N-1}; m, g, \lambda),$$

[20]This is to avoid the discontinuities of the Green's functions for timelike p, which in our case are easily seen to occur for $p^2 \geqslant m^2$.

[21]We will henceforth drop the index u from the fields, except when necessary to avoid confusion.

where

$$\Gamma(p_1, \ldots, p_{N-1}; m, g, \lambda)\delta(\sum p)$$
$$= K_1(p_1) \ldots K_N(p_N) \int d^4x_1 \ldots d^4x_N \, e^{i\sum x_k \cdot p_k}$$
$$\times \langle T\Phi_1(x_1) \ldots \Phi_N(x_N)\rangle_0 ; \tag{9.8}$$

here the K_k are the appropriate inverse propagators, $iK(p) = S_R^{-1}(p)$ for fermion fields, $iK(p) = D_R^{-1}(p)$ for gluons, etc.[22] Next, we calculate

$$\Gamma_{uD}(p_1, \ldots, p_{N-1}; m, g, \lambda),$$

by using $\mathscr{L}^{\xi}_{uD,\text{int}}$ (cf. Equation [9.2]). Then, Γ_R is obtained from Γ_{uD} as

$$\Gamma_R(p_1, \ldots, p_{N-1}; m, g, \lambda)$$
$$= Z_{\Phi_1}^{-1/2} \ldots Z_{\Phi_N}^{-1/2}\Gamma_{uD}(p_1, \ldots, p_{N-1}; Z_m m, Z_g g, Z_\lambda \lambda). \tag{9.9}$$

This equation takes on a more transparent appearance if we define the bare couplings[23]:

$$m_{uqD} = Z_{mq} m_q , \qquad \lambda_{uD} = Z_\lambda \lambda, \qquad g_{uD} = Z_g g, \tag{9.10}$$

for then (9.9) reads

$$\Gamma_R(p_1, \ldots, p_{N-1}; m, g, \lambda)$$
$$= Z_{\Phi_1}^{1/2} \ldots Z_{\Phi_N}^{1/2}\Gamma_{uD}(p_1, \ldots, p_{N-1}; m_{uD}, g_{uD}, \lambda_{uD}). \tag{9.11}$$

To see how this works, let us consider the quark propagator. According to the general discussion, and with $\alpha_g \equiv g^2/4\pi$,

$$S_R(p; g_R, m_R, \lambda_R) = Z_F^{1/2} Z_F^{1/2} S_{uD}(p; Z_g g, Z_m m, Z_\lambda \lambda).$$

We shall work to second order only. Hence, we may replace Z_g, Z_λ by unity, since the corrections will be of higher order in α_g. With this, and using the expression we calculated in (7.4), (7.5),

$$S_R(p; g, m, \alpha) = Z_F^{-1} \frac{i}{(\not p - Z_m m)}$$
$$= iZ_F^{-1} \frac{1 - C_F g^2 A_{D\xi}(p^2)}{\not p - Z_m m\{1 - C_F g^2 B_{D\xi}(p^2)\}} .$$

As stated, to determine the Z, we have to specify S_R at a given $p = \bar{p}$. We will do so by requiring that at this point, S_R equal the free propagator:

$$S_R(\bar{p}; g, m, \alpha) = \frac{i}{\bar{\not p} - m} . \tag{9.12}$$

[22] The amputation simply removes the poles associated with the external legs of a Feynman diagram. Since S_R, D_R are *renormalized*, Γ will contain a factor $Z_\Phi^{-1/2}$ for each field *and* a factor Z_Φ for each K_Φ, hence, an effective factor $Z_\Phi^{1/2}$ for each field Φ.

[23] It is at times convenient to think of masses and gauge parameter as coupling constants.

Thus we find, with $\bar{p}^2 = -\mu^2$:

$$Z_F \equiv Z_{FD}^{\xi}(\mu^2, m^2)$$

$$= 1 - C_F \frac{\alpha_g}{4\pi} \Bigg\{ (1-\xi)N_\epsilon - 1 - \int_0 dx [2(1-x) - \xi]$$

$$\times \log \frac{xm^2 + x(1-x)\mu^2}{\nu_0^2} + \xi(\mu^2 + m^2) \int_0 dx \frac{x}{m^2 + \mu^2 x} \Bigg\},$$

$$(9.13)$$

$$Z_m \equiv Z_m(\mu^2, m^2)$$

$$= 1 - C_F \frac{\alpha_g}{4\pi} \Bigg\{ 3N_\epsilon - 1 - 2\int_0 dx (1+x) \log \frac{xm^2 + x(1-x)\mu^2}{\nu_0^2}$$

$$+ \xi(\mu^2 + m^2) \int_0 dx \frac{x}{m^2 + \mu^2 x} \Bigg\}. \qquad (9.14)$$

An important fact that should be noted is that while the divergent part of Z_F depends on the gauge, that of Z_m is gauge-independent, although, in this scheme, the finite parts of Z_m are still gauge-dependent. The gauge dependence of Z_F means that one may find gauges in which it is *finite*. From (9.13), it is clear that, to second order, this will be the case for the Landau gauge,[24] $\xi = 1$.

In QED, there is a natural renormalization scheme: we take electrons and photons on their mass shells (i.e., choose $\bar{p}^2 = m^2$ for S and $\bar{q}^2 = 0$ in D). Because it is likely that confinement is a feature of QCD, no such natural scheme exists in our case. Therefore, we are free to choose renormalization schemes that simplify the calculation as much as possible. These are the minimal subtraction schemes, which will be discussed below.

ii The Minimal Subtraction Scheme

It was noted by 't Hooft (1973) that the simplest way to eliminate divergences in Green's functions is to chop off the poles $1/\epsilon$ that appear in dimensional regularization (*minimal subtraction*). Subsequently, Bardeen, *et*

[24] The Landau gauge is useful in zero-mass calculations; here, not only Z_F is finite, but $\Sigma^{(2)}$ actually vanishes.

al. (1978) realized that these poles always appear in the combination

$$N_\epsilon = \frac{2}{\epsilon} - \gamma_E + \log 4\pi. \tag{9.15}$$

Therefore, if one simply cancels the $2/\epsilon$, one introduces, somewhat artificially, the transcendentals γ_E, $\log 4\pi$. These, it will be recalled, appear due to the specific way we continued to the arbitrary dimension $D = 4 - \epsilon$, which yielded the terms

$$(4\pi)^{\epsilon/2}\Gamma(\epsilon/2) = N_\epsilon + O(\epsilon).$$

It seems natural to also eliminate these by introducing the *modified minimal subtraction scheme* ($\overline{MS}$, henceforth) in which the entire N_ϵ is subtracted.[25] In this scheme, we find

$$\bar{Z}_F = 1 - C_F \frac{\alpha_g}{4\pi}(1 - \xi)N_\epsilon, \tag{9.16}$$

$$\bar{Z}_m = 1 - C_F \frac{3\alpha_g}{4\pi} N_\epsilon. \tag{9.17}$$

We will mainly work with the $\overline{MS}$ scheme and shall therefore drop the bar over the Zs that characterizes their value in this scheme (In this scheme, Z_m is fully gauge independent. This has been checked to two loops by Tarrach (1981) and is likely true to all orders due to the gauge independence of the mass terms $m\bar{q}q$ themselves.) From (9.16) and (9.17), we see that, writing $C = cN_\epsilon$,

$$c_F^{(1)} = -C_F(1 - \xi) \tag{9.18}$$

$$c_m^{(1)} = -3C_F. \tag{9.19}$$

The second-order coefficients have also been calculated.[26]

Next we will calculate, in the $\overline{MS}$ scheme, the other renormalization constants. We begin with the gluon propagator. Its transverse part may be written as:

$$\begin{aligned} D^{\mu\nu}_{\text{utr};ab}(q, g_u, m_u, \lambda_u) = {} & i\frac{-g^{\mu\nu} + q^\mu q^\nu/q^2}{q^2}\delta_{ab} \\ & + \sum \frac{-g^{\mu\mu'} + q^\mu q^{\mu'}/q^2}{q^2}\delta_{aa'}\Pi^{a'b'}_{\mu'\nu'}\delta_{b'b} \\ & \times i\frac{-g^{\nu'\nu} + q^{\nu'}q^\nu/q^2}{q^2} + \cdots, \end{aligned} \tag{9.20}$$

where, to second order, the renormalization of the gauge parameter, mass, or g do not intervene. The part of Π originating from gluons and ghosts

[25] The $\overline{MS}$ scheme may be reduced to the MS scheme by replacing the continuation $d^D\hat{k} = \nu_0^{4-D} d^Dk/(2\pi)^D$ by $d^D\hat{k} = \{\nu_0^{4-D}/(2\pi)^D\}/\{(4\pi)^{(4-D)/2}\Gamma(3 - D/2)\}$.

[26] The calculations were performed by Nanopoulos and Ross (1979); Tarrach (1981) checked and corrected a trivial mistake in the original calculations.

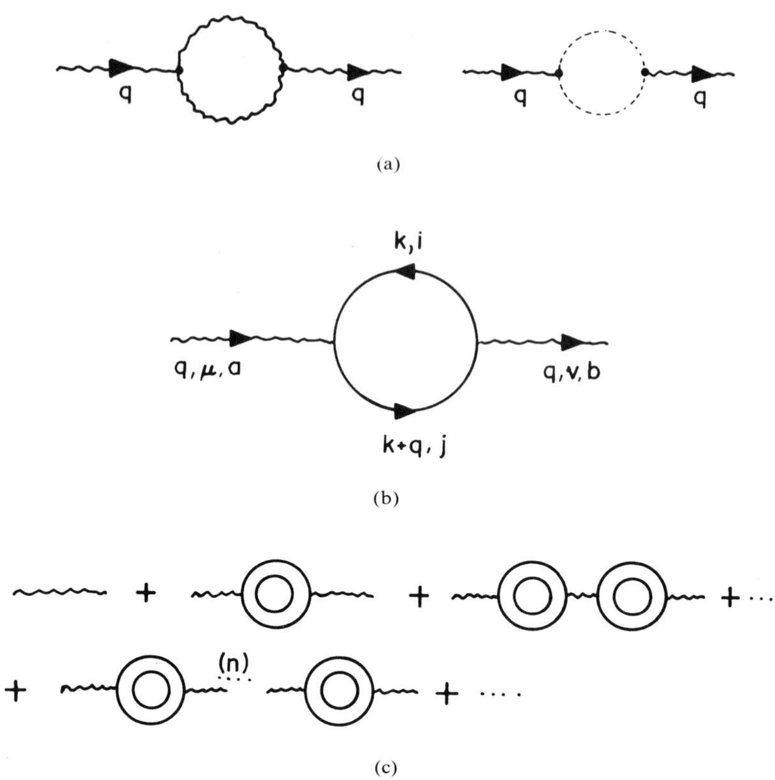

Figure 6. The gluon propagator.

(Figure 6a) was calculated earlier in Equation (5.9).[27] The part arising from a quark loop (Figure 6b) is, for each flavor f of quark

$$\Pi^{\mu\nu}_{f\,\text{quark};ab} = -ig^2 \sum_{ij} t^a_{ij} t^b_{ji} \int \frac{d^D k}{(2\pi)^D} \nu_0^{4-D} \frac{\operatorname{Tr}(\not k + m_f)\gamma^\mu(\not k + \not q + m_f)\gamma^\nu}{(k^2 - m_f^2)\left[(k+q)^2 - m_f^2\right]}.$$

The calculation may be carried out using standard techniques. The result is exactly the same as that for the photon vacuum polarization, apart from the factor $\operatorname{Tr} t^a t^b$. The result is, with n_f the total number of quark flavors:

$$\Pi^{\mu\nu}_{\text{all quarks};ab} = \delta_{ab} \frac{-2T_F g^2}{16\pi^2}(-g^{\mu\nu}q^2 + q^\mu q^\nu) \times \left\{ \frac{2}{3} N_\epsilon n_f - 4\int_0^1 dx \cdot x(1-x) \sum_{f=1}^{n_f} \log \frac{m_f^2 - x(1-x)q^2}{\nu_0^2} \right\}. \tag{9.21}$$

[27] Equation (5.9) was calculated without taking into account the factor ν_0^{4-D}. If we include it, the only difference is that $\log(-q^2)$ is replaced by $\log(-q^2/\nu_0^2)$.

To second order, we can sum all the graphs of Figure 6c, where the ring means any gluon, ghost, or quark loop. Noting that

$$\Pi^{\mu'\nu'}_{a'b'} = -\delta_{a'b'}(-g^{\mu'\nu'}q^2 + q^{\mu'}q^{\nu'})\Pi, \tag{9.22a}$$

we obtain the analog of Equation (7.5),

$$D^{\mu\nu}_{u\,\mathrm{tr};ab}(q) = i\delta_{ab}\frac{-g^{\mu\nu} + q^{\mu}q^{\nu}/q^2}{(1-\Pi)q^2}\,. \tag{9.22b}$$

Let us introduce the notation

$$f \overset{\mathrm{div}}{=} g,$$

which means that the N_ϵ coefficients of f and g are equal. The renormalized D is

$$D^{\mu\nu}_{R\,\mathrm{tr};ab} = Z_B^{-1}D^{\mu\nu}_{u\,\mathrm{tr};ab}\,;$$

using Equations (5.9), (9.20) and (9.21), we see that

$$1 - \Pi \overset{\mathrm{div}}{=} 1 + \frac{g^2}{32\pi^2}\left\{\frac{10C_A}{3} - \frac{8T_F n_f}{3}\right\}N_\epsilon\,,$$

and, therefore, in the $\overline{MS}$ scheme and in the Fermi–Feynman gauge,

$$Z_B = 1 + \frac{\alpha_g}{8\pi}\left\{\frac{10C_A}{3} - \frac{8T_F n_f}{3}\right\}N_\epsilon\,. \tag{9.23}$$

In an arbitrary gauge, Z_B has been calculated by Gross and Wilczek (1973a) and Politzer (1973). The corresponding $c^{(1)}_{B\xi}$ is

$$c^{(1)}_{B\xi} = \frac{1}{2}\left\{10 + 3\xi - \frac{4n_f}{3}\right\}. \tag{9.24}$$

We shall not calculate Z_λ explicitly, but shall just give the result[28]

$$c^{(1)}_{\lambda\xi} = c^{(1)}_{B\xi}. \tag{9.25}$$

It should be noted that, for the Landau gauge, ξ does not become renormalized at the one-loop level. Actually, the Slavnov–Taylor identities indicate that this is true to all orders, as we showed in Section 6.

To complete this section, let us calculate Z_g. This we will do from the $\bar{q}qB$ vertex. With the choice of momenta of Figure 7, we may write the vertex to second order as (cf., [9.7])

$$V^{\mu}_{uij,a} = ig\gamma^{\mu}t^a_{ij} + i\Gamma^{(2)\mu}_{uij,a}\,, \tag{9.26a}$$

[28] See, for example, de Rafael (1979) and work quoted there. The Slavnov–Taylor identity, which we proved in Section 6, implies that to all orders $Z_B = Z_\lambda$.

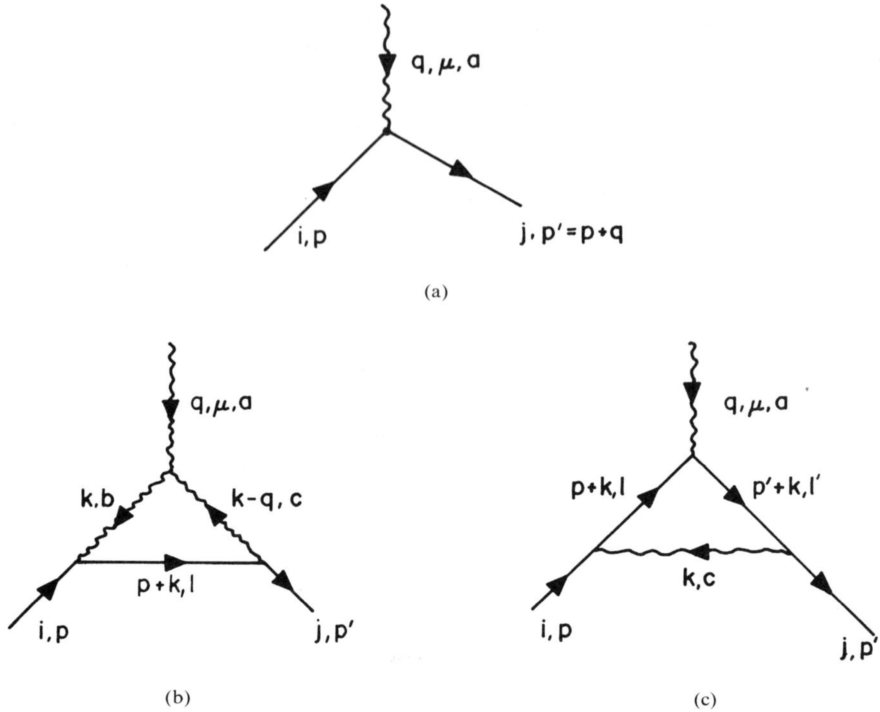

Figure 7. The quark-gluon vertex.

with

$$\Gamma^{(2)\mu}_{uij,a}(p, p') = \{\Gamma^{(b)} + \Gamma^{(c)}\}^{\mu}_{uij,a}. \tag{9.26b}$$

The indices in $\Gamma^{(b)}$ and $\Gamma^{(c)}$ refer to the contribution of Figures 7*b* and 7*c*. Diagram *a* gives the $ig\gamma t$ term in (9.26*a*). As should be obvious from the previous examples, the quark masses play no role in Z (except, of course, in Z_m), so we will simplify the calculation by taking $m = 0$. Also, only the divergent parts of the Γ shall be taken into account. Then, in the Fermi–Feynman gauge

$$i\Gamma^{(b)\mu}_{uij,a} \overset{\text{div}}{=} ig \int d^D\hat{k}\, \frac{\gamma^{\beta}\left[(2k-q)^{\mu} g_{\alpha\beta} - (k+q)_{\beta} g^{\mu}_{\alpha} + (2q-k)_{\alpha} g^{\mu}_{\beta}\right](\not{p} + \not{k})\gamma^{\alpha}}{\left[(p+k)^2 + i0\right]\left[(k-q)^2 + i0\right](k^2 + i0)} C^a_{ij}$$

$$\overset{\text{div}}{=} igC^a_{ij}\gamma^{\mu} \lim_{\eta \to 0} \int d^D\hat{k}\, \frac{2(2-D)/D - 2}{(k^2 - i\eta)^2} \overset{\text{div}}{=} g^3 \frac{3N_{\epsilon} C^a_{ij}}{16\pi^2}. \tag{9.27a}$$

We have used the notations

$$d^D\hat{k} \equiv \frac{d^Dk}{(2\pi)^D} \nu_0^{4-D},$$

and

$$C_{ij}^{a} \equiv -g^2 \sum t_{jl}^{b} t_{li}^{c} f^{abc} = \frac{1}{2} g^2 [t^b, t^c]_{ji} f^{bca}$$
$$= g^2 \frac{i}{2} C_A t_{ij}^{a} = \frac{3}{2} i t_{ij}^{a} g^2 :$$

we have profited from the antisymmetry of the f to replace $t^b t^c \to \frac{1}{2}[t^b, t^c]$. Likewise,

$$i\Gamma_{uij,a}^{(c)\mu} \overset{\text{div}}{=} -i^2 g \int d^D \hat{k} \frac{\gamma_\beta (\not{p} + \not{k}) \gamma^\mu (\not{p}' + \not{k}) \gamma_\alpha g^{\alpha\beta}}{[(p+k)^2 + i0][(p'+k)^2 + i0](k^2 + i0)} C_{ij}^{\prime a}$$
$$\overset{\text{div}}{=} ig \frac{N_\epsilon}{16\pi^2} \gamma^\mu C_{ij}^{\prime a} . \tag{9.27b}$$

Here,

$$C_{ij}^{\prime a} = g^2 \sum_c (t^c t^a t^c)_{ij} = g^2 \sum_c ([t^c, t^a] t^c)_{ij} + g^2 \Big(t^a \sum_c t^c t^c \Big)_{ij}$$
$$= g^2 t_{ij}^{a} \Big\{ -\frac{1}{2} C_A + C_F \Big\}, \tag{9.27c}$$

and we have repeatedly used the formulas of Appendix C. Thus,

$$\Gamma_{uij,a}^{(2)\mu} \overset{\text{div}}{=} \frac{N_\epsilon g^3}{16\pi^2} \{ C_A + C_F \} i t_{ij}^{a} \gamma^\mu \tag{9.28}$$

Now, the renormalization of the vertex involves Z_g *and* Z_F, Z_B,

$$V_{Rij,a}^{\mu} = Z_F^{-1} Z_B^{-1/2} Z_g V_{uij,a}^{\mu} ; \tag{9.29}$$

using the Z_F, Z_B values that have already been determined, and the expression just calculated for the divergent part of $\Gamma_u^{(2)}$, we find

$$Z_g = 1 - \frac{\alpha_g}{4\pi} \Big\{ \frac{11 C_A}{6} - \frac{2}{3} T_F n_f \Big\} N_\epsilon . \tag{9.30}$$

Therefore,

$$c_g^{(1)} = -\Big\{ \frac{11}{2} - \frac{n_f}{3} \Big\}.$$

It is interesting to watch the cancellation of the C_F terms. This is necessary because we could have calculated Z_g for a pure Yang–Mills theory (see below). We see that the cancellation is due to the gauge structure; cf., Equation (9.27c). To the next order, β has been calculated by Caswell (1974) and Jones (1974). The original calculation of the $c_g^{(1)}$ was performed by Gross and Wilczek (1973a) and Politzer (1973), who, instead of the $\bar{q}qB$ vertex, considered the three-gluon vertex. This involves the diagrams of Figure 8. One can also use the $\bar{\omega}\omega B$ vertex. Of course, the result is the same, which is just a reflection of gauge invariance.

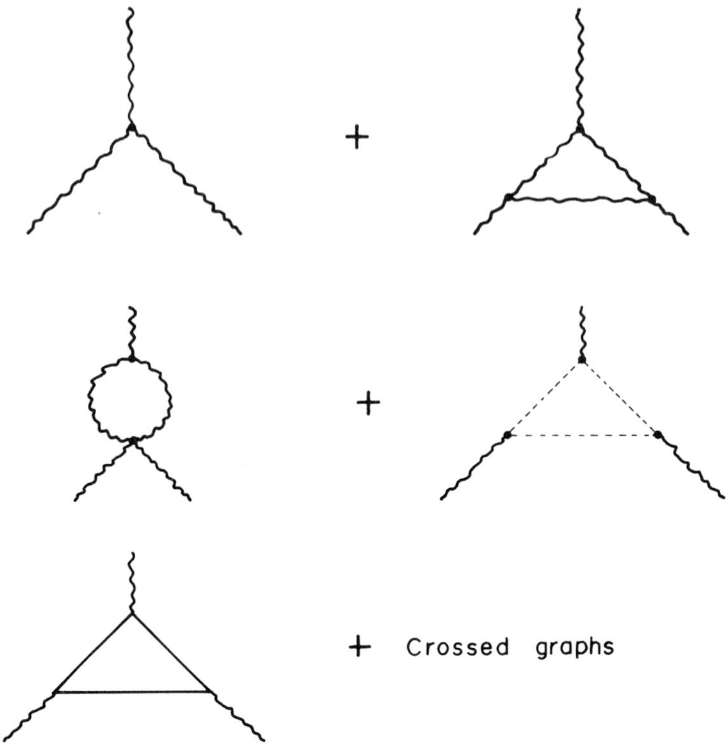

Figure 8. The three-gluon vertex.

A result that is worth noting is that, to all orders, the $c_g^{(n)}$ are gauge invariant in the $\overline{MS}$ scheme [Caswell and Wilczek (1974)].

10 Global Symmetries of the QCD Lagrangian: Conserved Currents

In this section we will discuss the *global* symmetries of the QCD Lagrangian. Since its form is unaltered by renormalization, we can neglect the distinction between the bare and renormalized $\mathscr{L}$.

Clearly, $\mathscr{L}$ is invariant under Poincaré transformations, $x \to \Lambda x + a$. The currents corresponding to Lorentz transformations Λ (generators of total spin) are not of great interest to us. Space-time translations generate the *energy-momentum tensor*. Its form is fixed by Noether's theorem, which yields

$$\Theta^{\mu\nu} = \sum_i \frac{\partial \mathscr{L}}{\partial(\partial_\mu \Phi_i)} \partial^\nu \Phi_i - g^{\mu\nu} \mathscr{L}, \tag{10.1}$$

and the sum over i runs over all the fields in the QCD Lagrangian. These currents are conserved,

$$\partial_\mu \Theta^{\mu\nu} = 0,$$

and, of course, the corresponding conserved "charges" are the components of four-momentum

$$P^\mu = \int d\vec{x}\,\Theta^{0\mu}(x).$$

The explicit expression for $\Theta^{\mu\nu}$ in QCD is[29]

$$\Theta^{\mu\nu} = i\sum_q \bar{q}\gamma^\mu D^\nu q - g^{\mu\nu} i \sum_q \bar{q}\not{D} q + g^{\mu\nu}\sum_q m_q \bar{q}q$$
$$- g_{\alpha\beta} G^{\mu\alpha} G^{\nu\beta} + \tfrac{1}{4} g^{\mu\nu} G^2 + \text{gauge fixing} + \text{ghost terms.} \quad (10.2)$$

Next, there are the currents and charges associated with the color rotations. We leave it to the reader to write them explicitly, since they are particular cases of color gauge transformations (with constant gauge parameters) and pass over to a different set of currents *not* associated with interactions of quarks and gluons among themselves.

If all the quark masses vanished, we would have invariance $\mathscr{L}$ under the transformations

$$q_f \to \sum_{f'=1}^{n_f} U_{ff'} q_{f'}, \qquad q_f \to \sum_{f'=1}^{n_f} V_{ff'} \gamma_5 q_{f'}, \quad (10.3)$$

provided U and V are $n_f \times n_f$ unitary matrices. This implies that the currents

$$V^\mu_{qq'}(x) = \bar{q}(x)\gamma^\mu q'(x), \qquad A^\mu_{qq'}(x) = \bar{q}(x)\gamma^\mu\gamma_5 q'(x) \quad (10.4)$$

are each separately conserved. When mass terms are taken into account, only the diagonal V^μ_{qq} are conserved; the others are what are called *quasi-conserved* currents, i.e., their divergences are proportional to masses. These divergences are easily calculated: since the transformations of (10.3) commute with the interaction part of $\mathscr{L}$, we may evaluate them with free fields, in which case use of the Dirac equation $i\not{\partial} q = m_{qq} q$ gives

$$\partial_\mu V^\mu_{qq'} = i(m_q - m_{q'})\bar{q}q'; \qquad \partial_\mu A^\mu_{qq'} = i(m_q + m_{q'})\bar{q}\gamma_5 q'. \quad (10.5)$$

In fact, there is a subtle point concerning the divergences of axial currents. Equation (10.5) is correct as it stands for the nondiagonal terms, $q \neq q'$; for

[29] In the quantum version, we understand that products are replaced by Wick ordered products. For a discussion of the arbitrariness in the definition of the energy momentum tensor, cf., Callan, Coleman, and Jackiw (1970) and Collins, Duncan, and Joglekar (1977).

Θ is not unique, and, as a matter of fact, from (10.1) above, we do not obtain a gauge-invariant tensor. The expression (10.2) is obtained by replacing derivatives with covariant derivatives. Alternatively, we may reformulate (10.1) by performing gauge transformations simultaneously to the translation $x^\mu \to x^\mu + \epsilon^\mu$. For example, writing $B^\mu_a \to B^\mu_a + (\epsilon_\alpha \partial^\alpha B^\mu_a \equiv D^\mu \epsilon_\alpha B^\alpha_a + \epsilon_\alpha G^{\alpha\mu}_a)$, the first term may be absorbed by a gauge transformation, so we get $B^\mu_a \to B^\mu_a + \epsilon_\alpha G^{\alpha\mu}_a$.

$q = q'$, however, one has instead

$$\partial_\mu A^\mu_{qq}(x) = i(m_q + m_q)\bar{q}(x)\gamma_5 q(x) + \frac{T_F g^2}{16\pi^2}\epsilon^{\mu\nu\rho\sigma}G_{\mu\nu}(x)G_{\rho\sigma}(x). \quad (10.6)$$

This is the so-called Adler–Bell–Jackiw triangle anomaly which will be discussed later in Sections 33, 37, and 38.

The *equal time commutation relations* of the V, A, and the fields can also be easily calculated. Using (10.3), one finds:

$$\begin{aligned}\delta(x^0 - y^0)\left[V^0_{qq'}(x), q''(y)\right] &= -\delta(x-y)\delta_{qq''}q'(x),\\ \delta(x^0 - y^0)\left[A^0_{qq'}(x), q''(y)\right] &= -\delta(x-y)\delta_{qq''}\gamma_5 q'^{(x)}, \quad \text{etc.,}\end{aligned} \quad (10.7)$$

for free fields. The V and A commute with gluon and ghost fields. The equal time commutations of the V, A among themselves, again for free fields, are best described by introducing the Gell-Mann λ^α matrices in flavor space (cf., Appendix C). For three flavors ($f = 1, 2, 3$), if we let

$$V^\mu_\alpha(x) = \sum_{ff'}\bar{q}_f(x)\lambda^\alpha_{ff'}\gamma^\mu q_{f'}(x), \qquad A^\mu_\alpha(x) = \sum_{ff'}\bar{q}(x)\lambda^\alpha_{ff'}\gamma^\mu\gamma_5 q_{f'}(x), \quad (10.8)$$

we find

$$\begin{aligned}\delta(x^0 - y^0)\left[V^0_\alpha(x), V^\mu_\beta(y)\right] &= 2i\delta(x-y)\Sigma f_{\alpha\beta\delta}V^\mu_\delta(x),\\ \delta(x^0 - y^0)\left[V^0_\alpha(x), A^\mu_\beta(y)\right] &= 2i\delta(x-y)\Sigma f_{\alpha\beta\delta}A^\mu_\delta(x),\\ \delta(x^0 - y^0)\left[A^0_\alpha(x), A^\mu_\beta(y)\right] &= 2i\delta(x-y)\Sigma f_{\alpha\beta\delta}V^\mu_\delta(x), \quad \text{etc.}\end{aligned} \quad (10.9)$$

Equations (10.7) and (10.9) have been derived only for free fields. However, they involve short distances; therefore, in QCD and because of asymptotic freedom, they will hold as they stand even in the presence of interactions.

The equal time commutation relations of conserved or quasi-conserved currents with the Hamiltonian (or Lagrangian) may also be easily obtained. If J^μ is conserved, then the corresponding charge

$$Q_J(t) = \int d\vec{x}J^0(t, \vec{x}), \qquad t = x^0$$

commutes with $\mathscr{H}$:

$$\left[Q_J(t), \mathscr{H}(t, \vec{y})\right] = 0.$$

Here $\mathscr{H}$ is the *Hamiltonian* (density), $\mathscr{H} = \Theta^{00}$. If J is quasi-conserved, let $\mathscr{H}'$ be the mass term in $\mathscr{H}$,

$$\mathscr{H}' = \sum_q m_q\bar{q}q.$$

Then,

$$\left[Q_J(t), \mathscr{H}'(t, \vec{y})\right] = i\partial_\mu J^\mu(t, \vec{y}).$$

Of course, Q_J still commutes with the rest of $\mathscr{H}$.

11 The Renormalization Group

Consider, for example, the renormalized quark propagator. In the μ-scheme and the Fermi–Feynman gauge,

$$S_R^{(\mu)}(p;\, g, m) = i\frac{1-(4/3)g^2A_R^{(\mu)}(p^2)}{\not p - m\{1-(4/3)g^2B_R^{(\mu)}(p^2)\}}, \tag{11.1a}$$

where

$$\begin{aligned} A_R^{(\mu)}(p^2) &= \frac{2}{16\pi^2}\int_0^1 dx(1-x)\frac{xm^2+x(1-x)\mu^2}{xm^2-x(1-x)p^2}, \\ B_R^{(\mu)}(p^2) &= \frac{-2}{16\pi^2}\int_0^1 dx(1+x)\log\frac{xm^2+x(1-x)\mu^2}{xm^2-x(1-x)p^2}. \end{aligned} \tag{11.1b}$$

In the $\overline{MS}$ scheme,

$$S_R^{(\nu)}(p;\, g, m) = i\frac{1-(4/3)g^2\overline{A}_R(p^2,\nu)}{\not p - m\{1-(4/3)g^2\overline{B}_R(p^2,\nu)\}}, \tag{11.2a}$$

and

$$\begin{aligned} \overline{A}_R &= \frac{1}{16\pi^2}\left\{-1-2\int_0^1 dx(1-x)\log\frac{xm^2-x(1-x)p^2}{\nu_0^2}\right\}; \\ \overline{B}_R &= \frac{1}{16\pi^2}\left\{1+2\int_0^1 dx(1+x)\log\frac{xm^2-x(1-x)p^2}{\nu_0^2}\right\}. \end{aligned} \tag{11.2b}$$

We see that the renormalization introduces an arbitrary mass parameter in the Green's functions: it is the renormalization point μ^2 in the μ-scheme or the mass scale ν^2 in the $\overline{MS}$ scheme.

Let us begin the discussion with the μ scheme. Suppose we change the renormalization point to μ'. If we only used Equations (11.1b) with μ replaced by μ', we would get a different value $S_R^{(\mu')}$ for the quark propagator. However, we want to have a unique theory; therefore, we must compensate for this change by allowing for a dependence on μ. So we should rewrite (11.1a) as

$$S_R^{(\mu)}(p;\, g, m(\mu)) = i\frac{1-(4/3)g^2A_R^{(\mu)}(p^2)}{\not p - m(\mu)\{1-(4/3)g^2B_R^{(\mu)}(p^2)\}}. \tag{11.3}$$

That such an $m(\mu)$ exists is clear from the expression of S_R in terms of the unrenormalized propagator:

$$\begin{aligned} S_R^{(\mu)}(p;\, g, m(\mu)) &= Z_F^{-1}(\mu)S_{uD}(p;\, g, m_{uD}); \\ m_{uD} &= Z_m(\mu)m(\mu). \end{aligned} \tag{11.4}$$

All we have to do is appropriately choose the dependence on μ of Z_F, Z_m. Otherwise stated, we can, once we know $S_R^{(\mu)}(p;\, g, m(\mu))$, calculate $S_R^{(\mu)}(p=\mu';\, g, m(\mu))$. Now, we fix $S_R^{(\mu')}$ by

$$S_R^{(\mu')}(\mu', g, m(\mu')) = \frac{i}{\not p - m(\mu')},$$

so we may fix $m(\mu')$, $Z_F(\mu')/Z(\mu)$ by requiring equality. This gives, for example:

$$m(\mu') = m(\mu)\left\{1 - \frac{2}{3}\frac{\alpha_g}{\pi}\int_0^1 dx(1+x)\log\frac{xm + x(1-x)\mu'^2}{xm + x(1-x)\mu^2}\right\}.$$

In the $\overline{MS}$ scheme, the argument is simpler, but also subtler.[30] We regularized in such a way that an arbitrary mass scale, ν_0, was introduced. If we want to obtain Green's functions independent of ν_0, we can make them so by cancelling not only the term $(4\pi)^{\epsilon/2}\Gamma(\epsilon/2)$, but the entire $(4\pi)^{\epsilon/2}\Gamma(\epsilon/2)\nu_0^{\epsilon}$. This can only be achieved at the expense of introducing a new mass scale, ν, so that $Z \rightarrow Z(\nu) = (\nu_0/\nu)^{\epsilon}Z$, which thus cancels $N_\epsilon^{(\nu)} = 2/\epsilon - \gamma_E + \log 4\pi + \log \nu_0$. The renormalized Green's functions will depend on ν, but no longer on ν_0. Now, we want a change of ν to have no physical effects. Then it will be sufficient to allow for a ν dependence of g, m, ξ (besides the Z). For amputated Γ,

$$\begin{aligned}&\Gamma_R(p_1,\ldots,p_{N-1};\, g(\nu), m(\nu), \xi(\nu); \nu)\\ &\quad = Z_{\Phi_1}^{1/2}(\nu)\ldots Z_{\Phi_N}^{1/2}(\nu)\Gamma_{uD}(p_1,\ldots,p_{N-1};\, g_{uD}, m_{uD}, \xi_{uD});\end{aligned} \tag{11.5}$$

$$\begin{aligned} g_{uD} &= Z_g(\nu)g(\nu), \qquad m_{uD} = Z_m(\nu)m(\nu),\\ \lambda_{uD} &= Z_\lambda(\nu)\lambda(\nu), \qquad \xi = 1 - \lambda^{-1}.\end{aligned} \tag{11.6}$$

It is very easy to see what ν dependence we need. We recall that the ν_0 entered in the combination

$$d^D\hat{k} = \frac{d^Dk}{(2\pi)^D}\nu_0^{4-D},$$

so the only dependence on ν_0 is in the divergent part:

$$\Gamma(2/\epsilon)(4\pi)^{\epsilon/2}(\nu_0^2)^{\epsilon/2}.$$

Therefore, all $Z(\nu)$ are of the form

$$Z_j(\nu) = 1 + C_j^{(1)}(\nu)\frac{g^2}{16\pi^2} + \cdots, \tag{11.7a}$$

$$C_j^{(1)}(\nu) = c_j^{(1)}\left\{\frac{2}{\epsilon} - \gamma_E + \log 4\pi + \log\frac{\nu_0^2}{\nu^2}\right\}: \tag{11.7b}$$

[30] Our version of the $\overline{MS}$ scheme is slightly different (although equivalent) to the standard one.

the coefficients of the $\log \nu^2$ terms are the same $c_j^{(1)}$ we have already calculated up to a sign. It is easy to show that the same is true for the coefficient of $\log \mu^2$ in the μ scheme, to lowest order.

The set of transformations $\mu \to \mu'$ (or $\nu \to \nu'$) constitute the *renormalization group*,[31] first introduced by Stückelberg and Peterman (1953); see also Gell-Mann and Low (1954) and Bogoliubov and Shirkov (1959). The invariance of physical quantities under this group may be used (as will be done here) to study the asymptotic behavior of Green's functions. This is best done by using the Callan (1970) and Symanzik (1970) equation, which will be the subject of the next section.

12 The Callan–Symanzik Equation

The CS equation is more simply derived by noting that the Γ_u, g_u, m_u, ξ_u are ν-independent (in, say, the $\overline{MS}$ scheme). Therefore, from (11.5) and (11.6) we immediately derive

$$\begin{gathered} \frac{\nu d}{d\nu}\Gamma_{uD}(p_1, \ldots, p_{N-1}; g_{uD}, m_{uD}, \lambda_{uD}) = 0, \quad \text{i.e.,} \\ \left\{ \frac{\nu\partial}{\partial\nu} + g\beta\frac{\partial}{\partial g} + (1-\xi)\lambda\delta\frac{\partial}{\partial\lambda} + \sum_q m_q\gamma_{m,q}\frac{\partial}{\partial m_q} - \gamma_\Gamma \right\} \\ \times \Gamma_R(p_1, \ldots, p_{N-1}; g(\nu), m(\nu), \lambda(\nu); \nu) = 0; \end{gathered} \tag{12.1}$$

we have defined the universal functions β, γ_k, δ:

$$\begin{aligned} \nu\frac{d}{d\nu}g(\nu) &= g(\nu)\beta, \\ \nu\frac{d}{d\nu}m_q(\nu) &= m_q(\nu)\gamma_{m,q}, \\ \nu\frac{d}{d\nu}\lambda(\nu) &= \{1-\lambda(\nu)\}\delta, \end{aligned} \tag{12.2}$$

and, moreover,

$$Z_\Gamma^{-1} = Z_{\Phi_1}^{1/2} \ldots Z_{\Phi_N}^{1/2}, \qquad Z_\Gamma^{-1}\nu\frac{d}{d\nu}Z_\Gamma = \gamma_\Gamma. \tag{12.3}$$

The functions β, γ, and δ can be calculated by recalling Equation (9.10) and that the g_u, m_u, ξ_u are independent of ν:

$$\begin{gathered} \beta = -Z_g^{-1}(\nu)\nu\frac{d}{d\nu}Z_g(\nu), \qquad \gamma_{m,q} = -Z_m^{-1}(\nu)\nu\frac{d}{d\nu}Z_m(\nu), \\ \delta = -Z_\lambda(\nu)\nu\frac{d}{d\nu}Z_\lambda^{-1}(\nu). \end{gathered} \tag{12.4}$$

Equation (12.1) is not very useful as it stands because it contains the partial derivative $\nu\partial/\partial\nu$. However, we can transform it into a more useful

[31] Actually, a group structure is obtained only within a given renormalization scheme. If we include the transformations $T(R_1 \to R)$ that change from one scheme to another, then we obtain a fiber bundle.

form by using dimensional analysis. Suppose ρ_Γ is the dimension of Γ_R; then $\nu^{-\rho_\Gamma}\Gamma_R$ is dimensionless,[32] and so it can only depend on ratios of dimensional parameters. We scale the momenta,

$$\nu^{-\rho_\Gamma}\Gamma_R(\lambda p_1, \ldots, \lambda p_{N-1}; g, m, a^{-1}; \nu)$$
$$= F(\lambda p_1/\nu, \ldots, \lambda p_{N-1}/\nu; g, m/\nu, a^{-1});$$

we replaced the gauge parameter by $a = \lambda^{-1}$ to avoid confusion with the scale λ. Now, we can replace the $\nu\partial/\partial\nu$ by $-\lambda\partial/\partial\lambda$ to obtain the Callan–Symanzik equation:

$$\left\{-\frac{\partial}{\partial\log\lambda} + g\beta\frac{\partial}{\partial g} + (a^{-1})\delta\frac{\partial}{\partial a^{-1}} + \sum_q m_q(\gamma_{m,q} - 1)\frac{\partial}{\partial m_q} + \rho_\Gamma - \gamma_\Gamma\right\}$$
$$\times\Gamma_R(\lambda p_1, \ldots, \lambda p_{N-1}; g, m, \xi; \nu) = 0. \tag{12.5}$$

To solve this equation, we define effective or "running" parameters given implicitly by the equations

$$\frac{d\bar{g}(\lambda)}{d\log\lambda} = \bar{g}(\lambda)\,\beta(\bar{g}(\lambda)), \qquad \frac{d\bar{m}(\lambda)}{d\log\lambda} = \bar{m}(\lambda)\gamma_{m,q}, \qquad \frac{d\bar{a}(\lambda)^{-1}}{d\log\lambda} = \bar{a}^{-1}\delta, \tag{12.6a}$$

with the boundary conditions

$$\bar{g}|_{\lambda=1} = g(\nu), \qquad \bar{m}|_{\lambda=1} = m(\nu), \qquad \bar{a}|_{\lambda=1} = a(\nu); \tag{12.6b}$$

then,

$$\Gamma_R(\lambda p_1, \ldots, \lambda p_{N-1}; g(\nu), m(\nu), \xi(\nu); \nu)$$
$$= \lambda^{\rho_\Gamma}\Gamma_R\left(p_1, \ldots, p_{N-1}; \bar{g}(\lambda), \bar{m}(\lambda), \bar{a}(\lambda)^{-1}; \nu\right)$$
$$\times\exp\left\{-\int_0^{\log\lambda} d\log\lambda'\,\gamma_\Gamma\left(\bar{g}(\lambda'), \bar{m}(\lambda'), \bar{a}(\lambda')^{-1}\right)\right\}. \tag{12.7}$$

We see that when we scale the momenta by λ, Γ_R does not scale simply as λ^{ρ_Γ}: corrections to this have developed (the exp term in the right-hand side of (12.7)). For this reason, γ_Γ is usually called the *anomalous dimension* of Γ_R. In this respect the renormalization group can be interpreted as the realization of scale invariance in local quantum field theory.[33] This realization is nontrivial due to the infinite character of renormalization, which introduces an extraneous mass scale.

[32]The dimension of a field is easily deduced, noting that the action $\mathscr{A} = \int d^4x\,\mathscr{L}(x)$ must be dimensionless. Hence, $[q] = [M]^{3/2}$, $[\omega] = [M]^1$, $[B] = [M]^1$. The dimension of Γ is obtained from those of the fields that it embodies: for example, $\rho_S = -1$ ($3/2 + 3/2$ for the fields, and -4 from d^4x), for the fermion propagator S.

[33]The development of this point of view can be seen in the papers of Callan, Coleman, and Jackiw (1970) and, for QCD, Collins, Duncan, and Joglekar (1977).

There is one additional point that has to be made about (12.7). In principle, it is valid independent of perturbation theory; but in practice, perturbation theory is essential to obtain real results from it.

13 Renormalization of Composite Operators

Because we probe hadronic structure with external currents, both weak and electromagnetic, we must discuss not only Green's functions, but matrix elements of composite operators as well. These operators may be classified into two categories: those that are conserved, or partially conserved, and those that are not conserved.

A conserved operator is the electromagnetic current of any number of quark flavors: $J^{\mu}_{\text{em}} = \sum Q_q V^{\mu}_q$ with

$$V^{\mu}_q(x) = :\bar{q}(x)\gamma^{\mu}q(x): ;$$

it satisfies

$$\partial_{\mu} V^{\mu}_q(x) = 0 \tag{13.1a}$$

to all orders in perturbation theory. A partially conserved current is, for example, the weak current

$$A^{\mu}_{qq'}(x) = :\bar{q}(x)\gamma^{\mu}\gamma_5 q'(x): .$$

Using the equations of motion (3.6), we see that it verifies

$$\partial_{\mu} A^{\mu}_{qq'}(x) = i(m_q + m_{q'})J^5_{qq'}(x), \qquad J^5_{qq'}(x) = :\bar{q}(x)\gamma_5 q'(x): , \tag{13.1b}$$

so it will be conserved in the high-energy limit when quark masses can be neglected.

In general, the matrix elements of any composite operator are divergent. However, if we take into account the counterterms in the QCD Lagrangian, then conserved and quasi-conserved currents have finite matrix elements.[34] Physically, this is obvious; a formal proof shall be given later.

Nonconserved operators, however, generally require renormalization. To see this, we begin with a simple example—the operator $\sum_i :\bar{q}^i(x)q^i(x): \equiv M(x)$. As discussed in Sections 8 and 9, we may either work with $\bar{q}_u q_u$ and calculate, taking counterterms into account, or use $Z_F^{-1}\bar{q}_u q_u$, replace $g \to g_u = Z_g g$, $m \to m_u = Z_m m$ and neglect counterterms. Nevertheless, this is generally insufficient to make M finite: to obtain finite matrix elements, we must still multiply by an extra Z_M, called the *operator renormalization*:

$$M_R(x) = Z_M M(x). \tag{13.2}$$

To prove this, we use the formulas of Section 3: with the subscript or

[34] Note that we are working to lowest order in weak and electromagnetic interactions. Otherwise, we would have to include weak and electromagnetic counterterms, Z_F^w, Z_F^{em}, etc.

superscript 0 meaning free fields, $q_0 \equiv q_u^0$, $B_0 \equiv B_u^0$,

$$M_R(x) = Z_M T :\bar{q}_0(x)q_0(x): \exp i \int d^4z \mathscr{L}_{\text{int}}^0(z).$$

To the lowest order in g, this is

$$M_R(x) = Z_M Z_F^{-1} :\bar{q}_0(x)q_0(x):$$
$$- \frac{g^2}{2!} Z_M \sum \int d^4z_1 d^4z_2 T :\bar{q}_o(x)q_0(x): :\bar{q}_0(z_1)t^a\gamma_\mu q_0(z_2):$$
$$\times :\bar{q}_0(z_2)t^b\gamma_\nu q_0(z_2): B_{0a}^\mu(z_1)B_{0b}^\nu(z_2). \tag{13.3}$$

Because we expect $Z_M = 1 + O(g^2)$, we can neglect the Z_M in the second term of the right-hand side of (13.3). We next consider divergent matrix elements; to be precise, matrix elements of M_R between quark states with equal momentum p: it is not difficult to see that, in our case, the divergence is the same for diagonal and off-diagonal terms. Denoting by $\langle M \rangle_p$, $\langle M_R \rangle_p \ldots$ to these matrix elements, eq. (13.3) gives, after simple manipulations and in the Fermi–Feynman gauge,

$$\langle M_R \rangle_p = Z_M Z_F^{-1} \langle M_0 \rangle_p$$
$$+ i\langle M_0 \rangle_p \left\{ g^2 C_F \int d^D\hat{k} \frac{-\gamma^\mu(\not{p}+\not{k})(\not{p}+\not{k})\gamma_\mu}{k^2(p+k)^4} + S_u(p) + S_u(p) \right\},$$
$$M_0 \equiv :\bar{q}_0 q_0: . \tag{13.4}$$

The relevant Feynman diagrams are shown in Fig. 9. The first term in the right-hand side is that corresponding to Figure 9a, the last two to Figure 9b, the integral to Figure 9c. We have performed the calculation, neglecting the quark mass; it is not difficult to check that this does not affect the divergences. Clearly, the divergent part of one of the S_u in the right-hand side of (13.4) is exactly cancelled by the Z_F; thus, we need only

$$-iC_F g^2 \int \frac{d^Dk}{(2\pi)^D} \nu_0^{4-D} \frac{\gamma^\mu\gamma_\mu}{k^2(p+k)^2} \overset{\text{div}}{=} \frac{4g^2 C_F}{16\pi^2} \Gamma(\epsilon/2)(4\pi)^{\epsilon/2}\nu_0^\epsilon;$$

therefore, adding the contribution of the remaining S_u, we find:

$$Z_M(\nu) = 1 - \frac{3C_F\alpha_g}{4\pi} \left\{ \frac{2}{\epsilon} + \log 4\pi - \gamma_E - \log \nu^2/\nu_0^2 \right\}. \tag{13.5}$$

We have calculated in the Fermi–Feynman gauge, but it is not difficult to verify that Z_M is actually gauge independent.

If we had carried over the calculation for $\bar{q}\gamma^\mu q'$ or $\bar{q}\gamma^\mu\gamma_5 q'$ instead of $\bar{q}q$, we would have obtained zero for the anomalous dimension. As stated, this is a special case of a general result, which we now prove. Let J^μ be a quasi-conserved operator, i.e., as the masses go to zero, $\partial_\mu J^\mu(x) = 0$.

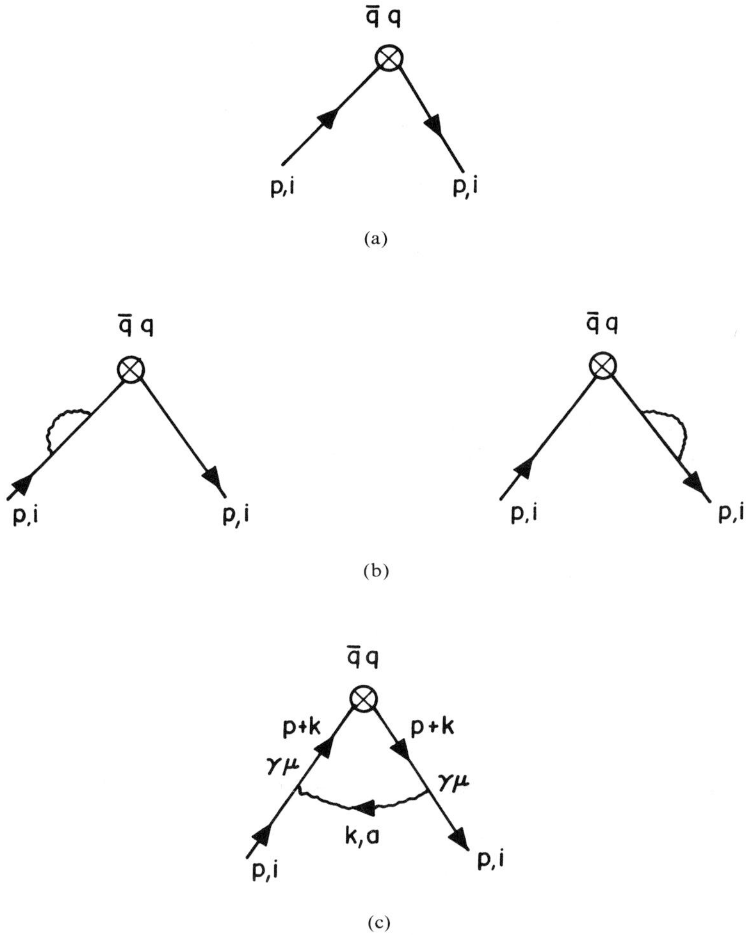

Figure 9. Renormalization of the operator $\bar{q}q$.

Consider any T-product with arbitrary fields Φ_i

$$TJ^{\mu}(x)\Phi_1(y_1)\ \dots\ \Phi_N(y_N).$$

Then, using $\partial_0\theta(x^0 - y^0) = \delta(x^0 - y^0)$, we have the Ward identity:

$$\begin{aligned}
&\partial_\mu TJ^{\mu}(x)\Phi_1(y_1)\ \dots\ \Phi_N(y_N)\\
&\qquad = T(\partial_\mu J^{\mu}(x))\Phi_1(y_1)\ \dots\ \Phi_N(y_N)\\
&\qquad\quad + \sum_{k=1}^{N}\delta(x^0 - y_k^0)T\Phi_1(y_1)\ \dots\ \left[J^0(x), \Phi_k(y_k)\right]\ \dots\ \Phi_N(y_N). \qquad (13.6)
\end{aligned}$$

Now let

$$\delta(x^0 - y_k^0)\left[J^0(x), \Phi_k(y_k)\right] = \Phi_k'(y_k)\delta(x - y_k);$$

if Z_J and Z_D are the anomalous dimensions of J^μ and $\partial_\mu J^\mu$, and γ_J and γ_D are the coefficients of $-(g^2/16\pi^2)N_\epsilon$ in Z_J and Z_D, we find, for (13.6), by differentiating $\nu d/d\nu$, that

$$\begin{aligned}\gamma_J\partial_\mu TJ^\mu(x)\Phi_1(y_1)\ldots\Phi_N(y_N)\\ = T\left\{\sum\gamma_m m\frac{\partial}{\partial m}\partial_\mu J^\mu(x)\right\}\Phi_1(y_1)\ldots\Phi_N(y_N)\\ +\gamma_D T(\partial_\mu J^\mu(x))\Phi_1(y_1)\ldots\Phi_N(y_N).\end{aligned}\tag{13.7}$$

This is possible only if $\gamma_J = 0$ and, moreover,

$$\gamma_D\partial_\mu J^\mu = -\sum\gamma_m m\frac{\partial}{\partial m}\partial_\mu J^\mu.\tag{13.8}$$

Equation (13.8) may be verified for the case $J^\mu = \bar{q}'\gamma^\mu q$ with the help of the previous calculation, because

$$\partial_\mu J^\mu = i(m - m')\bar{q}'q;$$

We may then use the γ_m to be determined in Section 14. Alternatively, we can take into account (9.17) and (11.6) to verify that to second order,

$$m_{uD}(\bar{q}q)_{uD} = m_R Z_m(\bar{q}q)_{uD} = M_R Z_M(\bar{q}q)_{uD} = M_R(\bar{q}q)_R,$$

for Z_m indeed equals the Z_M we have just calculated.

14 The Running Coupling Constant and the Running Mass in QCD: Asymptotic Freedom

Let us now turn to Equations (12.6) and (12.7). To solve (12.6), we assume that for some ν the renormalized coupling constant is sufficiently small that we can expand the functions β, γ, δ in a power series in $g(\nu)$:

$$\left.\begin{aligned}\beta &= -\left\{\beta_0\frac{g^2(\nu)}{16\pi^2} + \beta_1\left(\frac{g^2(\nu)}{16\pi^2}\right)^2 + \beta_2\left(\frac{g^2(\nu)}{16\pi^2}\right)^3 + \ldots\right\},\\ \gamma_m &= \gamma_m^{(0)}\frac{g^2(\nu)}{16\pi^2} + \gamma_m^{(1)}\left(\frac{g^2(\nu)}{16\pi^2}\right)^2 + \ldots,\\ \delta &= \delta^{(0)}\frac{g^2(\nu)}{16\pi^2} + \delta^{(1)}\left(\frac{g^2(\nu)}{16\pi^2}\right)^2 + \ldots.\end{aligned}\right\}\tag{14.1}$$

The value of β_0 can be read off from (9.30), (12.4):

$$\beta_0 = \tfrac{1}{3}\{11C_A - 4n_f T_F\} = \tfrac{1}{3}(33 - 2nf).\tag{14.2a}$$

Using the calculations of Z_g to the second order [Caswell (1974); Jones (1974)] and to the third order [Tarasov, Vladimirov, and Zharkov (1980)], we also have[35]:

$$\beta_1 = \tfrac{34}{3} C_A^2 - \tfrac{20}{3} C_A T_F n_f - 4 C_F T_F n_f = 102 - \tfrac{38}{3} n_f ;$$

$$\begin{aligned}\beta_2 &= \tfrac{2857}{54} C_A^3 - \tfrac{1415}{27} C_A^2 T_F n_f \\ &\quad + \tfrac{158}{27} C_A T_F^2 n_f^2 - \tfrac{205}{9} C_A C_F T_F n_f + \tfrac{44}{9} C_F T_F^2 n_f^2 + 2 C_F^2 T_F n_f \\ &= \tfrac{2857}{2} - \tfrac{5033}{18} n_f + \tfrac{325}{54} n_f^2 . \end{aligned} \tag{14.2b}$$

We calculate the lowest order expression for $\bar{g}$. With the standard notation $\alpha_s \equiv \bar{g}^2/4\pi$, Equations (12.6$a$) yield

$$\frac{d\bar{g}}{d\log\lambda} = -\beta_0 \frac{\bar{g}^3}{16\pi^2},$$

i.e., with $\lambda^2 = Q^2/\nu^2$,

$$\int_{\alpha_g(\nu)}^{\alpha_s(Q^2)} \frac{d\alpha_s}{\alpha_s^2} = \frac{-\beta_0}{2\pi} \int_0^{(1/2)\log Q^2/\nu^2} d\log\lambda',$$

with the solution

$$\alpha_s(Q^2) = \frac{\alpha_g(\nu)}{1 + \alpha_g(\nu)\beta_0(\log Q^2/\nu^2)/4\pi.} \tag{14.3}$$

It is convenient to reexpress this in terms of an invariant mass parameter, Λ, choosing it so that (14.3) becomes

$$\alpha_s(Q^2) = \frac{4\pi}{\beta_0 \log Q^2/\Lambda^2}; \qquad \Lambda^2 = \nu^2 e^{-4\pi/\beta_0\alpha_g(\nu)}. \tag{14.4a}$$

In explicit terms,

$$\alpha_s(Q^2) = \frac{12\pi}{(33 - 2n_f)\log Q^2/\Lambda^2}. \tag{14.4b}$$

If we had also taken into account β_1, we would have obtained the second-order expression:

$$\alpha_s^{(1)}(Q^2) = \frac{12\pi}{(33-2n_f)\log Q^2/\Lambda^2}\left\{1 - 3\frac{153 - 19n_f}{(33-2n_f)^2} \cdot \frac{\log\log Q^2/\Lambda^2}{\frac{1}{2}\log Q^2/\Lambda^2}\right\}. \tag{14.4c}$$

We see that $\alpha_s^{(2)}(Q^2)/\alpha_s(Q^2) \to 1$ and that each of them vanishes logarithmically[36] as the momentum scale $Q^2 \to \infty$. This is the celebrated property of *asymptotic freedom*, first discussed by Gross and Wilczek (1973a) and Politzer (1973). Recalling (12.7), it means that, at large

[35] β_0, β_1 are scheme independent; the value given for β_2 is in the $\overline{MS}$ scheme.

[36] Provided $n_f < 16$, a bound comfortably satisfied: only five flavors have been found experimentally, and current theories require a sixth one.

spacelike momenta $\lambda p_i \sim q$, $q^2 = -Q^2$, the theory will behave as a free-field theory modulo logarithmic corrections. What is more, as $Q^2 \to \infty$, $\alpha_s \to 0$. Hence these corrections will be calculable in a perturbative series in powers of α_s.

The running mass is also calculated. To lowest order, we require (12.2), (12.6), and (9.14). We then have

$$\frac{1}{\overline{m}} \cdot \frac{d\overline{m}}{d\log\lambda} = \gamma_m^{(0)} \frac{g^2}{16\pi^2} = \frac{\gamma_m^{(0)}}{2\beta_0 \log\lambda}.$$

Using (14.4a) with $\log Q^2/\Lambda^2 = 2\log\lambda$, and introducing the integration constant $\hat{m}$ (which is the mass analog of Λ), this gives

$$\overline{m}(Q^2) = \frac{\hat{m}}{\left(\frac{1}{2}\log Q^2/\Lambda^2\right)^{-\gamma_m^{(0)}/\beta_0}}, \qquad \gamma_m^{(0)} = -3C_F; \tag{14.5a}$$

explicitly,

$$\overline{m}(Q^2) = \frac{\hat{m}}{\left(\frac{1}{2}\log Q^2/\Lambda^2\right)^{d_m}}, \qquad d_m = \frac{12}{33 - 2n_f}. \tag{14.5b}$$

d_m is sometimes called the *anomalous dimension* of the mass.

The running gauge parameter can be similarly calculated. The details may be found in Narison (1982). One finds

$$\bar{\xi}(Q^2) = 1 - \frac{1}{\hat{\lambda}\left(\frac{1}{2}\log Q^2/\Lambda^2\right)^{d_\xi}} \left\{1 + \frac{9}{39 - 4n_f} \cdot \frac{1}{\hat{\lambda}(\log Q^2/\Lambda^2)^{d_\xi}}\right\}^{-1},$$

$$d_\xi = \frac{1}{2} \cdot \frac{39 - 4n_f}{33 - 2n_f}.$$

We end this section with the two-loop expression for $\overline{m}$ (Tarrach, 1981):

$$\overline{m}^{(2)}(Q^2) = \frac{\hat{m}}{\left(\frac{1}{2}\log Q^2/\Lambda^2\right)^{d_m}} \left\{1 \frac{-\gamma_m^{(0)}\beta_1}{\beta_0^2} \cdot \frac{\log\log Q^2/\Lambda^2}{2\log Q^2/\Lambda^2} + \frac{1}{2\beta_0^2}\left(\gamma_m^{(1)} - \gamma_m^{(0)}\frac{\beta_1}{\beta_0}\right)\frac{1}{\log Q^2/\Lambda^2}\right\}, \tag{14.5c}$$

$$\gamma_m^{(1)} = 3\left(\frac{n_c^2 - 1}{2n_c}\right)^2 + \frac{97}{6} \cdot \frac{n_c^2 - 1}{4} - \frac{5n_f(n_c^2 - 1)}{3n_c},$$

where n_c = number of colors = 3, and, as an example of the techniques, we give an evaluation of the behavior of the quark propagator, at large momentum:

$$S_R(p, g(\nu), m(\nu), \xi(\nu); \nu), \qquad p^2 = -Q^2 \gg \Lambda^2.$$

The naive dimension of S_R is $\rho_S = -1$. Hence, Equation (12.7) yields with $p = \lambda n$, $n^2 = -\Lambda^2$, and noting that $Z = Z_F$ (S_R is *not* amputated; the amputated S_R would simply be equal to S_R^{-1}):

$$S_R(p, g(\nu), m(\nu), \xi(\nu); \nu) = S_R\left(n, \bar{g}(\lambda), \bar{m}(\lambda), \bar{\xi}(\lambda); \nu\right)\left(\frac{Q^2}{\Lambda^2}\right)^{-1/2} \times \exp\left\{-\int_0^{\log Q/\Lambda} d\log\lambda' \frac{1-\xi}{3\pi}\alpha_g(\lambda')\right\}.$$

To leading order in α_s,

$$S_R\left(n, \bar{g}(\lambda), \bar{m}(\lambda), \bar{\xi}(\lambda); \nu\right) \underset{Q^2 \to \infty}{\simeq} \frac{i}{\not{n}};$$

so, using (14.4*a*),

$$S_R(p, g(\nu), m(\nu), \xi(\nu); \nu) \underset{Q^2 \gg \Lambda^2}{\simeq} \frac{i}{\not{p}} \cdot \frac{1}{\left(\frac{1}{2}\log Q^2/\Lambda^2\right)^{d_{F\xi}}}, \tag{14.6a}$$

where the anomalous quark dimension is

$$d_{F\xi} = \frac{3}{2} \cdot \frac{(1-\xi)C_F}{11C_A - 4T_F n_f} = 2\frac{1-\xi}{33-2n_f}: \tag{14.6b}$$

S_R behaves as a free propagator, except for the logarithmic correction $(\log Q/\Lambda)^{-d_{F\xi}}$. Note that $d_{F\xi}$ depends on the gauge parameter, as expected, and vanishes in the Landau gauage where S has canonical dimension.

CHAPTER III

Deep Inelastic Processes

"Is there any point to which you would wish to draw my attention?"
"To the curious incident of the dog in the night-time."
"The dog did nothing in the night-time!"
"That was the curious incident," remarked Sherlock *Holmes.*

ARTHUR CONAN DOYLE, 1892

15 e^+e^- Annihilation into Hadrons

The Lagrangian for strong and electromagnetic interactions of quarks may be written as

$$\mathscr{L}_{\text{QCD+em}} = \sum_q \{i\bar{q}\not{D}q - m_q\bar{q}q\} - \frac{1}{4}(D\times B)^2 + e\sum_q Q_q\bar{q}\gamma_\mu qA^\mu - \frac{1}{4}F_{\mu\nu}F^{\mu\nu}, \tag{15.1}$$

where Q_q is the charge of quark q in units of the proton charge, e. We have omitted gauge fixing and ghost terms from (15.1). The electromagnetic current is:

$$J^\mu = \sum_q Q_q :\bar{q}\gamma^\mu q: .$$

Let us consider a generic hadron state, Γ. The (unpolarized) hadron annihilation cross section of e^+e^- is defined as the sum of all the cross sections for $e^+e^- \to \Gamma$, averaged over the spins of e^+e^-. To calculate it, we consider the matrix element

$$\langle\Gamma|S_{\text{QCD+em}}|e^+e^-\rangle = \langle\Gamma|T\exp i\int d^4x\{\mathscr{L}_{\text{int, QCD}}(x) + \mathscr{L}_{\text{int, em}}(x)\}|e^+e^-\rangle.$$

We will calculate to lowest order in electromagnetic interactions; we obtain

$$\langle\Gamma|S_{\mathrm{QCD+em}}|e^+e^-\rangle = \frac{-e^2}{2!}\langle\Gamma|\int d^4x_1 d^4x_2 T\mathscr{L}^0_{\mathrm{int,\,em}}(x_1)\mathscr{L}^0_{\mathrm{int,\,em}}(x_2)$$
$$\times \exp i\int d^4x \mathscr{L}^0_{\mathrm{int,\,QCD}}(x)|e^+e^-\rangle.$$

Using the Feynman rules for QED, we then find, with the kinematics of Figure 10*a*,

$$F(e^+e^-\to\Gamma) = \frac{2\pi e^2}{q^2}\bar{v}(p_1,\sigma_1)\gamma_\mu u(p_2,\sigma_2)\langle\Gamma|J^\mu(0)|0\rangle;$$

hence,

$$\sigma_h(s) = \sum_\Gamma \sigma\left(e^+e^-\to\Gamma, s=(p_1+p_2)^2\right)$$
$$= \frac{2\alpha^2}{s^3}4\pi^2 l_{\mu\nu}\sum_\Gamma(2\pi)^4\delta(p_1+p_2-p_\Gamma)\langle\Gamma|J^\nu(0)|0\rangle\langle\Gamma|J^\mu(0)|0\rangle^*,$$
$$(15.2)$$

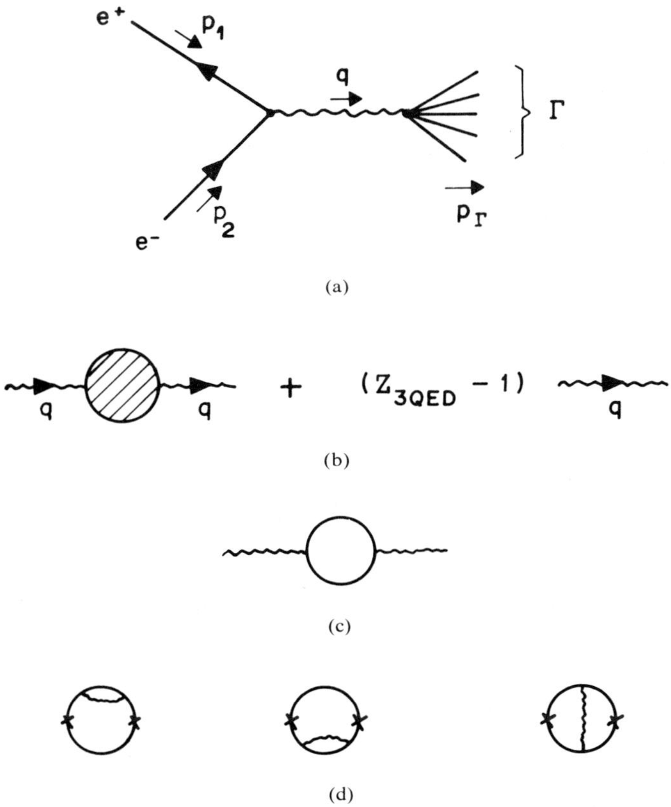

Figure 10. Diagrams for $e^+e^-\to$hadrons.

where $l_{\mu\nu}$ is the tensor

$$l_{\mu\nu} = \frac{1}{4}\sum_{\sigma_1\sigma_2} \bar{v}(p_1,\sigma_1)\gamma_\mu u(p_2,\sigma_2)\left[\bar{v}(p_1,\sigma_1)\gamma_\nu u(p_2,\sigma_2)\right]^*$$
$$= \frac{1}{2}\left\{q_\mu q_\nu - q^2 g_{\mu\nu} - (p_1-p_2)_\mu(p_1-p_2)_\nu\right\},$$

and we have neglected the electron mass. We see that we are led to consider the quantity

$$\Delta^{\mu\nu} = \sum_\Gamma (2\pi)^4\delta(p_1+p_2-p_\Gamma)\langle 0|J^\mu(0)|\Gamma\rangle\langle\Gamma|J^\nu(0)|0\rangle.$$

Using completeness, $\sum_\Gamma|\Gamma\rangle\langle\Gamma| = 1$ (and, that, because of energy-momentum conservation, the term with reversed order of Js gives zero), one can rewrite $\Delta^{\mu\nu}$ as

$$\Delta^{\mu\nu} = \int d^4x\, e^{iq\cdot x}\langle[J^\mu(x),J^\nu(0)]\rangle_0. \tag{15.3}$$

It is convenient to define

$$\Pi^{\mu\nu}(q) = i\int d^4x\, e^{iq\cdot x}\langle TJ^\mu(x)J^\nu(0)\rangle_0, \tag{15.4a}$$

with $p_1+p_2 = q$, as indeed one can see[37] that $\Delta^{\mu\nu} = 2\,\mathrm{Im}\,\Pi^{\mu\nu}$: the e^+e^- annihilation cross section is related to the imaginary part of the photon propagator.

A slight complication is introduced here due to the interplay of electromagnetic and strong interactions. Because $\Pi^{\mu\nu}$ is the coefficient of a second-order term in e, we must also consider the renormalization of the electric charge; that is to say, both diagrams in Figure 10*b* have to be included. The simplest solution to this is to neglect the problem, which does not arise for the physically relevant quantity, $\mathrm{Im}\,\Pi^{\mu\nu}$.

The electromagnetic currents are conserved, and hence their anomalous dimensions vanish. If we explicitly extract the tensor $-g^{\mu\nu}q^2 + q^\mu q^\nu$ from $\Pi^{\mu\nu}$,

$$\Pi^{\mu\nu}(q) = (-g^{\mu\nu}q^2 + q^\mu q^\nu)\Pi(q), \tag{15.4b}$$

then the general theory gives us

$$\mathrm{Im}\,\Pi_R(q; m(\nu), g(\nu); \nu) = \mathrm{Im}\,\Pi_R(\nu n; \bar{m}(Q^2), \bar{g}(Q^2); \nu),$$
$$Q^2 = -q^2 = s, \qquad n^2 = 1. \tag{15.5}$$

Therefore, all we have to do is calculate $\mathrm{Im}\,\Pi_R(q; m(\nu), g(\nu); \nu)$ and then replace $q = \nu$, $m(\nu)\to\bar{m}(Q^2)$, $g(\nu)\to\bar{g}(Q^2)$. To zero order, we have the diagram of Figure 10*c*, which gives, neglecting quark masses that give corrections $\bar{m}^2/s$,

$$\mathrm{Im}\,\Pi_R^{(0)} = \frac{1}{12\pi}3\sum_{f=1}^{n_f} Q_f^2. \tag{15.6}$$

[37]The simple, but somewhat tedious procedure, is to apply unitarity equations (2.8) and (2.9) to the forward $e^+e^-\to e^+e^-$ scattering to order α^2.

This justifies the old parton model result [Feynman (1972); Cabibbo, Parisi, and Testa (1970)] in which quarks were considered to be free: because of this, it is customary to define the quantity

$$R(s) = \frac{\sigma_h(s)}{\sigma^{(0)}_{e^+e^-\to\mu^+\mu^-}(s)}, \tag{15.7}$$

where $\sigma^{(0)}_{e^+e^-\to\mu^+\mu^-}(s)$ is the lowest order (in α) cross section for $e^+e^- \to \mu^+\mu^-$. We see that we have obtained

$$R^{(0)}(s) = 3\sum_{f=1}^{n_f} Q_f^2 . \tag{15.8}$$

The next correction involves the diagrams of Figure 10*d*. One may profit from the fact that they are the same as for QED with the gluon replaced by a photon, except for the group-theoretic factor

$$\sum_{a,k} t^a_{ik} t^a_{kj} = C_F \delta_{ij},$$

and that, in QED, they were calculated long ago by Jost and Luttinger (1950). So we find [Appelquist and Georgi (1973); Zee (1973)]

$$R^{(1)}(s) = 3\sum_{f=1}^{n_f} Q_f^2 \left\{1 + \frac{\alpha_s(Q^2)}{\pi}\right\}. \tag{15.9}$$

The following correction has also been calculated elsewhere [Chetyrkin, Kataev, and Tkachov (1979); Dine and Sapiristein (1979). In the $\overline{MS}$ scheme,

$$R^{(2)}(s) = 3\sum_{f=1}^{n_f} Q_f^2 \left\{1 + \frac{\alpha_s(Q^2)}{\pi} + r_2\left(\frac{\alpha_s(Q^2)}{\pi}\right)^2\right\},$$

$$r_2 = \left[\frac{2}{3}\zeta(3) - \frac{11}{12}\right] n_f + \frac{365}{24} - 11\zeta(3) \simeq 2.0 - 0.12 n_f . \tag{15.10}$$

The symbol ζ stands for Riemann's zeta function and the second order expression should be used for α_s.

There is yet another point to be discussed: how many quark flavors are to be taken into account? This is related to the problem of the quark masses. If we have a quark of mass m_q, then for $s \gg m_q^2$, the corrections introduced by $m_q \neq 0$ are obviously

$$0\left(\frac{m_q^2}{s}\right).$$

This can be neglected, compared to any $[\alpha_s]^n$ correction as $s \to \infty$. The situation is different, however, if $m_q^2 \gg s$, for then there is insufficient energy to create $\bar{q}q$ pairs. We will discuss this in detail later on; for the moment, we adopt the heuristic recipe of taking the sums $\sum_1^{n_f}$ to go up to

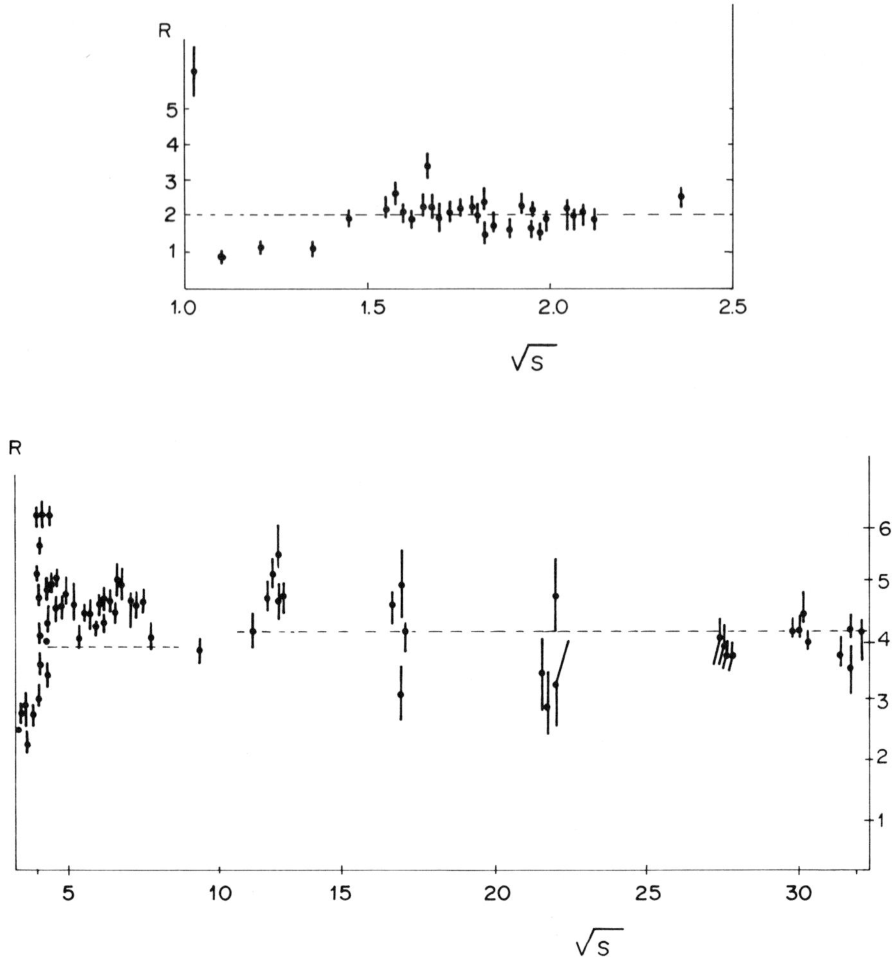

Figure 11. Plot of R vs. s. The dotted line is the (leading) QCD prediction. The data are from the compilation of Wiik and Wolf (1979).

the number of flavors that can be excited at each s, i.e., to include only those quarks whose mass is $m_q^2 \ll s$. The troubled regions $4m_q^2 \approx s$ (thresholds) shall be omitted. In fact, it may be shown that perturbative QCD is not directly applicable there.

With this, the theoretical predictions are seen to agree well with data, thus furnishing our first experimental test of QCD (Figure 11).[38] However, because of the systematic errors in the data, it is difficult to go beyond the leading order in QCD.

[38] More refined analyses may be found in the paper of Barnett, Dine, and McLerran (1980) and work quoted therein.

16 Digression on the Renormalization Scheme Dependence of Calculations and Parameters

In QED there is a natural renormalization scheme: one renormalizes with photons and electrons on their mass shells. This is useful because of Thirring's (1950) theorem which says that at zero photon energy, the Compton amplitude is exactly given (to all orders in α) by the classical approximation; so we may use results from classical physics to determine the fundamental parameters α, m_e. In QCD there is no such preferred scheme, at least not based on physical grounds. Therefore, a discussion on what happens when we change the scheme is necessary. We will neglect masses and gauge parameters in the discussion; their introduction would not cause problems different from the ones we shall consider now.

Take a physical observable, P. Clearly, P must be independent of the renormalization scheme we use to calculate it. However, when we write

$$P = \sum_n C_n(R)\left[\alpha_s(R)\right]^n, \tag{16.1}$$

both C_n and α_s depend upon the scheme R in which we are calculating. The relation with a new scheme, R', is found by writing

$$P = \sum_n C_n(R')\left[\alpha_s(R')\right]^n, \tag{16.2}$$

expanding $\alpha_s(R')$ in terms of $\alpha_s(R)$, and equating. The expansion of $\alpha_s(R')$ in terms of $\alpha_s(R)$ will be

$$\alpha_s(R') = \alpha_s(R)\{1 + a_1(R', R)\alpha_s(R) + \cdots\}.$$

It is clear that the expansion must begin with unity, because to zero order, $\alpha_s = g_u^2/4\pi$, which is independent of the scheme. This means that $C_{0,1}(R) = C_{0,1}(R')$. It is clear, however, that all other C_n will vary:

$$C_2(R) = C_2(R') + a_1(R', R)C_1(R'), \text{ etc.}$$

As an example,[39] consider the quantity R studied in the previous section. If we had calculated it in the minimal scheme (in which, it will be remembered, one only cancels the $2/\epsilon$ divergences instead of the full $N_\epsilon = 2/\epsilon - \gamma_E + \log 4\pi$), we would have obtained that (15.10) is replaced by

$$R_{\text{ms}}^{(2)}(s) = 3\sum_{f=1}^{n_f} Q_f^2\left\{1 + \frac{\alpha_{s,\text{ms}}(Q^2)}{\pi} + r_{2,\text{ms}}\left(\frac{\alpha_{s,\text{ms}}(Q^2)}{\pi}\right)^2\right\},$$

$$r_{2,\text{ms}} = r_2 + (\log 4\pi - \gamma_E)\frac{33 - 2n_f}{12}. \tag{16.3}$$

[39] A comprehensive discussion of this, in particular, for deep inelastic scattering, may be found in Bardeen *et al.* (1978).

The expression for $\alpha_{s,\text{m.s.}}$ also differs from (14.4c). One finds

$$\alpha_{s,\text{ms}}(Q^2) = \frac{12\pi}{(33-2n_f)\log Q^2/\Lambda^2} \times \left\{1 - 3\frac{153-19n_f}{(33-2n_f)^2}\cdot\frac{\log\log Q^2/\Lambda^2}{\frac{1}{2}\log Q^2/\Lambda^2} - \frac{\log 4\pi - \gamma_E}{\log Q^2/\Lambda^2}\right\}, \tag{16.4}$$

as expected. The form of Equation (14.4c) could be maintained if we defined a new parameter Λ_{ms},

$$\Lambda_{\text{ms}}^2 = e^{\gamma_E - \log 4\pi}\Lambda^2, \tag{16.5}$$

for then (16.4) would read:

$$\alpha_{s,\text{ms}}(Q^2) = \frac{12\pi}{(33-2n_f)\log Q^2/\Lambda_{\text{ms}}^2}\left\{1 - 3\frac{153-19n_f}{(33-2n_f)^2}\cdot\frac{\log\log Q^2/\Lambda_{\text{ms}}^2}{\frac{1}{2}\log Q^2/\Lambda_{\text{ms}}^2}\right\}, \tag{16.6}$$

up to terms $O([\alpha_s]^3)$.

A simple point, which, unfortunately, is not always remembered, is that the parameters of the theory can be obtained only with second-order calculations: at leading order, Λ and Λ_{ms} can be interchanged, because the error committed is of second order. Also, when quoting a value for, say, Λ (the same is true for $\hat{m}$), one has to specify the renormalization scheme: Λ or $\hat{m}$ are invariant under the renormalization group, but they vary from one scheme to another. The $\overline{MS}$ scheme is preferred in this book because of its simplicity; no unnecessary transcendentals (like $-\gamma_E + \log 4\pi$) are introduced. Generally speaking, it also gives reasonably small second-order corrections. For example, in ms,

$$r_{2\,\text{ms}} \approx 7.4 - 0.44 n_f,$$

compared with the value 2.0–0.12 n_f that we had in $\overline{MS}$.

In this scheme, the preferred experimental value for Λ is

$$\Lambda \approx 0.13\,{}^{+0.07}_{-0.05}\ \text{GeV}.$$

This corresponds to $\Lambda_{\text{ms}} \approx 0.05$ GeV. The values of the $\hat{m}$ are

$$10 \gtrsim \hat{m}_n \gtrsim 5\ \text{MeV}, \qquad 20 \gtrsim \hat{m}_d \gtrsim 10\ \text{MeV}, \qquad 400 \gtrsim \hat{m}_s \gtrsim 200\ \text{MeV}.$$

The method for obtaining Λ could be by fitting R to experiment, but the data are not good enough (cf., Figure 11): one uses other processes like electron or neutrino deep inelastic scattering, and ψ, Υ decays. How the $\hat{m}$ can be estimated will be discussed in Section 32.

17 Kinematics of Deep Inelastic Scattering: The Parton Model

We consider the process $l + h \to l' + \text{all}$, where l and l' are leptons, h is a hadron target, and "all" means that we sum over all possible final states Γ (Figure 12a). If $l = l' = e$ or μ (Figure 12b), we are probing h with lowest order electromagnetic interactions, and the relevant operator will be the electromagnetic current,

$$J^{\mu}_{\text{em}} = \sum_q Q_q \bar{q}\gamma^{\mu} q; \qquad \mathscr{L}_{\text{int, em}} = eJ^{\mu}_{\text{em}} A_{\mu} .$$

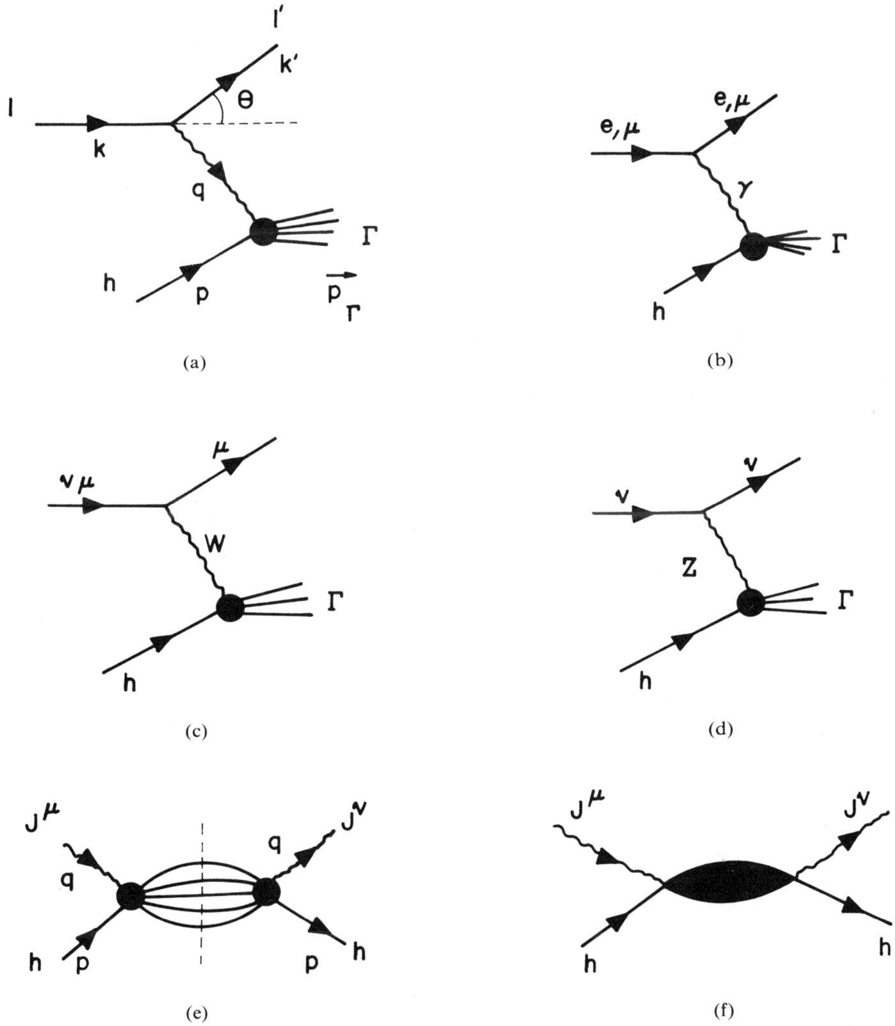

Figure 12. Diagrams for deep inelastic scattering.

If $l = \nu_\mu$, $l' = \mu$ (Figure 12c), we have (charged) weak interactions and

$$J_w^\mu = \bar{u}\gamma^\mu(1-\gamma_5)d_\theta + \bar{c}\gamma^\mu(1-\gamma_5)s_\theta + \cdots, \qquad \mathscr{L}_{\text{int},w} = \frac{1}{2\sqrt{2}}\, g_w J_w^\mu W_\mu,$$

$$g^2/M_w^2 = 4\sqrt{2}\,G_F, \qquad G_F = 1.027 m_{\text{proton}}^{-2}$$

with d_θ, s_θ the Cabibbo-rotated quarks:

$$d_\theta = d\cos\theta_C + s\sin\theta_C, \qquad s_\theta = -d\sin\theta_C + s\cos\theta_C.$$

If $l = l' = \nu_\mu$, we have weak neutral interactions (Figure 12d); then, in the standard model,

$$J_Z^\mu = \left(\frac{1}{2} - \frac{4\sin^2\theta_w}{3}\right)\bar{u}\gamma^\mu u + \left(-\frac{1}{2} + \frac{2\sin^2\theta_w}{3}\right)\bar{d}\gamma^\mu d$$
$$+\frac{1}{2}\bar{u}\gamma^\mu\gamma_5 u - \frac{1}{2}\bar{d}\gamma^\mu\gamma_5 d;$$

$$\mathscr{L}_{\text{int},Z} = \frac{e}{2\cos\theta_w\sin\theta_w} J_Z^\mu Z_\mu$$

$\sin^2\theta_w \simeq 0.22$.

Let us introduce the Bjorken variables

$$Q^2 = -q^2, \qquad \nu = p\cdot q, \qquad x = Q^2/2\nu;$$

note that, in terms of these,

$$s = p_\Gamma^2 = -Q^2 + m_h^2 + 2\nu = 2\nu\{1 + m_h^2/2\nu - x\}.$$

The deep inelastic, or *Bjorken limit* will be defined as Q^2, $\nu \gg \Lambda^2$; $x = Q^2/2\nu$ fixed. Using the standard rules, the scattering amplitude may be written for, say, the e/μ case, as

$$\mathscr{T}_{e+h\to e+\Gamma} = \frac{2\alpha}{q^2}\bar{u}(k',\sigma')\gamma^\mu u(k,\sigma)$$
$$\times(2\pi)^2\delta(p+q-p_\Gamma)\langle\Gamma|J_\mu(0)|p,\tau\rangle. \tag{17.1}$$

Here σ/σ' are the spins of the incoming/outgoing electron, and τ that of the target h. Note that we normalize the states covariantly (cf., Appendix G):

$$\langle p',\tau'|p,\tau\rangle = 2p^0\delta_{\tau\tau'}\delta(\vec{p}-\vec{p}\,').$$

The unpolarized cross section for $e + h \to e +$ all will thus involve the tensors (we systematically neglect lepton masses)

$$L^{\mu\nu} = \frac{1}{2}\sum_{\sigma\sigma'}\bar{u}(k',\sigma')\gamma^\mu u(k,\sigma)\left[\bar{u}(k',\sigma')\gamma^\nu u(k,\sigma)\right]^*$$
$$= 2(k^\mu k'^\nu + k^\nu k'^\mu - k.k' g^{\mu\nu})$$

and[40]

$$W^{\mu\nu}(p,q) = \frac{1}{2}\,\frac{1}{2}\sum_{\tau}\sum_{\Gamma}(2\pi)^6\delta(p+q-p_\Gamma)\langle p,\tau|J^\mu(0)^+|\Gamma\rangle \times \langle\Gamma|J^\nu(0)|p,\tau\rangle. \tag{17.2a}$$

Of course, $J^{\nu+} = J^\nu$, but we have written the general expression that also holds for weak currents. (17.2*a*) may be recast[41] in the form

$$W^{\mu\nu}(p,q) = \frac{1}{2}(2\pi)^2\int d^4z e^{iq\cdot z}\langle p|[J^\mu(z)^+, J^\nu(0)]|p\rangle, \tag{17.2b}$$

where the average over the target spin τ is understood.

Let us consider the general case of weak or electromagnetic currents. The general expression for $W^{\mu\nu}$ in terms of invariants is

$$W^{\mu\nu}(p,q) = (-g^{\mu\nu} + q^\mu q^\nu/q^2)W_1 + \frac{1}{m_h^2}(p^\mu - \nu q^\mu/q^2)(p^\nu - \nu q^\nu/q^2)W_2 + i\epsilon^{\mu\nu\alpha\beta}\frac{p_\alpha q_\beta}{2m_h^2}W_3. \tag{17.3}$$

Other terms give zero when contracted with the lepton tensor $L_{\mu\nu}$. The corresponding cross sections are,[42] in the lab system (h at rest) and with θ = angle of $\vec{k}$, $\vec{k}'$, $d\Omega = d\cos\theta d\phi$,

$$\frac{d\sigma^e}{d\Omega dk_0'} = \frac{\alpha^2}{4m_h k_0^2\sin^4(\theta/2)}\left\{W_2^e\cos^2\frac{\theta}{2} + 2W_1^e\sin^2\frac{\theta}{2}\right\}, \tag{17.4a}$$

and

$$\frac{d\sigma^{\nu/\bar\nu}}{d\Omega dk_0'} = \frac{G_F^2 k_0'^2}{2\pi^2 m_h}\left\{W_2^{\nu\pm}\cos^2\frac{\theta}{2} + 2W_1^{\nu\pm}\sin^2\frac{\theta}{2} \mp \frac{k^0 + k'^0}{2m_h}W_3^{\nu\pm}\right\}; \tag{17.4b}$$

in (17.4*b*), ($\pm$) refers to ν, $\bar\nu$, and G_F is the Fermi constant,

$$G_F = \sqrt{2}\,g_w^2/8M_w^2.$$

The W_i are invariant and depend on Q^2 and ν. It is convenient to define the *structure functions*[43]

$$f_1^a(x,Q^2) = 2xW_1^a, \qquad f_2^a(x,Q^2) = \frac{\nu}{m_h^2}W_2^a, \qquad f_3^a(x,Q^2) = \frac{Q^2}{2m_h}W_3^a, \tag{17.5}$$

[40]The factors 1/2 in (17.2) are introduced to average over the spin of the initial nucleon and the "helicity" of the virtual photon.

[41]The equivalence may be seen by inserting a complete sum of states $\sum_\Gamma|\Gamma\rangle\langle\Gamma|$ in (17.2*b*), noting that the second term of the commutator does not contribute because of energy-momentum conservation.

[42]We write all formulas for e scattering; for μ scattering, they are identical. Only the charged-current processes will be discussed here for ν scattering.

[43]These f_i are slightly different from the standard ones; to be precise, $f_1 = 2xF_{1\ \text{standard}}$, $f_2 = F_{2\ \text{st}}$, $f_3 = xF_{3\ \text{st}}$. Our definition aims at unifying the QCD equations that will be written later.

where superscript a indicates the process $(e/\mu h, \nu h, \bar{\nu} h)$. The *longitudinal* structure function,

$$f_L^a(x, Q^2) = f_2^a(x, Q^2) - f_1^a(x, Q^2), \tag{17.6}$$

is also used in lieu of f_1^a. It is useful to rewrite (17.3) in terms of the f_i^a, neglecting q^μ, q^ν terms (which will yield zero when contracted with the leptonic tensor $L_{\mu\nu}$)[44]:

$$\frac{1}{2}(2\pi)^2 \int d^4z\, e^{iq\cdot z} \langle p | [J_a^\mu(z)^+, J_a^\nu(0)] | p \rangle$$

$$= \frac{\nu}{q^2} g^{\mu\nu} f_1^a + \frac{p^\mu p^\nu}{\nu} f_2^a + i\epsilon^{\mu\nu\alpha\beta} \frac{q_\alpha p_\beta}{q^2} f_3^a$$

$$= -\frac{\nu g^{\mu\nu}}{q^2} f_L^a + \left(\frac{\nu}{q^2} g^{\mu\nu} + \frac{p^\mu p^\nu}{\nu} \right) f_2^a + i\epsilon^{\mu\nu\alpha\beta} \frac{q_\alpha p_\beta}{q^2} f_3^a. \tag{17.7}$$

As for e^+e^- annihilations, we will find it convenient to consider a T-product:

$$T_a^{\mu\nu}(p,q) = i(2\pi)^3 \int d^4z\, e^{iq\cdot z} \langle p | T J_a^\mu(z)^+ J_a^\nu(0) | p \rangle; \tag{17.8a}$$

note that if we write

$$T_a^{\mu\nu} = \frac{\nu}{q^2} g^{\mu\nu} T_1^a(x, Q^2) + \frac{p^\mu p^\nu}{\nu} T_2^a(x, Q^2)$$

$$+ i\epsilon^{\mu\nu\alpha\beta} \frac{q_\alpha p_\beta}{q^2} T_3^a(x, Q^2), \tag{17.8b}$$

then, as shown diagramatically in Figures 12*e* and 12*f*,

$$f_i^a = \frac{1}{2\pi} \operatorname{Im} T_i^a. \tag{17.8c}$$

Let us now consider the Bjorken limit in the so-called *infinite momentum frame*,

$$p = (p^0, 0, 0, p^0); \qquad q = \left(\nu/2p^0, \sqrt{Q^2}, 0, \nu/2p^0\right); \qquad p^0 \approx \nu^{1/2} \to \infty. \tag{17.9}$$

Rewriting $q \cdot z$ as

$$q \cdot z = \frac{1}{2}(q^0 - q^3)(z^0 + z^3) + \frac{1}{2}(q^0 + q^3)(z^0 - z^3) - q^1 z^1,$$

we see that $z \cdot q \approx 0$ corresponds, in the Bjorken limit, to

$$z^0 \pm z^3 \approx 1/\nu^{1/2}, \qquad z^1 \approx 1/\nu^{1/2}.$$

[44] Throughout this section, we use z for the position vector to distinguish it from the Bjorken variable, x.

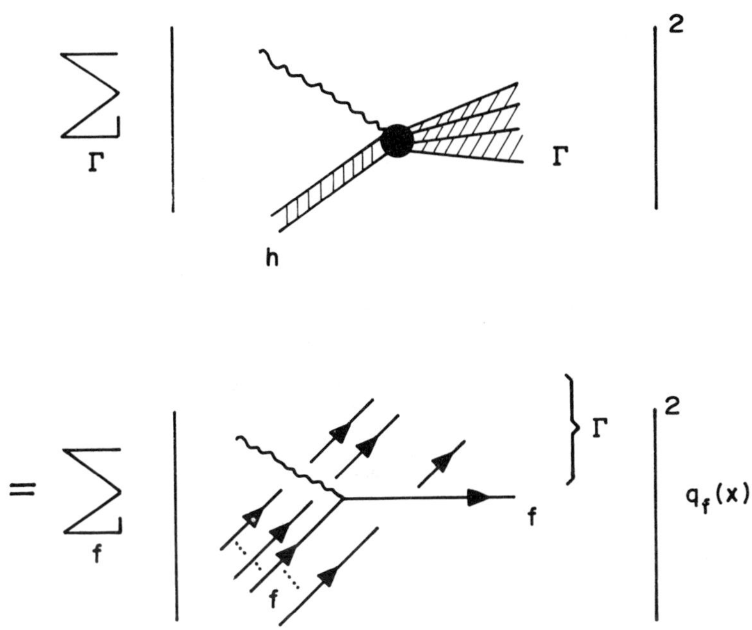

Figure 13. The parton model.

More simply stated, $z^2 \to 0$.[45] By virtue of a well-known property of Fourier transforms, it follows that the fixed x, large q behavior of the transform of the commutator in Equation (17.2*b*) or of the time-ordered product in (17.8*a*) is given by the z with $z^2 \approx O(1/q^2)$: that is to say, by the light-cone behavior of

$$[J^\mu(z)^+, J^\nu(0)] \quad \text{or} \quad TJ^\mu(z)^+J^\nu(0). \tag{17.10}$$

Because of asymptotic freedom, we expect that, up to logarithmic corrections, we could calculate these commutators or time-ordered products, neglecting quark interactions and treating the hadron target as a group of free quarks. This is the parton model of Feynman (1969). To see some of its implications, consider deep inelastic *ep* scattering. Let $q_f(x)$ be the probability for finding a quark flavor f with momentum fraction x. The total cross section, $e + p \to e +$ all, is then obtained by incoherently adding (the quarks are free) the cross sections $e + f \to e + f$, weighted with $q_f(x)$ (Figure 13). The cross sections $e + f \to e + f$ are, of course, trivial, and thus we immediately find $f_2^{ep}(x, Q^2) = f_1^{ep}(x, Q^2)$ and

$$f_2^{ep}(x, Q^2) \underset{Q^2 \to \infty}{=} x \sum_f Q_f^2 q_f(x). \tag{17.11}$$

[45] Actually, we could have z_2 as large as we wished. However, this corresponds to $z^2 < 0$ where, by locality, the commutator $[J(z), J(0)]$ vanishes and we get no contribution except when $z_2^2 \sim z_0$, i.e., $z^2 \sim 0$.

It is worth noting that the sum over f also runs over antiquarks, for it is to be expected that the probability for finding anti-quarks inside a proton be nonzero. Later we will rewrite (17.11) in a more detailed form, specifying some properties of the various quark densities $q_f(x)$.

A remarkable feature of (17.11) is *scaling*. Scaling was proposed by Bjorken (1969) before the parton model—which, indeed, was devised to explain it. Scaling means that as $Q^2 \to \infty$, the structure functions $f_i^a(x, Q^2)$ should become independent of Q^2:

$$f_i^a(x, Q^2) \underset{\substack{Q^2 \to \infty \\ x \text{ fixed}}}{\to} f_i^a(x). \tag{17.12}$$

We shall see that QCD justifies this in the sense that it predicts scaling up to logarithmic $(\log Q^2/\Lambda^2)^d$ corrections; what is more, these corrections can be calculated, and the predictions are confirmed by the experimental measurements of scaling violations.

18 The Operator Product Expansion (OPE)

The tool for rigorously analyzing the product of operators at short or lightlike distances is the *operator product expansion* (OPE).[46] To discuss it, we begin with free fields and the simplest possible case of the time-ordered product of two massless scalar fields:

$$T\phi(x)\phi(y).$$

As $x \to y$, this is singular; but the singularity is a c-number. We may separate it and write

$$T\phi(x)\phi(y) = \Delta(x-y)1 + :\phi(x)\phi(y):,$$

where Δ is the propagator and 1 stands for the unit operator:

$$\Delta(x) = \frac{i}{(2\pi)4} \int d^4k\, e^{-ik\cdot x} \frac{1}{k^2 + i0} = \frac{1}{(2\pi)^2} \cdot \frac{1}{x^2 - i0}.$$

The operator $:\phi(x)\phi(y):$ is regular as $x \to y$, and so is, of course, the unit operator. In general, we can write the product of the local (elementary or composite) operators A and B as the *short distance*, or *Wilson expansion*

$$TA(x)B(y) = \sum_t C_t(x-y)N_t(x, y), \tag{18.1}$$

[46]The operator product expansion was first introduced by Wilson (1969) and further developed (for short distances) by Zimmermann (1970), Wilson and Zimmermann (1972), and others. For the light cone, cf., Brandt and Preparata (1971), Fritzsch and Gell–Mann (1971). The general use of this tool in deep inelastic scattering was developed by Christ, Hasslacher, and Muller (1972); its application to QCD is discussed by Gross and Wilczek (1973*b*, 1974) and Georgi and Politzer (1974).

where the C_t are c-numbers (*Wilson coefficients*), in general, singular functions of $x-y$, and $N_t(x, y)$ are bilocal operators, regular as $x \to y$; the use of the letter N for them is a reminder that they will be normal-ordered composite operators. The expansion (18.1) is nothing but a generalization of the free-field case. We write,

$$TA(x)B(y) = \sum \frac{i^n}{n!} \int dz_1 \ldots dz_n \, TA^0(x)B^0(y)\mathcal{L}^0_{\text{int}}(z_1) \ldots \mathcal{L}^0_{\text{int}}(z_n),$$

where the superscript 0 means that all fields are to be taken free. Systematic application of Wick's theorem then produces (18.1). However, it is seldom necessary to write the above expression in complete generality; if we are interested in the behavior as $x \to y$ only, there is a simpler way to proceed: one considers a basis formed by all the operators with the same quantum numbers and transformation properties as the product AB (in particular, if A and B are scalars, and gauge invariant, only scalar, gauge-invariant operators have to be considered). In this case, we have the operators

$$1, :\bar{q}(x)q(y):, :\bar{q}(x)\not{D}q(y):, \ldots, :(\bar{q}(x)q(y))^2:, \ldots, :G(x)G(y):, \ldots, \tag{18.2}$$

in fact, an infinite array: but, in the limit $x \to y$, only a few (at times only one for the leading behavior) are required. This may be seen as follows: let ρ_N be the (naive) mass dimension of the operator N; then, the lowest dimensional operators in (18.2) are 1, $\rho_1 = 0$, $:\bar{q}q:$ with $\rho_{\bar{q}q} = 3$, $:\bar{q}\not{D}q:$ with $\rho_{\bar{q}Dq} = 4$ and $:G^2:$ with $\rho_{G^2} = 4$. If we suppose that the dimension of each A, B is 3, simple counting tells us that the Wilson coefficient C_1 has dimension 6, $C_{\bar{q}q}$ has dimension 3, and $C_{\bar{q}Dq}$, C_{G^2} both have dimension 4. Therefore, extracting explicitly a mass from $C_{\bar{q}q}$,

$$\begin{aligned} C_1(x-y) &\approx (x-y)^{-6}, & C_{\bar{q}q}(x-y) &\approx m(x-y)^{-2}, \\ C_{\bar{q}Dq}(x-y) &\approx (x-y)^{-2}, & C_{G^2}(x-y) &\approx (x-y)^{-2}, \end{aligned} \tag{18.3}$$

where x^6 means $(x \cdot x)^3$, x^{-2} is $1/x^2$, etc. Clearly, this will be exactly true only in free-field theory; in QCD, asymptotic freedom guarantees that the corrections to (18.3) will merely be logarithmic, which does not substantially alter the discussion.

The coefficients of other operators will be finite as $x \to 0$. If we now take any matrix element of (18.1),

$$\begin{aligned} \langle \Phi | TA(x)B(0) | \Psi \rangle \underset{x \to 0}{=} & \, C_1(x)\langle \Phi | \Psi \rangle + C_{\bar{q}q}(x)\langle \Phi | :\bar{q}(0)q(0): | \Psi \rangle \\ & + C_{\bar{q}Dq}(x)\langle \Phi | :\bar{q}(0)\not{D}q(0): | \Psi \rangle \\ & + C_{G^2}(x)\langle \Phi | :G^2(0): | \Psi \rangle + \cdots, \end{aligned} \tag{18.4}$$

then, because the operators N_t are regular, we find that the $x \to 0$ behavior of the left-hand side in (18.4) is given by that of the Wilson coefficients up to the finite constants $\langle \Phi | N_t | \Psi \rangle$. Thus, the leading behavior as $x \to 0$ of $TA(x)B(0)$ is given by $C_1(x)$, and the subleading one by $C_{\bar{q}q}$, $C_{\bar{q}Dq}$, and C_{G^2}.

Let us return to (18.1). Since the operators $N_t(x, y)$ are regular, we can expand them in $x - y$. With $y = 0$, we then write

$$N_t(x,0) = \sum_n x_{\mu_1} \ldots x_{\mu_n} N_t^{(n)\mu_1 \ldots \mu_n}(0,0).$$

For example,

$$:\bar{q}(0)q(-x):= \sum_n x_{\mu_1} \ldots x_{\mu_n} \frac{(-1)^n}{n!} :\bar{q}(0)\partial^{\mu_1} \ldots \partial^{\mu_n} q(0):. \quad (18.5)$$

In a gauge theory such as QCD, we should replace the derivatives in (18.5) by covariant derivatives.[47] So we get the expansion:

$$TA(x)B(0) \underset{x \to 0}{\simeq} C_1(x)1 + C_{\bar{q}q}(x) \sum_n x_{\mu_1} \ldots x_{\mu_n} \frac{(-1)^n}{n!} \times :\bar{q}(0)D^{\mu_1} \ldots D^{\mu_n} q(0): + \cdots . \quad (18.6)$$

As $x \to 0$, the derivatives in (18.6) are (in general) subleading because of the extra x_μ factors. This is, however, *not* true in the *light-cone expansion*. In this expansion, we are interested in the behavior as $x^2 \to 0$, but not necessarily as $x \to 0$. Because of this, in the light-cone limit, all derivatives in the right-hand side of (18.6) contribute equally.

Let us now apply this to the T-product of currents that appear in deep inelastic scattering and whose light-cone behavior controls the Bjorken limit of structure functions. Before proceeding, let us set up some notation. First of all, we consider vector and axial currents of the form described in Section 17. These may be written as combinations of the eighteen currents

$$\begin{aligned} V_a^\mu(x) &= \sum_{ff'} :q_f(x)\lambda_{ff'}^a \gamma^\mu q_{f'}(x):, \\ A_a^\mu(x) &= \sum :q_f(x)\lambda_{ff'}^a \gamma^\mu \gamma_5 q_{f'}(x):, \\ V_0^\mu(x) &= \sum :\bar{q}_f(x)\gamma^\mu q_f(x):, \\ A_0^\mu(x) &= \sum :\bar{q}_f(x)\gamma^\mu \gamma_5 q_f(x): . \end{aligned} \quad (18.7)$$

This may be unified by letting $\lambda_{ff'}^0 = \delta_{ff'}$, and allowing the index a to run from 0 to 8. For example, the electromagnetic current is given by

$$J_{\text{em}}^\mu = \frac{1}{2}\left\{ V_3^\mu + \frac{1}{\sqrt{3}} V_8^\mu \right\}. \quad (18.8)$$

[47]This is intuitively obvious. A formal proof is obtained by noting that an operator $\bar{q}(0)q(-x)$ is *not* gauge invariant. A gauge-invariant is obtained by inserting the exponential $P \exp \int_{-x}^0 dy_\mu \sum t^a B_a^\mu$. See, for example, Wilson (1975) and Appendix I.

Note that the matrices λ act in flavor space; we consider a three-flavor model, $q_1 = u$, $q_2 = d$, and $q_3 = s$. The introduction of extra flavors presents no difficulty. Of course, sums over implicit color indices are understood.

We start with the free-field case. Using Wick's theorem, we can write for the T-product of two vector currents:

$$
\begin{aligned}
TV_a^\mu(x)V_b^\nu(y) &= \sum T :\bar{q}_{i\alpha}(x)\lambda_{ik}^a\gamma_{\alpha\beta}^\mu q_{k\beta}(x): :\bar{q}_{j\delta}(y)\lambda_{jl}^b\gamma_{\delta\rho}^\nu q_{l\rho}(y): \\
&\underset{z^2\to 0}{=} \frac{2n_c\delta_{ab}(g^{\mu\nu}z^2 - 2z^\mu z^\nu)}{\pi^4(z^2 - i0)^4}\cdot 1 \\
&\quad + \sum(if_{abc} + d_{abc})\gamma_{\alpha\beta}^\mu S_{\beta\delta}(x-y)\gamma_{\delta\rho}^\nu :\bar{q}_\alpha(x)\lambda^c q_\rho(y): \\
&\quad + \sum(-if_{abc} + d_{abc})\gamma_{\alpha\beta}^\nu S_{\beta\delta}(y-x)\gamma_{\delta\rho}^\mu \\
&\quad \times :\bar{q}_\alpha(y)\lambda^c q_\rho(x): + \cdots,
\end{aligned}
\tag{18.9}
$$

where $z = x - y$ and n_c = number of colors = 3, and the colons indicate operators with four quarks $:\bar{q}\, q\, \bar{q}\, q:$. As explained, they will give subleading contributions in the light cone, and for the moment we are only interested in leading effects. To obtain (18.9), we have repeatedly used the relation

$$
Tq_\beta(x)\bar{q}_\delta(y) = -:\bar{q}_\delta(y)q_\beta(x): + S_{\beta\delta}(x-y),
$$

and properties of the λ, γ matrices (Appendix A and Appendix C). Next, we substitute S by its light-cone behavior

$$
S(z) \underset{z^2\to 0}{\simeq} \frac{2i\not{z}}{(2\pi)^2(z^2 - i0)^2},
$$

easily obtained from the explicit form

$$
S(z) = \frac{i}{(2\pi)^4}\int d^4p\, e^{-ip\cdot z}\frac{\not{p} + m}{p^2 - m^2 + i0}
$$

(Appendix F). After some manipulation of γ matrices (Appendix A), Equation (18.9) can be simplified to

$$
\begin{aligned}
&TV_a^\mu(x)V_b^\nu(y) \\
&\underset{z^2\to 0}{=} 2i\sum(if_{abc} + d_{abc})\Bigg\{ S^{\mu\alpha\nu\beta}\frac{z_\alpha}{(2\pi)^2(z^2 - i0)^2} :\bar{q}(x)\lambda^c\gamma_\beta q(y): \\
&\qquad + i\epsilon^{\mu\alpha\nu\beta}\frac{z_\alpha}{(2\pi)^2(z^2 - i0)^2} :\bar{q}(x)\lambda^c\gamma_\beta\gamma_5 q(y): \Bigg\} \\
&\quad + (x\leftrightarrow y,\, a\leftrightarrow b,\, \mu\leftrightarrow\nu) \\
&\quad + \text{constant term,}
\end{aligned}
\tag{18.10}
$$

and we have not explicitly written the constant term

$$\frac{6\delta_{ab}(g^{\mu\nu}z^2 - 2z^\mu z^\nu)}{\pi^4(z^2 - i0)^4}1$$

because it does not contribute to the commutator that appears in $W^{\mu\nu}$ (it would have been the leading term in other cases like, e.g., $\langle TV^aV^b\rangle_0$). Next, taking $y = 0$ and expanding the regular operators $:\bar{q} \ldots q:$ in powers of z, we find the light-cone expansion

$$\begin{aligned} TV_a^\mu(z)V_b^\nu(0) \underset{z^2\to 0}{=} & -i \sum_{n=\text{odd}} d_{abc} S^{\mu\alpha\nu\beta} \frac{z_\alpha}{\pi^2(z^2 - i0)^2} \cdot \frac{z_{\mu_1} \ldots z_{\mu_n}}{n!} \\ & \times :\bar{q}(0)\lambda^c\gamma_\beta D^{\mu_1} \ldots D^{\mu_n}q(0): \\ & + i \sum_{n=\text{odd}} f_{abc}\epsilon^{\mu\alpha\nu\beta} \frac{z_\alpha}{\pi^2(z^2 - i0)^2} \cdot \frac{z_{\mu_1} \ldots z_{\mu_n}}{n!} \\ & \times :\bar{q}(0)\lambda^c\gamma_\beta\gamma_5 D^{\mu_1} \ldots D^{\mu_n}q(0): \\ & + \text{constant term} + \text{gradient terms} \\ & + \text{odd terms under } (\mu \leftrightarrow \nu, a \leftrightarrow b). \end{aligned} \tag{18.11}$$

We have brought all the derivatives to act on the right by adding, if necessary, a gradient; also, we have not written explicitly odd terms in $(\mu \leftrightarrow \nu, a \leftrightarrow b)$. When plugged into $W^{\mu\nu}$ all these terms yield zero because we take diagonal matrix elements, $\langle p|TJJ|p\rangle$.[48]

It is convenient to rearrange (18.11). We will not do this in the general case, but shall simply exemplify the method with the product of two electromagnetic currents. For this, Equations (18.8) and (18.11) give (we drop the index "em," and will not make constant or gradient terms or odd terms in $\mu \leftrightarrow \nu$ explicit)

$$TJ^\mu(z)J^\nu(0)$$

$$\underset{z\,0}{=} i \sum_{n=\text{odd}} S^{\mu\alpha\nu\beta} \frac{-z_\alpha}{\pi^2(z^2 - i0)^2} \cdot \frac{z_{\mu_1} \ldots z_{\mu_n}}{n!} :\bar{q}(0)\, Q_e^2\gamma_\beta D^{\mu_1} \ldots D^{\mu_n}q(0):,$$

where Q_e is the electric charge operator in flavor space:

$$Q_e = \begin{bmatrix} 2/3 & & 0 \\ & -1/3 & \\ 0 & & -1/3 \end{bmatrix} = \frac{1}{2}\left(\lambda^3 + \frac{1}{\sqrt{3}}\lambda^8\right).$$

Next, we separate a term proportional to $g^{\mu\nu}$ (which will thus be identified with f_1) and another that will yield f_2. This is easily done with the help of the explicit expression for $S^{\mu\alpha\nu\beta}$. We then obtain, after some slight changes

[48] For processes that involve nondiagonal matrix elements, gradient terms have to be taken into account. An example of the last situation will be found in Section 27*iii*.

of notation,

$$TJ^{\mu}(z)J^{\nu}(0)$$

$$\underset{z^2\to 0}{=} i\left\{ g^{\mu\nu}\frac{1}{\pi^2(z^2-i0)^2}\sum_{n=\text{even}} z_{\mu_1}\dots z_{\mu_n}\frac{1}{(n-1)!}\right.$$

$$\times :\bar{q}(0)Q_e^2\gamma^{\mu_1}D^{\mu_2}\dots D^{\mu_n}q(0):$$

$$+\frac{-1}{2\pi^2(z^2-i0)}\sum_{n=\text{even}} z_{\mu_1}\dots z_{\mu_n}\frac{1}{n!}$$

$$\left.\times\left[:\bar{q}(0)Q_e^2\gamma^{\mu}D^{\nu}D^{\mu_1}\dots D^{\mu_n}q(0):+(\mu\leftrightarrow\nu)\right]\right\}, \qquad (18.12)$$

and (for the second term in the right-hand side) we have written $z_\alpha/(z^2-i0)^2=-\frac{1}{2}\partial_\alpha/(z^2-i0)$, transferring the ∂_α to act on the z_{μ_1}. As a final step, we decompose Q_e^2 into a piece proportional to the unit matrix (and, hence, singlet under flavor $SU_F(3)$ transformations) and a component along Q_e, hence nonsinglet:

$$Q_e^2=c_{eNS}Q_e+c_{eF}=\frac{1}{6}\lambda^3+\frac{1}{6\sqrt{3}}\lambda^8+\frac{2}{9};$$

$$c_{eNS}=1/3,\qquad c_{eF}=2/9. \qquad (18.13)$$

In this way, we finally obtain

$$TJ^{\mu}(z)J^{\nu}(0)=-g^{\mu\nu}\frac{i}{\pi^2(z^2-i0)^2}\sum_{n=\text{even}} z_{\mu_1}\dots z_{\mu_n}\frac{i^{n-1}}{n-1}$$

$$\times\left\{N_{NS,3}^{(e)\mu_1\dots\mu_n}(0)+\frac{1}{6\sqrt{3}}N_{NS,8}^{(e)\mu_1\dots\mu_n}(0)+\frac{2}{9}N_F^{(e)\mu_1\dots\mu_n}(0)\right\}$$

$$+\frac{i}{2\pi^2(z^2-i0)}\sum_{n=\text{even}} z_{\mu_1}\dots z_{\mu_n}i^{n-1}$$

$$\times\left\{\frac{1}{6}N_{NS,3}^{(e)\mu\nu\mu_1\dots\mu_n}(0)+\frac{1}{6\sqrt{3}}N_{NS,8}^{(e)\mu\nu\mu_1\dots\mu_n}(0)\right.$$

$$\left.+\frac{2}{9}N_F^{(e)\mu\nu\mu_1\dots\mu_n}(0)+(\mu\leftrightarrow\nu)\right\}, \qquad (18.14a)$$

and we have defined

$$N_{NS,a}^{(e)\mu_1\dots\mu_n}=\frac{i^{n-1}}{(n-2)!}:\sum_{ff'}\bar{q}_f(0)\gamma^{\mu_1}D^{\mu_2}\dots D^{\mu_n}\lambda^a_{ff'}q_{f'}(0):,$$

$$N_F^{(e)\mu_1\dots\mu_n}=\frac{i^{n-1}}{(n-2)!}:\sum_{f}\bar{q}_f(0)\gamma^{\mu_1}D^{\mu_2}\dots D^{\mu_n}q_f(0):, \qquad (18.14b)$$

$$a=1,\dots,8.$$

To end this section, we will re-derive scaling from the light-cone expansion in the free-field case (parton model), Equations (18.12) and (18.14). We have [cf., (17.8)]

$$T_{\rm em}^{\mu\nu}(p,q)$$

$$\overset{\rm Bj}{=} (2\pi)^3\left\{\frac{-g^{\mu\nu}}{\pi^2}\int d^4z\, e^{iq\cdot z}\sum_{n={\rm even}}\frac{iz_{\mu_1}\dots iz_{\mu_n}}{(z^2-i0)^2(n-1)}A_n^{\mu_1\dots\mu_n}(p)\right.$$

$$\left.-\frac{1}{2\pi^2}\int d^4z\, e^{iq\cdot z}\sum_n\frac{iz_{\mu_1}\dots iz_{\mu_n}}{z^2-i0}\left[A_n^{\mu\nu\mu_1\dots\mu_n}(p)+(\mu\leftrightarrow\nu)\right]\right\},\tag{18.15a}$$

where "Bj" means that equality holds in the Bjorken limit, and

$$A_n^{\mu_1\dots\mu_n}(p)=i^n\langle p|\frac{1}{(n-2)!}:\bar{q}(0)\,Q_e^2\gamma^{\mu_1}D^{\mu_2}\dots D^{\mu_n}q(0):|p\rangle.\tag{18.15b}$$

We may write the A in terms of invariants

$$A_n^{\mu_1\dots\mu_n}(p)=-ip^{\mu_1}\dots p^{\mu_n}a_n+\text{trace terms},$$

where the trace terms (containing $g^{\mu_i\mu_j}$) yield terms proportional to p^2, and hence need not be considered now. Thus,

$$T_{\rm em}^{\mu\nu}(p,q)\overset{\rm Bj}{=}i(2\pi)^3\left\{\frac{g^{\mu\nu}}{\pi^2}\int d^4z\, e^{iq\cdot z}\frac{1}{(z^2-i0)^2}\sum_{n={\rm even}}(iz\cdot p)^n a_n\frac{1}{n-1}\right.$$

$$\left.+\frac{p^\mu p^\nu}{\pi^2}\int d^4z\, e^{iq\cdot z}\frac{1}{z^2-i0}\sum_{n={\rm even}}(iz\cdot p)^n a_{n+2}\right\}.$$

Comparing this with (17.8b), we find

$$T_1^{\rm em}(x,Q^2)\overset{\rm Bj}{=}i\frac{q^2}{\nu}\cdot\frac{(2\pi)^3}{\pi^2}\int d^4z\, e^{iq\cdot z}\frac{1}{(z^2-i0)^2}\sum_{n={\rm even}}(iz\cdot p)^n\frac{a_n}{n-1},$$

$$T_2^{\rm em}(x,Q^2)\overset{\rm Bj}{=}i\nu\frac{(2\pi)^3}{\pi^2}\int d^4z\, e^{iq\cdot z}\frac{1}{z^2-i0}\sum_{n={\rm even}}(iz\cdot p)^n a_{n+2}.\tag{18.16}$$

The final formula we will require is

$$\frac{\partial}{\partial q_{\mu_1}}\cdots\frac{\partial}{\partial q_{\mu_n}}=2^n q^{\mu_1}\dots q^{\mu_n}\left(\frac{\partial}{\partial q^2}\right)^n+\text{trace terms},\tag{18.17}$$

for then, replacing the iz_{μ_j} by $\partial/\partial q^{\mu_j}$, Equations (18.16) can be written as follows:

$$T_1^{\text{em}}(x, Q^2) \overset{\text{Bj}}{=} i\frac{q^2}{\nu} \cdot \frac{(2\pi)^3}{\pi^2} \sum \frac{2^n a_n}{n-1} q_{\mu_1} \cdots q_{\mu_n} p^{\mu_1} \cdots p^{\mu_n} \left(\frac{\partial}{\partial q^2}\right)^n$$

$$\times \int d^4z \frac{e^{iq\cdot z}}{(z^2 - i0)^2}$$

$$\overset{\text{Bj}}{=} -(2\pi)^3 \frac{q^2}{\nu} \sum (2\nu)^n \frac{a_n}{n-1} \left(\frac{\partial}{\partial q^2}\right)^n \cdot \log q^2$$

$$= 2(2\pi)^3 \sum \frac{(n-2)!\, a_n}{x^{n-1}} = t(x)/x, \tag{18.18a}$$

$$T_2^{\text{em}}(x, Q^2) \overset{\text{Bj}}{=} i\nu \frac{(2\pi)^3}{\pi^2} \sum a_{n+2} (2\nu)^n \left(\frac{\partial}{\partial q^2}\right)^n \int d^4z \frac{e^{iq\cdot z}}{z^2 - i0}$$

$$\overset{\text{Bj}}{=} -4\nu(2\pi)^3 \sum (2\nu)^n a_{n+2} \left(\frac{\partial}{\partial q^2}\right)^n \frac{1}{q^2}$$

$$= 2(2\pi)^3 \sum \frac{n!\, a_{n+2}}{x^{n+1}} = T_1^{\text{em}}(x, Q^2). \tag{18.18b}$$

We have used Fourier transforms of Appendix F, and defined

$$t(x) \equiv 2(2\pi)^3 \sum_n n!\, a_{n+2} \frac{1}{x^n}. \tag{18.18c}$$

Taking the imaginary part, we therefore obtain Bjorken scaling with $f_1(x) = f_2(x)$. This last relation, which implies that the longitudinal structure function vanishes, is known as the Callan-Gross relation (Callan and Gross (1969); see also Bjorken and Paschos (1969)). Another derivation that makes it apparent that $f_2(x)/x$ is the probability that the quark has fraction x of total momentum p in the infinite momentum frame may be found in Gross (1976).

19 The OPE for Deep Inelastic Scattering in QCD: Moments

Throughout the discussions of the previous section, the underlying field theory was unspecified except when it was free-field theory. Now we shall add substance to that earlier discussion.

We again consider a two-current T-product:

$$TJ_p^{\mu}(x)^{+}J_p^{\nu}(y), \tag{19.1}$$

where p labels any current or combination of currents, among those of (18.7), but now we want to take interactions into account. We still will neglect terms that are suppressed by powers of M^2/Q^2, where M is any mass. The OPE may be written by specifying a basis consisting of operators that give leading contributions, in powers of M^2/Q^2, for free fields: in QCD this will only be modified by extra logarithmic corrections. If one classifies operators according to their *twist* τ, where $\tau = \rho - j$ with ρ the (free field) dimension and j is the spin of the operator, it is not difficult to see, by mere dimensional analysis, that the leading operators are those of twist 2. Operators of twist $\tau = 2n + 2$ are suppressed by $(M^2/Q^2)^n$ with respect to the former.

Now, the only operators of twist 2 that can be formed, and which can be connected to (19.1), are[49]

$$N_{NS,a\pm}^{\mu_1 \ldots \mu_n} = \frac{1}{2}\,\frac{i^{n-1}}{(n-2)!}\,\mathscr{S}\,:\bar{q}(0)\lambda^a\gamma^{\mu_1}(1 \pm \gamma_5)D^{\mu_1}\ldots D^{\mu_n}q(0):\,,$$

$$a = 1, \ldots, 8;$$

$$N_{F\pm}^{\mu_1 \ldots \mu_n} = \frac{1}{2}\,\frac{i^{n-1}}{(n-2)!}\,\mathscr{S}\,:\bar{q}(0)\lambda^0\gamma^{\mu_1}(1 \pm \gamma_5)D^{\mu_2}\ldots D^{\mu_n}q(0):\,;$$

$$N_{V}^{\mu_1 \ldots \mu_n} = \frac{i^{n-2}}{(n-2)!}\,\mathscr{S}\,\mathrm{Tr}\,:G^{\mu_1\alpha}(0)D^{\mu_2}\ldots D^{\mu_{n-1}}G_{\alpha}{}^{\mu_n}(0):, \tag{19.2}$$

where $\mathscr{S}$ stands for symmetrization, viz., $\mathscr{S}a_{i_1 \ldots i_n} = (1/n!)\sum_{\text{permutations }\pi} a_{\pi(i_1, \ldots, i_n)}$, the trace refers to the color indices and

$$D_\mu G^a_{\alpha\beta} \equiv \sum_c \left\{ \partial_\mu \delta^{ac} + g \sum f^{abc} B^b_\mu \right\} G^c_{\alpha\beta}\,.$$

Of the operators of (19.2), we have already encountered the first two types [Equation (18.14)] with $N^e = N^+ + N^-$; obviously, the only way in which currents consisting of quarks can have nonzero projection on purely gluon operators is to take interactions into account which is the reason why the third term only appears now.

If we work with a gauge that requires ghosts, there are other operators as well as (19.2) that have to be considered; they are made up of ghosts. However, it can be proved that, *for twist 2 operators*, ghosts can be entirely neglected. This can be done because the mixing matrix is triangular [cf., Kluberg–Stern and Zuber (1975); Dixon and Taylor (1974)]. We will return

[49] The labels F/V denote fermion/vector boson singlet operators.

to this point later on. We write the OPE of (19.1) as follows:

$$\begin{aligned} TJ_p^\mu(z)^+J_p^\nu(0) = &-\sum_{j,n}\bar{C}^n_{1pj}(z^2)g^{\mu\nu}i^{n-1}z_{\mu_1}\ldots z_{\mu_n}N_j^{\mu_1\ldots\mu_n}(0)\\ &-\sum_{j,n}\bar{C}^n_{2pj}(z^2)i^{n-1}z_{\mu_1}\ldots z_{\mu_n}N_j^{\mu\nu\mu_1\ldots\mu_n}(0)\\ &+\sum_{j,n}\bar{C}^n_{3pj}(z^2)\epsilon^{\mu\nu\alpha\beta}i^{n-2}z_\beta z_{\mu_1}\ldots z_{\mu_n}N_j^{\alpha\mu_1\ldots\mu_n}(0), \end{aligned} \tag{19.3}$$

where j indicates all the operators among those in (19.2) that have quantum numbers connected with those of $J_p^+J_p$. In this context, it is worth noting that the flavor symmetries are preserved by the QCD interaction, and therefore the λ-flavor algebra can be carried over as in the free-field case. For example, for two electromagnetic currents, (19.3) is

$$\begin{aligned} &iTJ^\mu_{\text{em}}(z)J^\nu_{\text{em}}(0)\\ &= g^{\mu\nu}\Big\{\sum_{n=\text{even}}\bar{C}^n_{1NS}(z^2)\Big[\frac{1}{6}N^{\mu_1\ldots\mu_n}_{NS,3}(0)+\frac{1}{6\sqrt{3}}N^{\mu_1\ldots\mu_n}_{NS,8}(0)\Big]\\ &\qquad+\frac{2}{9}\bar{C}^n_{1F}(z^2)N_F^{\mu_1\ldots\mu_n}(0)\Big\}i^nz_{\mu_1}\ldots z_{\mu_n}\\ &\quad+\sum_{n=\text{even}}\Big\{\bar{C}^n_{2NS}(z^2)\Big[\frac{1}{6}N^{\mu\nu\mu_1\ldots\mu_n}_{NS,3}(0)+\frac{1}{6\sqrt{3}}N^{\mu\nu\mu_1\ldots\mu_n}_{NS,8}(0)\Big]\\ &\qquad+\frac{2}{9}\bar{C}^n_{2F}(z^2)N_F^{\mu\nu\mu_1\ldots\mu_n}(0)\Big\}i^nz_{\mu_1}\ldots z_{\mu_n}\\ &\quad+\Big\{g^{\mu\nu}\sum_{n=\text{even}}\bar{C}^n_{1V}(z^2)\frac{2}{9}N_V^{\mu_1\ldots\mu_n}(0)\\ &\qquad+\sum_{n=\text{even}}\bar{C}^n_{2V}(z^2)\frac{2}{9}N_V^{\mu\nu\mu_1\ldots\mu_n}(0)\Big\}i^{n-1}z_{\mu_1}\ldots z_{\mu_n}. \end{aligned} \tag{19.4}$$

We have symmetrized the N; this is permissible if, as occurs in our case, only diagonal matrix elements are required and terms in m_N^2/Q^2 are neglected; cf. Equations (18.15b,c). In fact, both (19.3) and (19.4) have been written rather sketchily. When taking into account interactions, renormalization will occur. This causes, among others, two important effects. First, because operators N_F, N_V have the same quantum numbers (those of a flavor singlet), they will mix under renormalization. Only the NS operators are renormalized by themselves. Second, renormalizations introduce a dependence of the C, N on a dimensional parameter that we will temporarily denote by μ to avoid confusion with Bjorken's variable, $\nu = p\cdot q$.

The currents J,

$$J^\mu(x) = aV^\mu(x) + bA^\mu(x), \tag{19.5}$$

do not require specific renormalization because the operators V, A are conserved or quasi-conserved (Section 13). However, and except in special instances, the operators N require renormalization, and so do the Wilson coefficients.

For the nonsinglet operators, which do not mix, renormalization simply reads[50]

$$N_{NS,a\pm R}^{\mu_1\ldots\mu_n} = Z_{n-2}^{a\pm}(\mu)N_{NS,a\pm}^{\mu_1\ldots\mu_n}. \tag{19.6a}$$

Actually, the Z are independent of $a\pm$.

For the singlet, however, we have matrix renormalization

$$\vec{N}_R^{\mu_1\ldots\mu_n} = \mathbf{Z}_{n-2}\vec{N}^{\mu_1\ldots\mu_n}; \tag{19.6b}$$

we have introduced the vector

$$\vec{N} = \begin{pmatrix} N_F \\ N_V \end{pmatrix}, \tag{19.6c}$$

and the matrix

$$\mathbf{Z} = \begin{pmatrix} Z_{FF} & Z_{FV} \\ Z_{VF} & Z_{VV} \end{pmatrix}. \tag{19.6d}$$

With this, we define the anomalous dimension and anomalous dimension matrices for the operators N,

$$\begin{aligned}\gamma_{NS}(n, g) &= -(Z_n(\mu))^{-1}\frac{\mu\partial}{\partial\mu}Z_n(\mu),\\ \gamma(n, g) &= -(\mathbf{Z}_n(\mu))^{-1}\frac{\mu\partial}{\partial\mu}\mathbf{Z}_n(\mu),\end{aligned} \tag{19.7}$$

and their expansions:

$$\begin{aligned}\gamma_{NS}(n, g) &= \sum_{k=0}^{\infty}\gamma_{NS}^{(k)}(n)\left(\frac{g^2}{16\pi^2}\right)^{k+1},\\ \gamma(n, g) &= \sum_{k=0}^{\infty}\gamma^{(k)}(n)\left(\frac{g^2}{16\pi^2}\right)^{k+1}.\end{aligned} \tag{19.8}$$

We will pursue this matter of renormalization later; for the moment we return to the formal machinery. Let us consider momentum space and write the part of the OPE that contributes to the nonsinglet piece of the structure function f_2 (i.e., to the part of f_2 that contains nonsinglet operators). We have, selecting the appropriate portion of (19.3),

$$\begin{aligned}&i\int d^4z\, e^{iq\cdot z}TJ^{\mu}(z)J^{\nu}(0)\Big|_{p^{\mu}p^{\nu}}^{NS}\\ &\quad = \sum_n \int d^4z\, e^{iq\cdot z}\bar{C}_{2NS}^{n}(z^2)i^n z_{\mu_1}\ldots z_{\mu_n}N_{NS}^{\mu\nu\mu_1\ldots\mu_n}(0),\end{aligned} \tag{19.9}$$

[50]Note that as in Section 13, the quark or gluon fields entering into the N_{NS}, N are assumed *renormalized*.

so if we take the matrix element relevant to deep inelastic scattering as in Equation (17.8*a*), we obtain

$$\frac{p^{\mu}p^{\nu}}{\nu} T_{2NS} = (2\pi)^3 \sum_n \int d^4z\, e^{iq\cdot z} \overline{C}^n_{2NS}(z^2) i^n z_{\mu_1} \dots z_{\mu_n} \langle p | N^{\mu\nu\mu_1\dots\mu_n}_{NS}(0) | p \rangle. \tag{19.10}$$

We can write, up to trace terms,

$$i\langle p | N^{\mu\nu\mu_1\dots\mu_n}_{NS}(0) | p \rangle = p^{\mu}p^{\nu}p^{\mu_1}\dots p^{\mu_n} \overline{A}^n_{NS} \tag{19.11}$$

and replace

$$z_{\mu_1}\dots z_{\mu_n} \to (-i)^n \frac{\partial}{\partial q_{\mu_1}} \dots \frac{\partial}{\partial q_{\mu_n}} = (-2i)^n q_{\mu_1}\dots q_{\mu_n}\left(\frac{\partial}{\partial q^2}\right)^2 + \text{trace terms}. \tag{19.12}$$

Therefore, (19.10) becomes

$$\begin{aligned} T_{2NS}(x, Q^2; g, \mu) &= (2\pi)^3 \nu \sum_{n=\text{even}} 2^n \overline{A}^n_{NS} \left(\frac{\partial}{\partial q^2}\right)^n \int d^4z\, e^{iq\cdot z} \frac{1}{i} \overline{C}^n_{2NS}(z^2)(q\cdot p)^n \\ &= \frac{1}{2}(2\pi)^3 \sum_{n=\text{even}} (2\nu)^{n+1} \overline{A}^n_{NS} \left(\frac{\partial}{\partial q^2}\right)^n \int d^4z\, e^{iq\cdot z} \frac{1}{i} \overline{C}^n_{2NS}(z^2). \end{aligned} \tag{19.13}$$

Because we know that in the free-field case $\overline{C}^n_{2NS}(z^2)$ behaves as (cf. Section 18)

$$i\overline{C}^n_{2NS}(z^2)|_{g=0} \underset{z^2\to 0}{=} \frac{1}{\pi^2(z^2 - i0)}, \tag{19.14}$$

we will define new coefficients in momentum space by

$$C^n_{2NS}(Q^2/\mu^2, g^2/4\pi) \equiv 4(Q^2)^{n+1}\left(\frac{\partial}{\partial q^2}\right)^n \int d^4z\, e^{iq\cdot z} \frac{1}{i} \overline{C}^n_{2NS}(z^2), \tag{19.15}$$

and arrive at the expression

$$T_{2NS}(x, Q^2; g, \mu) = 2\sum \frac{1}{x^{n+1}} A^n_{NS} C^n_{2NS}(Q^2/\mu^2, g^2/4\pi); \qquad A \equiv (2\pi)^3 \overline{A}. \tag{19.16}$$

As we will see later, asymptotic freedom allows us to calculate the Wilson coefficients C in (19.16); but, in general, the A are unknown constants. To be able to extract physical information, we have to single out the individual terms in (19.16). This is done by using the known analyticity properties of the T to write a dispersion relation[51] for T_2 at fixed Q^2, in the variable ν:

$$T_{2NS}(x, Q^2; g, \mu) = \frac{1}{\pi}\left\{\int_{Q^2/2}^{\infty} \frac{d\nu'}{\nu' - \nu} \operatorname{Im} T_{2NS}\left(\frac{Q^2}{2\nu'}, Q^2; g, \mu\right) - \int_{-\infty}^{-Q^2/2} \frac{d\nu'}{\nu' - \nu} \operatorname{Im} T_{2NS}\left(\frac{Q^2}{2\nu'}, Q^2; g, \mu\right)\right\}. \tag{19.17}$$

We can only relate this to the physical structure functions if T has a definite signature, i.e., is even/odd under the exchange $q \to -q$. This is the case for T_2 in electroproduction, as $T_2(x, \ldots) = T_2(-x, \ldots)$. Then we can change variables, $\nu' \to x' = Q^2/2\nu'$ and rewrite (19.17) as

$$T_{2NS}(x, Q^2; g, \mu) = \frac{1}{\pi}\int_0^1 \frac{dx'}{x'(1 - x'^2/x^2)} \operatorname{Im} T_{2NS}(x', Q^2; g, \mu).$$

It only remains to expand in powers of x'/x to obtain [Cornwall and Norton (1969)]

$$T_{2NS}(x, Q^2; g, \mu) = 2\sum_n \frac{1}{x^n} \mu_{2NS}(n+1, Q^2; g, \mu), \tag{19.18}$$

where the *moments* μ_{2NS} are defined by

$$\mu_{2NS}(n, Q^2; g, \mu^2) = \int_0^1 dx'\, x'^{n-2} f_{2NS}(x', Q^2; g, \nu). \tag{19.19}$$

Comparing with (19.16), we immediately obtain the expression for the moments:

$$\mu_{2NS}(n, Q^2; g, \mu^2) = A_{NS}^n C_{2NS}^n(Q^2/\mu^2, g^2/4\pi). \tag{19.20}$$

It should be kept in mind that we have derived Equations (19.19) and (19.20) under the assumption of evenness for T: otherwise we cannot replace the integral $\int_{-1}^{0} dx'$ by $\int_0^1 dx'$. Therefore, Equations (19.19) and (19.20) are only valid for $n =$ even (if T is even, as is the case for electroproduction) or $n =$ odd, if T is odd, as occurs for T_3 in ν scattering. The corresponding equations for other n have to be obtained by analytic (Regge–Carlson) continuation. This is trivial for the leading order calculations (see Section 20) but presents some tricky points for second order ones. Another point is that, as already noted, we have to restrict above equations to values of n such that (19.19) converges. From Regge theory we expect that this will occur at least when $\operatorname{Re} n \geqslant 1$ for the nonsinglet, and for $\operatorname{Re} n \geqslant 2$ for the singlet (see also Section 23*ii*).

[51] In principle, the dispersion relation should be written with subtractions, but it may be seen that these alter nothing, provided that (19.19) is convergent. For information on dispersion relations, see the treatise of Eden *et al.* (1966).

20 Renormalization Group Analysis: The QCD Equations for the Moments

We will now write a renormalization equation for the moments. Since these are integrals over the structure functions, they are physical observables and hence, independent of the renormalization point. As a result of Equations (19.6), (19.11) and (19.20), it follows that the renormalization constant of the C is precisely the inverse of that of the operators N_R. Thus we obtain the Callan–Symanzik equation

$$\left\{\mu\frac{\partial}{\partial\mu}+\beta(g)g\frac{\partial}{\partial g}-\gamma_{NS}(g,n)\right\}C_{2NS}^{n}(Q^2/\mu^2,\,g^2/4\pi)=0, \tag{20.1}$$

with solution

$$C_{2NS}^{n}(Q^2/\mu^2,\,g^2/4\pi)=e^{-\int_0^t d\log(Q'/\mu)\gamma_{NS}(g(Q'^2),n)}C_{2NS}^{n}\left[1,\alpha_s(Q^2)\right],$$
$$t=\tfrac{1}{2}\log Q^2/\mu^2. \tag{20.2}$$

For the singlet case, there are some complications due to the coupled character of the equations. It is necessary to introduce an extra structure function,

$$f_V(x,\,Q^2),$$

whose physical interpretation is that it describes the gluons one finds in a nucleon. With the vector notation

$$\vec{f}=\begin{pmatrix} f_F\\ f_V\end{pmatrix},\qquad \vec{C}^n=\begin{pmatrix} C_F^n\\ C_V^n\end{pmatrix},$$
$$\vec{\mu}_2(n,\,Q^2)=\int_0^1 dx\,x^{n-2}\vec{f}_2(x,\,Q^2), \tag{20.3}$$

the analog of (20.2) is

$$\vec{C}_2^n(Q^2/\mu^2,\,g^2/4\pi)$$
$$=Te^{-\int_0^t d\log(Q'/\mu)\gamma(g(Q'^2),n)}\vec{C}_2^n(1,\alpha_s(Q^2)). \tag{20.4}$$

The T operator is formally identical to time ordering except that here it orders the variable $t=\frac{1}{2}\log Q^2/\mu^2$. [For details, see Gross and Wilczek (1974); Gross (1976).] Because of asymptotic freedom, we can use perturbation theory, and from Equations (20.2) and (20.4), calculate the Wilson coefficients; but since the $\bar{A}^n$ are still unknown, we will only be able to predict the *evolution* of the moments with Q^2. To see this, consider (20.2) to lowest order. We obtain

$$C_{2NS}^{n}(Q^2/\mu^2,\,g^2(\mu)/4\pi)=C_{2NS}^{n}(1,0)\left(\frac{\log Q^2/\Lambda^2}{\log\mu^2/\Lambda^2}\right)^{d(n)}, \tag{20.5}$$

where the anomalous dimension $d(n)$ is

$$d(n) = -\gamma_{NS}^{(0)}(n)/2\beta_0. \tag{20.6a}$$

$C_{2NS}^n(1,0)$ is merely the free-field value of the Wilson coefficient, which we calculated in Section 18. One can eliminate the unknown constant $\overline{A}^n$, and μ^2 by normalizing to a given Q_0^2 sufficiently large so that $\alpha_s(Q_0^2)$ will be small enough to justify perturbative expansions. Then we obtain the QCD equations for the evolution of the moments to leading order: dropping unnecessary labels,

$$\mu_{NS}(n, Q^2) = \left[\frac{\alpha_s(Q_0^2)}{\alpha_s(Q^2)}\right]^{d(n)} \mu_{NS}(n, Q_0^2). \tag{20.6b}$$

For the singlet,

$$\begin{gathered}\vec{\mu}(n, Q^2) = \left[\frac{\alpha_s(Q_0^2)}{\alpha_s(Q^2)}\right]^{\mathbf{D}(n)} \vec{\mu}(n, Q_0^2);\\ \mathbf{D}(n) = -\gamma^{(0)}(n)/2\beta_0.\end{gathered} \tag{20.7}$$

It only remains for us to calculate the $\gamma_{NS}^{(0)}$, $\gamma^{(0)}(n)$. First, one has to deduce the Feynman rules for the operators N. This is straightforward (see Section 42); they are collected in Appendix E. Then we have to calculate the renormalization constants for the N. The singlet case may be found in the paper of Gross and Wilczek (1974); here we will concentrate on $N_{NS}^{\mu_1 \cdots \mu_n}$, which involves the diagrams of Figure 14. In the Feynman gauge,

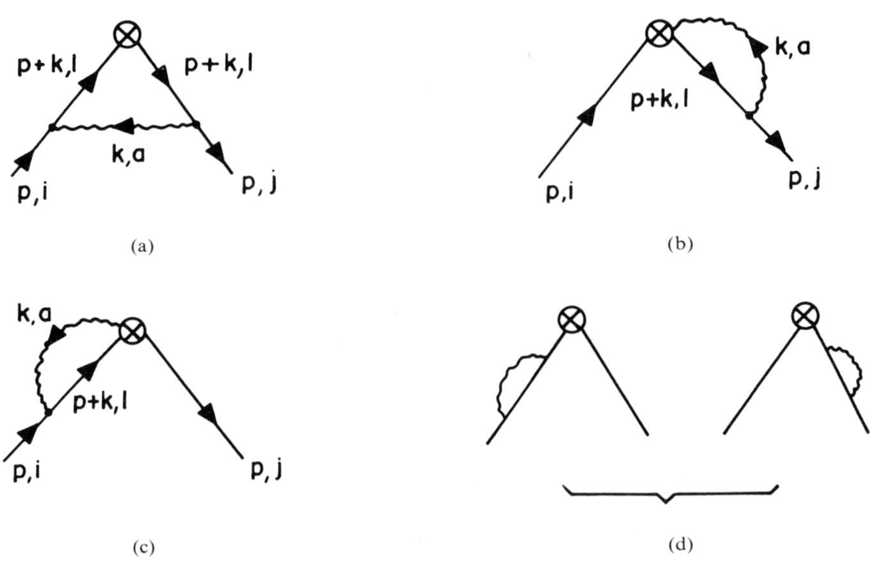

Figure 14. Diagrams involved in the calculation of Z_n^{NS}.

diagram a yields

$$V_{Aij} = i^5 g^2 \int d^D\hat{k}\, \frac{\gamma^\mu \not{k} \not{\Delta} (\Delta \cdot k)^{n-1} \not{k} \gamma^\nu (-g_{\mu\nu})}{k^4 (k-p)^2} \sum_{a,l} t^a_{il} t^a_{lj}\, .$$

To calculate Z, we only require the divergent part of the coefficient of $(\Delta \cdot p)^{n-1}\not{\Delta}$. Let us use the notation $a \overset{\text{eff}}{=} b$ to mean that a and b have equivalent divergent parts. And after standard manipulations,

$$V_{Aij} = ig^2 C_F \delta_{ij}$$

$$\times \int_0^1 dx\,(1-x) \int d^D\hat{l}\, \frac{-2\gamma^\alpha (\not{l} + x\not{p}) \not{\Delta} (\not{l} + x\not{p}) \gamma_\alpha \left[\Delta \cdot (l + xp)\right]^{n-1}}{\left(l^2 + x(1-x)p^2\right)^3}\, .$$

The divergent part of the term proportional to $(\Delta \cdot p)^{n-1}\not{\Delta}$ is easily identified:

$$V_{Aij} \overset{\text{eff}}{=} ig^2 \delta_{ij} C_F \int_0^1 dx\,(1-x) \int \frac{d^D\hat{l}}{\left[l^2 + x(1-x)p^2\right]^3}$$

$$\times \left\{ -\frac{2l^2}{D} \gamma^\alpha \gamma^\beta \not{\Delta} \gamma_\beta \gamma_\alpha x^{n-1} \right\} \not{\Delta} (\Delta \cdot p)^{n-1}$$

$$= \frac{g^2}{16\pi^2} N_\epsilon C_F \frac{2}{n(n+1)} (\Delta \cdot p)^{n-1} \not{\Delta} \delta_{ij}\, . \tag{20.8}$$

As for Figure 14b,

$$V_{Bij} = -i^3 g^2 C_F \delta_{ij} \int d^D\hat{k}\, \frac{\Delta^\mu \not{\Delta} \left\{ \sum_{l=0}^{n-2} (\Delta \cdot p)^l \left[\Delta \cdot (p+k)\right]^{n-l-2} (\not{p} + \not{k}) \gamma_\mu \right.}{k^2 (k+p)^2}\, .$$

Here, also, we have to extract the coefficient of the $(\Delta \cdot p)^{n-1}\not{\Delta}$ term; so

$$V_{Bij} \overset{\text{eff}}{=} 2ig^2 C_F \delta_{ij} \not{\Delta} \int_0^1 dx \int d^D\hat{q}\, \frac{\sum_{l=0}^{n-2} (\Delta \cdot p)^l \left[\Delta \cdot q + x\Delta \cdot p\right]^{n-1-l}}{\left(q^2 + x(1-x)p^2\right)^2}$$

$$\overset{\text{eff}}{=} -2 \frac{g^2 N_\epsilon}{16\pi^2} C_F \delta_{ij} (\Delta \cdot p)^{n-1} \not{\Delta} \int_0^1 dx \sum_{l=1}^{n-1} x^l$$

$$= \frac{g^2}{16\pi^2} N_\epsilon C_F \delta_{ij} \left(-2 \sum_{l=2}^{n} \frac{1}{l} \right) (\Delta \cdot p)^{n-1} \not{\Delta}. \tag{20.9}$$

Figure 14c gives the same result as Figure 14b, Equation (20.9). Figure 14d merely gives terms equivalent to Z_F. To obtain γ_{NS}, we have to add the counterterm contribution, to obtain $Z_n Z_F^{-1} N$ finite. Therefore, using the value of Z_F calculated in Section 9, we find

$$Z_n^{NS} = 1 + \frac{g^2 N_\epsilon}{16\pi^2} C_F \left\{ 4S_1(n) - 3 - \frac{2}{n(n+1)} \right\}, \tag{20.10}$$

$$S_1(n) = \sum_{j=1}^{n} \frac{1}{j}, \tag{20.11}$$

and thus

$$\gamma_{NS}^{(0)}(n) = 2C_F \left\{ 4S_1(n) - 3 - \frac{2}{n(n+1)} \right\}, \tag{20.12}$$

$$d(n) = \frac{16}{33 - 2n_f} \left\{ \frac{1}{2n(n+1)} + \frac{3}{4} - S_1(n) \right\}. \tag{20.13}$$

Likewise, for the singlet case, we obtain the matrix

$$\mathbf{D}(n) = \frac{16}{33 - 2n_f} \times \begin{bmatrix} \frac{33 - 2n_f}{16} d(n) & \frac{3n_f}{8} \cdot \frac{n^2 + n + 2}{n(n+1)(n+2)} \\ \frac{n^2 + n + 2}{2n(n^2 - 1)} & \frac{33 - 2n_f}{16} + \frac{9}{4} \left\{ \frac{1}{n(n-1)} + \frac{1}{(n+1)(n+2)} - S_1(n) \right\} \end{bmatrix}. \tag{20.14}$$

The expression we have given for $S_1(n)$ can be continued analytically to complex n. Due to Carlson's theorem [see, e.g., Titchmarsh (1939)], there is only one such continuation with the property that Equations (19.19), (20.3), (20.6), and (20.7) remain valid for complex n; it is

$$S_1(n) = n \sum_{k=1}^{\infty} \frac{1}{k(k+n)}. \tag{20.15a}$$

Note that with this definition [which coincides with (20.11) for integer n], one has

$$S_1(n) = \psi(n+1) + \gamma_E, \qquad \psi(z) \equiv \frac{d \log \Gamma(z)}{dz}. \tag{20.15b}$$

To this order, there is no problem with even/odd structure functions nor with the corresponding validity of the original equations for only even/odd values of n because the continuations of $\gamma^{(0)}(n)$ starting from even or odd values of n coincide.

21 QCD Equations for the Moments to Second Order

In the previous section, we derived the QCD equations for the moments to leading order. Now we will turn to the second-order corrections.

From Equations (20.2) and (20.4), we see that to calculate next-to-leading contributions, we have to consider two separate effects.[52] First, we have the effect of the anomalous dimension to second order, $\gamma_{NS}^{(1)}(n)$, $\gamma^{(1)}(n)$. Then we must calculate the next term in the expansion of the Wilson coefficients:

$$C_{NS}^{n}\left(1,\alpha_s(Q^2)\right)=C_{NS}^{n}(1,0)\left\{1+C_{NS}^{n(1)}(1,0)\frac{\alpha_s(Q^2)}{4\pi}+\cdots\right\}. \quad (21.1)$$

The calculation of the anomalous dimensions was carried out by Floratos, Ross, and Sachrajda (1977), for the NS, and (1979), for the singlet. Their results were simplified and expressed analytically by González–Arroyo *et al.* (1979) for the NS, and by González–Arroyo and López (1980), for the singlet. These results have been checked recently by Curci, Furmanski, and Petronzio (1980) and Furmanski and Petronzio (1980); only in the coefficient of the term for $\gamma_{VV}^{(1)}(n)$ is there disagreement with the previous calculation of *FRS*.[53] We let $\gamma_{NS}^{(1)\pm}(n)$ refer to even/odd structure functions. Then,

$$\begin{aligned}\gamma_{NS}^{(1)\pm}(n)={}&\frac{32}{9}S_1(n)\left[67+8\frac{2n+1}{n^2(n+1)^2}\right]-64S_1(n)S_2(n)\\&-\frac{32}{9}\left[S_2(n)-S_2^{\pm}(n/2)\right]\left\{2S_1(n)-\frac{1}{n(n+1)}\right\}\\&-\frac{128}{9}\tilde{S}^{\pm}(n)+\frac{32}{3}S_2(n)\left[\frac{3}{n(n+1)}-7\right]+\frac{16}{9}S_3^{\pm}\left(\frac{n}{2}\right)\\&-28-16\frac{151n^4+260n^3+96n^2+3n+10}{9n^3(n+1)^3}\\&\pm\frac{32}{9}\cdot\frac{2n^2+2n+1}{n^3(n+1)^3}+\frac{32n_f}{27}\\&\times\left\{6S_2(n)-10S_1(n)+\frac{3}{4}+\frac{11n^2+5n-3}{n^2(n+1)^2}\right\},\end{aligned} \quad (21.2a)$$

$$\begin{aligned}&S_l^{+}(x/2)=S_l(x/2),\qquad S_l^{-}(x/2)=S_l\left(\frac{x-1}{2}\right),\\&\tilde{S}^{\pm}(x)=-\frac{5}{8}\zeta(3)\mp\sum_{k=1}^{\infty}\frac{(-1)^k}{(k+x)^2}S_1(k+x).\end{aligned} \quad (21.2b)$$

[52] Besides, of course, using the second-order expression for $\alpha_s(Q^2)$, Section 14, and taking into account the finite parts of the first-order diagrams.

[53] The Furmanski–Petronzio result has recently been checked independently.

The $\gamma_{ij}^{(1)}(n)$ may be found collected in López and Ynduráin (1981) with the Furmanski–Petronzio result for $\gamma_{VV}^{(1)}$.

Let us now turn to the Wilson coefficients. Since these are constants, they can be calculated by taking matrix elements of $TJ^{\mu}J^{\nu}$ between arbitrary states. We are at liberty to take whichever states make the calculation simplest, and, of course, we choose quark and gluon states. A point to be kept in mind is that, unlike the anomalous dimensions, the Wilson coefficients depend on the process and structure function under consideration. The collected values[54] of the $C_{NS}^{n(1)}(1,0)$, $\vec{C}^{n(1)}(1,0)$ may be found in the paper of Bardeen *et al.* (1978) or the review of Buras (1980). Here we will present a sample calculation for the longitudinal structure function.

To leading order, the two structure functions f_1, f_2 are equal, and hence $f_L = 0$. This was shown in Section 18 for the free-field case: but since leading order corrections only multiply $C_L^n(1,0)$ by a factor $(\log Q^2/\Lambda^2)^{\delta(n)}$, with $\delta = d$ or $\mathbf{D}$, it follows that all moments of f_L vanish to this order, as claimed. This means that for the longitudinal case, (21.1) should actually read

$$C_L^n(1,\alpha_s) = C_L^{n(1)}(1,0)\frac{\alpha_s}{4\pi} + \cdots . \tag{21.3}$$

This measures the perturbative violation of the Callen–Gross relation. It is convenient to extract a factor that depends upon the process and a process-independent part, writing

$$C_{PL}^{n(1)}(1,0) = \delta_P B_L^{n(1)}. \tag{21.4}$$

The factors δ_P are, with $N = p$ or n, I an "isoscalar" nucleon,

$$\begin{aligned} \delta_{PNS} &= \begin{cases} \frac{1}{6}, & \text{for } f_2^{eN} \\ 1, & \text{for } f_2^{\nu \pm I} \end{cases} \\ \delta_{PF} &= \begin{cases} \frac{5}{18}, & \text{for } f_{2F}^{eN} \\ 1, & \text{for } f_2^{\nu \pm I}. \end{cases} \end{aligned} \tag{21.5}$$

Let us then consider f_L^{NS}. The calculation involves the diagrams of Figure 15, for all other diagrams contribute equally to the terms of which f_L^{NS} is the

[54] Some of the C were calculated previously by De Rújula, Georgi, and Politzer (1977*a*); Calvo (1977); Altarelli, Ellis, and Martinelli (1978); Kubar-André and Paige (1979); Abad and Humpert (1978); Zee, Wilczek, and Treiman (1974); Kingsley (1973); Walsh and Zerwas (1973); Hinchliffe and Llewellyn Smith (1977); Floratos, Ross, and Sachrajda (1979); Witten (1976), etc. The values reported by Bardeen *et al.* (1978) or Buras (1980) have all been checked by at least two independent calculations.

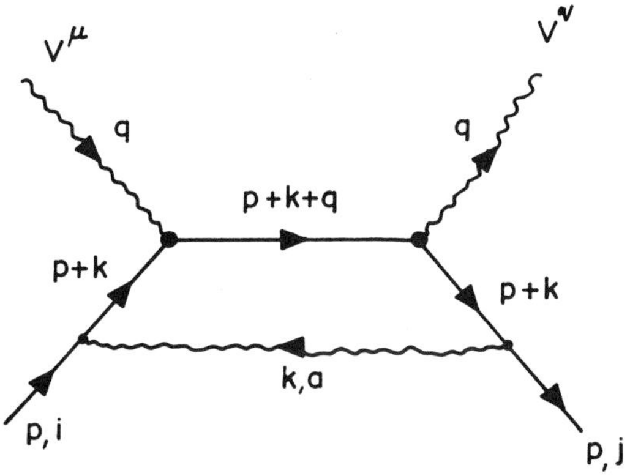

Figure 15. Diagram contributing to f_L (nonsinglet).

difference, or to the singlet part.[55] Moreover, because f_L begins at order α_s, we do not have to worry about the contribution of the renormalization of the N operators which will, in this case, yield terms $O(\alpha_s^2)$. The calculation is further simplified by noting that if we keep the q^μ, q^ν terms in $T^{\mu\nu}$, then f_L is the only invariant amplitude which is multiplied by $q^\mu q^\nu$: for, say, vector currents,

$$T^{\mu\nu} = (g^{\mu\nu} - q^\mu q^\nu/q^2)T_L + \left(g^{\mu\nu} - p^\mu p^\nu \frac{q^2}{\nu} + \frac{p^\mu q^\nu + p^\nu p^\mu}{\nu}\right)T_2,$$
$$f_L = \frac{1}{2\pi}\,\mathrm{Im}\,T_L\,. \tag{21.6}$$

In general, we have to carry out the calculation for $p^2 < 0$ to regulate infrared divergences; but, again, this is unnecessary for f_L which, to the order we are working, remains finite when $p^2 \to 0$.

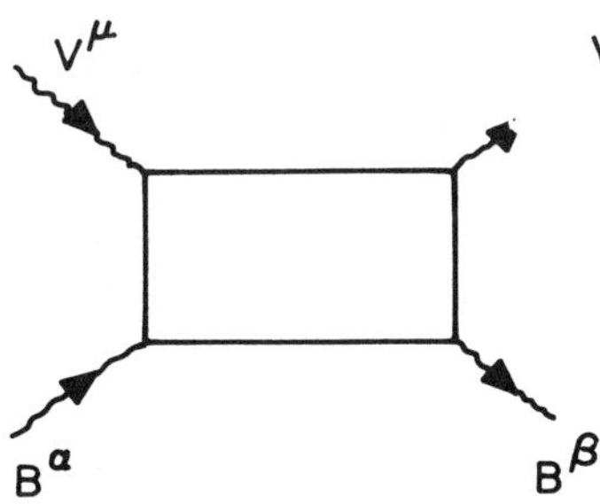

Figure 16. Diagram contributing to the singlet part of f_L.

[55] For the singlet part, the diagrams of Figure 16 would also have to be considered.

The amplitude for the diagram of Figure 15 is

$$\frac{i}{2}(2\pi)^3\sum_\sigma\int d^4z\, e^{iq\cdot z}\langle p,\sigma|TJ^\mu(z)J^\nu(0)|p,\sigma\rangle$$

$$=\left[T_{ij}'^{\mu\nu}=-iC_F\delta_{ij}g^2\frac{1}{4}\sum_\sigma\bar{u}(p,\sigma)\right.$$

$$\left.\times\int d^D\hat{k}\,\frac{\gamma_\alpha(\not{p}+\not{k})\gamma^\mu(\not{p}+\not{k}+\not{q})\gamma^\nu(\not{p}+\not{k})\gamma^2}{(p+k)^4(p+k+q)^2k^2}u(p,\sigma)\right]$$

$$+\text{ crossed term.}$$

Using

$$\sum_\sigma\bar{u}(p,\sigma)\mathcal{M}u(p,\sigma)=\mathrm{Tr}(\not{p}\mathcal{M}),$$

extracting the term proportional to $q^\mu q^\nu$, and introducing Feynman parameters, we find

$$T_L'^{NS}=\frac{g^2}{16\pi^2}C_F\frac{8}{x}\int_0^1 d\alpha\cdot\alpha\int_0^1 d\beta\frac{(1-u_2)u_1}{\left[1-u_2-(1-(u_1+u_2)/x)\right]^2},$$

where $u_1=\alpha\beta$ and $u_2=1-\alpha$. Expanding in powers of $1/x$ and integrating,

$$T_L'^{NS}=\frac{g^2}{16\pi^2}4C_F\sum_{n=1}^\infty\frac{1}{n+1}\left(\frac{1}{x}\right)^n.$$

The crossed diagram doubles even and cancels odd powers of $1/x$, so

$$T_L^{NS}=\frac{2g^2}{16\pi^2}C_F\sum_{n=\text{even}}^\infty\frac{4}{n+1}\left(\frac{1}{x}\right)^n;\tag{21.7}$$

writing the analog to Equation (19.18), we therefore find

$$\begin{aligned}B_L^{n(1)NS}&=\frac{4}{n+1}C_F,\qquad n=\text{even},\\ \mu_L^{NS}(n,Q^2)&=\delta_L^{NS}\frac{\alpha_s(Q^2)}{\pi}\cdot\frac{C_F}{n+1}\mu_2^{NS}(n,Q^2).\end{aligned}\tag{21.8}$$

For details about the calculation of the other B, we refer to the very comprehensible paper of Bardeen *et al.* (1978); and here we will merely give the results for electroproduction on proton targets:

$$C_{NS}^{(1)}(n) = C_F^{(1)}(n)$$
$$= C_F\left\{2[S_1(n)]^2 + 3S_1(n) - 2S_2(n) - \frac{2S_1(n)}{n(n+1)} + \frac{3}{n} + \frac{4}{n+1} + \frac{2}{n^2} - 9\right\}, \tag{21.9a}$$

$$C_V^{(1)}(n) = 4T_F n_f\left\{-\frac{1}{n} + \frac{1}{n^2} + \frac{6}{n+1} - \frac{6}{n+2} - S_1(n)\frac{n^2+n+2}{n(n+1)(n+2)}\right\}. \tag{21.9b}$$

Once we have explained the general methods, we can write explicitly the QCD equations for the moments to second order. For the nonsinglet, we have

$$\mu_{NS}(n, Q^2) = \left[\frac{\alpha_s(Q_0^2)}{\alpha_s(Q^2)}\right]^{d(n)} \times \frac{1 + C_{NS}^{(1)}(n)\alpha_s(Q^2)/4\pi}{1 + C_{NS}^{(1)}(n)\alpha_s(Q_0^2)/4\pi}\left\{\frac{1 + \beta_1\alpha_s(Q^2)/4\pi\beta_0}{1 + \beta_1\alpha_s(Q_0^2)/4\pi\beta_0}\right\}^{p(n)} \times \mu_{NS}(n, Q_0^2);$$
$$p(n) = \tfrac{1}{2}\left\{\gamma_{NS}^{(1)}(n)/\beta_1 - \gamma_{NS}^{(0)}(n)/\beta_0\right\}. \tag{21.10}$$

For the singlet, we have some extra complications. We have to begin by defining the matrix $\mathbf{C}^{(1)}(n)$ with $C_{12}^{(1)}(n) = C_V^{(1)}(n)$, $C_{11}^{(1)}(n) = C_F^{(1)}(n)$,

$$C_{21}^{(1)}(n) = \frac{D_{21}(n)}{D_{12}(n)}C_{12}^{(1)}(n); \quad C_{22}^{(1)}(n) = C_{11}^{(1)}(n) + \frac{D_{22}(n) - D_{11}(n)}{D_{12}(n)}C_{12}^{(1)}(n).$$

With these definitions, $\mathbf{C}^{(1)}$ and $\mathbf{D}$ commute. Writing α for $\alpha_s(Q^2)$ and α_0 for $\alpha_s(Q_0^2)$, the equations are [González–Arroyo and López (1980)]

$$\vec{\mu}(n, Q^2) = \mathbf{C}(n,\alpha)\mathbf{C}^{-1}(n,\alpha_0)\mathbf{M}(n;\alpha,\alpha_0)\,\vec{\mu}(n, Q_0^2) \tag{21.11}$$

with $\mathbf{C} = 1 + \mathbf{C}^{(1)}\alpha/4\pi$,

$$\mathbf{R}(n;\alpha,\alpha_0) = 1 - \frac{\alpha - \alpha_0}{4\pi}\cdot\frac{\beta_1}{2\beta_0^2}\gamma^{(0)}(n) + \mathbf{\Delta}(n;\alpha,\alpha_0),$$

$$\mathbf{\Delta}(n;\alpha,\alpha_0) = \frac{-3}{32}\cdot\frac{\alpha_0}{4\pi}\int_0^r dr'\, e^{-3\beta_0 r'/16}\left[\mathbf{M}^{(0)}(n,r')\right]^{-1}\gamma^{(1)}(n)\mathbf{M}^{(0)}(n,r'),$$

$$\mathbf{M}(n;\alpha,\alpha_0) = \left(\frac{\alpha_0}{\alpha}\right)^{\mathbf{D}(n)}\mathbf{R}(n;\alpha,\alpha_0),$$

and, finally,

$$r=\frac{16}{3\beta_0}\log\frac{\alpha_0}{\alpha},\qquad \mathbf{M}^{(0)}(n,r')=e^{-3r'\gamma^{(0)}(n)/32}.$$

The equations for the singlet can be rewritten in another form that is useful for some applications. Let $\mathbf{S}(n)$ be the matrix that diagonalizes $\mathbf{D}(n)$,

$$\mathbf{S}^{-1}(n)\mathbf{D}(n)\mathbf{S}(n)=\hat{\mathbf{D}}(n)\equiv\begin{pmatrix} d_+(n) & 0\\ 0 & d_-(n)\end{pmatrix},\qquad d_+(n)>d_-(n);$$

we can take this such that $\det\mathbf{S}=S_{11}=1$:

$$\mathbf{S}(n)=\begin{bmatrix} 1 & \dfrac{D_{12}(n)}{d_-(n)-d_+(n)}\\[2ex] \dfrac{d_+(n)-D_{11}(n)}{D_{12}(n)} & \dfrac{d_-(n)-D_{11}(n)}{d_-(n)-d_+(n)}\end{bmatrix}. \tag{21.12}$$

Define $\bar{\gamma}$ to be the transform of $\gamma^{(1)}$ under $\mathbf{S}$:

$$\mathbf{S}^{-1}(n)\gamma^{(1)}(n)\mathbf{S}(n)=\bar{\gamma}(n). \tag{21.13}$$

We then have

$$\begin{aligned}\alpha^{\hat{\mathbf{D}}(n)}&\left\{1+\frac{\alpha}{4\pi}\boldsymbol{\Gamma}(n)\right\}\mathbf{S}^{-1}(n)\mathbf{C}^{-1}(n,\alpha)\vec{\mu}(n,Q^2)\\ &=\alpha_0^{\hat{\mathbf{D}}(n)}\left\{1+\frac{\alpha_0}{4\pi}\boldsymbol{\Gamma}(n)\right\}\mathbf{S}^{-1}(n)\mathbf{C}^{-1}(n,\alpha_0)\vec{\mu}(n,Q_0^2)\\ &\equiv\vec{b}(n)\quad\text{independent of }Q^2.\end{aligned} \tag{21.14}$$

Here[56]

$$\boldsymbol{\Gamma}(n)=\frac{-1}{2\beta_0}\begin{bmatrix}\bar{\gamma}_{11}(n)+2\beta_1 d_+(n) & \dfrac{\bar{\gamma}_{12}(n)}{d_+(n)-d_-(n)+1}\\[2ex] \dfrac{\bar{\gamma}_{21}(n)}{d_-(n)-d_+(n)+1} & \bar{\gamma}_{22}(n)+2\beta_1 d_-(n)\end{bmatrix}.$$

Equations (2.10) and (2.11) apply to the moments of functions f_2, f_3. For f_L, we may express it in terms of f_2, using (21.8):

$$f_L=f_L^{NS}+f_L^F+f_L^V, \tag{21.15a}$$

$$f_L^{NS}(x,Q^2)=\frac{4\alpha_s}{3\pi}\int_x^1 dy\,\frac{x^2}{y^3}\,f_2^{NS}(y,Q^2), \tag{21.15b}$$

$$f_L^{F}(x,Q^2)=\frac{4\alpha_s}{3\pi}\int_x^1 dy\,\frac{x^2}{y^3}\,f_2^{F}(y,Q^2), \tag{21.15c}$$

$$f_L^{V}(x,Q^2)=\frac{4\alpha_s}{3\pi}\,\delta_L\int_x^1 dy\,\frac{x^2}{y^3}\left(1-\frac{x}{y}\right)f_2^{V}(y,Q^2), \tag{21.15d}$$

[56]The equations are somewhat modified for the two values $n_\pm$ such that $d_-(n_\pm)-d_+(n_\pm)+1=0$, where the next-to-leading correction is not $O(\alpha_s)$, but $O(\alpha_s\log\alpha_s)$.

where, for electroproduction on protons,

$$\delta_L = \frac{3n_f}{2}. \tag{21.16}$$

22 The Altarelli–Parisi Method

The OPE method for analysis of deep inelastic scattering is fairly rigorous and not too difficult to use; but it does not, perhaps, appeal to physical intuition. In particular, its connection with the parton model is not particularly transparent. This is one of the reasons for the success of the Altarelli–Parisi method [Altarelli and Parisi (1977); see also Dokshitzer, Dyakonov, and Troyan (1980)] in which close contact with the parton model is maintained at each step.

Before we discuss the partonic interpretation, let us further elaborate the equations we have. For the sake of definiteness, we will consider the nonsinglet part of $f_2(x, Q^2)$; in fact, we will take the contribution of a given quark flavor f to f_2^{NS}, q_f. In the free-parton model [recall Equation (17.11)], q_f is independent of Q^2, but when interactions are taken into account it will acquire a Q^2 dependence. If we let μ^2 be a fixed reference momentum and define $t = \frac{1}{2}\log Q^2/\mu^2$, then we generalize (17.11) to

$$f_2^{NS}(x, Q^2) = \sum \delta_f^{NS} x q_f(x,t), \tag{22.1}$$

where the δ_f are known coefficients.

What the QCD equations give us is the *evolution* of moments with t; thus, we recast (20.6) in differential form, which for the q_f reads

$$\frac{d\tilde{q}_f(n,t)}{dt} = \frac{\gamma_{NS}^{(0)}(n)\alpha_g(t)}{4\pi}\tilde{q}_f(n,t); \tag{22.2}$$

we have used (14.3) with ν replaced by μ and defined the moments

$$\tilde{q}_f(n,t) = \int_0^1 dx\, x^{n-1} q_f(x,t). \tag{22.3}$$

Of course, Equations (22.1)–(22.3) are completely equivalent to (20.6). Next we invert the Mellin transform (22.3). If we define the function $P_{NS}^{(0)}(z)$ by

$$\int_0^1 dz\, z^{n-1} P_{NS}^{(0)}(z) = \gamma_{NS}^{(0)}(n), \tag{22.4a}$$

then the convolution theorem[57] tells us that

$$\frac{dq_f(x,t)}{dt} = \frac{\alpha_g(t)}{4\pi}\int_x^1 \frac{dy}{y} q_f(y,t) P_{NS}^{(0)}\left(\frac{x}{y}\right). \tag{22.4b}$$

This is the Altarelli–Parisi equation; it can be rewritten in infinitesimal

[57] A simple way to obtain this theorem is to relate it to the standard one for Laplace transforms by changing variables, $z \to \log z = \zeta$. Alternatively, one may integrate (22.5) to get (22.2).

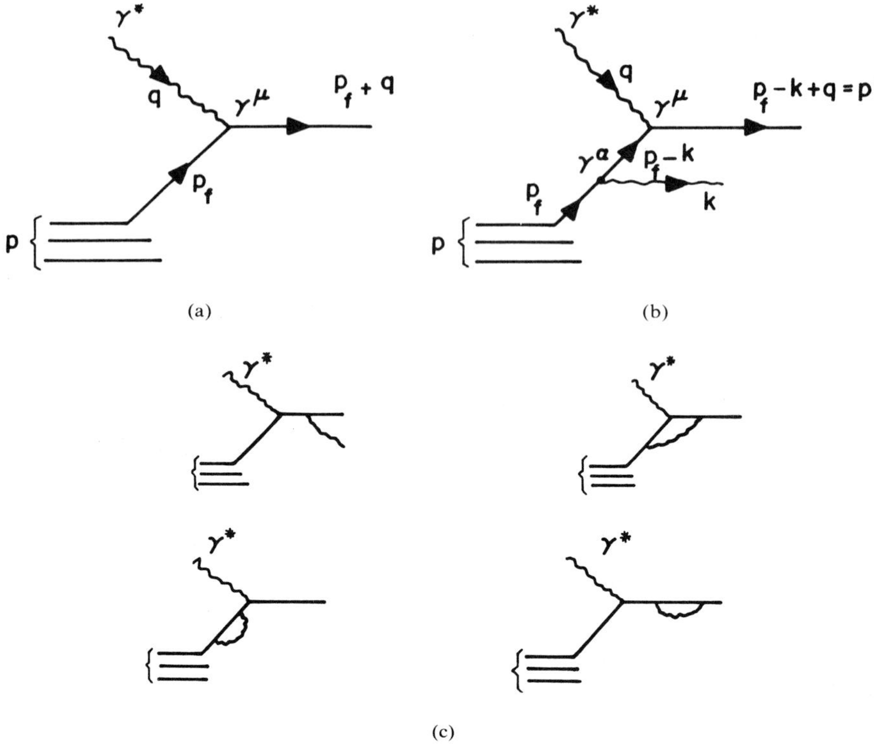

Figure 17. Elementary processes contributing to $\gamma^* + p \to$ all.

form as

$$q_f(x,t) + dq_f(x,t) = \int_0^1 dy \int_0^1 dz\, \delta(zy - x) q_f(y,t)$$
$$\times \left\{ \delta(z-1) + \frac{\alpha_y(t)}{4\pi} P_{NS}^{(0)}(z)\, dt \right\}. \qquad (22.5)$$

We see that $P_{NS}^{(0)}(z)$ may be interpreted as governing the rate of change of the parton distribution probability with t. Nevertheless, we will present a more attractive interpretation immediately.

Consider the scattering of an off-shell probe (say, a photon) off a parton. In the parton model, quarks are assumed free, with a certain probability of having a fraction of the proton momentum, $q_f(x)$. We now allow for a dependence on t, which is due to the fact that the quark may radiate gluons (Figure 17). If we work with an axial gauge, the calculation simplifies enormously, since only Figure 17*b* has to be considered (it is the only one that yields a term in t). Moreover, all the asymptotic freedom corrections are taken into account by replacing $g^2/4\pi$ by α_s.[58] So we will work with this gauge.

[58] This is easily understood if we recall our calculation of Equation (5.18) and compare it to (9.29) and (9.30): the entire contribution to Z_g comes from the gluon propagator in this gauge.

To zero order in g, we only have to consider Figure 17a. Let us take quarks as massless, and work in the reference frame (infinite-momentum frame) where

$$q = (0, \mathbf{0}, -Q), \qquad p = \frac{Q}{2x}(1, \mathbf{0}, q);$$

here $\mathbf{0}$ is the zero vector in the xy plane. The structure function f_2 is a cross-section and will be the sum of the pointlike cross-sections on each quark weighted with the q_f, as we saw in Section 17. With the obvious changes of notation and with w proportional to the cross-section,

$$\frac{1}{x} f_2^{NS}(x,t) = \sum \delta_f^{NS} \int_0^1 \frac{dy}{y} q_f(y,t) w_{\text{pointlike}}(p_f, q). \tag{22.6}$$

The origin of each term is clear. We have defined y by $p_f = yp$. Now, because quarks are massless, we must have $(p_f + q)^2 = 0$, and therefore

$$w_{\text{point}}(p_f, q) = \delta(y/x - 1).$$

We recover (22.1), as could have been expected. We will write (22.6) identically as

$$q_f(x,t) = \int_0^1 \frac{dy}{y} \delta(y/x - 1) q_f(x,t). \tag{22.7}$$

Of course, (22.7) is valid only to zero order in g (free-parton model). We require the corrections to it due to the gluon interactions. These can be split into two parts: vertex and propagator corrections to Figure 17a, given by the diagrams of Figure 17c, and corrections due to the possibility of emitting a real gluon (Figure 17b). First, we consider the latter effect. Its amplitude is[59]

$$\mathscr{A}^{\mu} = (2\pi)^{-2} \bar{u}(p_f - k + q, \sigma') \gamma^{\mu} \frac{i}{\not p_f - \not k} i\gamma^{\alpha} g t_{ij}^{a} u(p_f, \sigma) \epsilon_{\alpha}^{*}(k, \lambda).$$

Therefore, the probability for this process is proportional to

$$\begin{aligned} w^{\mu\nu} &= \frac{1}{2} \int \frac{d\vec{k}}{2k^0} \cdot \frac{d\vec{p}'}{2p'^0} \delta(p_f + q - k - p') \sum_{\text{spins}} \mathscr{A}^{\mu *} \mathscr{A}^{\nu} \\ &= \frac{1}{2} \sum_{\sigma,\sigma',\lambda} \sum_{a,j} \int d^4k \theta(k^0) \\ &\quad \times \delta(k^2) \theta(p_f^0 - k^0 + q^0) \delta\left[(p_f - k + q)^2\right] \mathscr{A}^{\mu *} \mathscr{A}^{\nu}. \end{aligned}$$

Note that the gluon is real; therefore, we have to take

$$\sum_{\lambda} \epsilon_{\alpha}(k,\lambda) \epsilon_{\beta}^{*}(k,\lambda) = -g_{\alpha\beta} + \frac{k_{\alpha} u_{\beta} + k_{\beta} u_{\alpha}}{k \cdot n},$$

[59] The equation is normalized so that if γ^* were real, the scattering amplitude would be $F(\gamma^* + q \to G + q') = \epsilon_{\mu} \mathscr{A}^{\mu}$.

and we have chosen a lightlike gauge:

$$k \cdot \epsilon = u \cdot \epsilon = 0, \qquad u^2 = 0.$$

Because of this, with $\delta_+(v^2) = \delta(v^2)\theta(v^0)$, we obtain

$$w^{\mu\nu} = \frac{1}{2(2\pi)^2} g^2 C_F \Phi^{\mu\nu}, \tag{22.8a}$$

$$\Phi^{\mu\nu} = \int d^4k\, \delta_+(k^2)\delta_+\big[(p_f - k + q)^2\big]\left(-g_{\alpha\beta} + \frac{k_\alpha u_\beta + k_\beta u_\alpha}{k \cdot u}\right)$$
$$\times \frac{\operatorname{Tr}(\not p_f - \not k)\gamma^\mu(\not p_f - \not k + \not q)\gamma^\nu(\not p_f - \not k)\gamma^\beta \not p_f \gamma^\alpha}{(p_f - k)^4}. \tag{22.8b}$$

Expression (22.8) is divergent for massless gluons and quarks, so it has to be regulated. One could use dimensional regularization for this, but it is simpler to consider the initial quark to be off-shell: $p_f^2 = -\mu^2$. Because the region of integration is compact, this is the only way we may obtain a logarithm which, we will see, is of the form $\log Q^2/p_f^2$. As a matter of fact, it is only the logarithmic term that is of interest to us; this will greatly simplify the calculation.

First of all, throughout (22.8), except in the denominator, we may take $p_f^2 = 0$: the corrections will be $O(\mu^2/Q^2)$. Thus,

$$\left(-g_{\alpha\beta} + \frac{k_\alpha u_\beta + k_\beta u_\alpha}{k \cdot u}\right)\operatorname{Tr}(\not p_f - \not k)\gamma^\mu(\not p_f - \not k + \not q)\gamma^\nu(\not p_f - \not k)\gamma^\beta \not p_f \gamma^\alpha$$
$$= -2(p_f - k)^2\Big\{\operatorname{Tr}\gamma^\mu(\not p_f - \not k + \not q)\gamma^\nu \not k + \operatorname{Tr}\gamma^\mu(\not p_f - \not k + \not q)\gamma^\nu$$
$$\times\big[(p \cdot u)(\not p_f - \not k) + (p_f - k)\cdot u \not p + 2k \cdot p_f \not u\big]\frac{1}{u \cdot k}\Big\}.$$

Since $p_f^2 = k^2 = 0$, $2k \cdot p_f = -(p_f - k)^2$; hence, the last term of the above equation is proportional to $(p_f - k)^4$ and it does not contribute to the logarithm. We may thus use "$\overset{\log}{=}$", meaning "equal $\log Q^2$ terms", and obtain,

$$\Phi^{\mu\nu} \overset{\log}{=} -2\int \frac{d\vec{k}}{2k^0}\delta_+\big[(p_f - k + q)^2\big]\frac{1}{(p_f - k)^2}$$
$$\times \operatorname{Tr}\big\{\gamma^\mu(\not p_f - \not k + \not q)\gamma^\nu \not k + \gamma^\mu(\not p_f - \not k + \not q)\gamma^\nu$$
$$\times\big[(\not p_f - \not k)(p_f \cdot u/k \cdot u) + \not p_f\big[(p_f - k)\cdot u/k \cdot u\big]\big]\big\} \tag{22.9}$$

Next, we write the denominator of (22.9) as

$$(p_f - k)^2 = -\mu^2 - 2k^0 p_f^0 + 2k^3 p_f^3 \cos\theta.$$

It may vanish only when $\cos\theta = 1$; that is, when k and p_f are colinear. (This, incidentally, indentifies the gluons that give corrections to scaling.) Thus we may take $\cos\theta = 1$ in all other cases, so that, in particular, the

$\delta[(p_f - k + q)^2]$ function in (22.9) becomes

$$\delta\left[(p_f - k + q)^2\right] = \delta(2\nu - Q^2 - 2Qk^0) = \delta\left[2\nu\left(1 - x - \frac{Qk^0}{\nu}\right)\right].$$

It is convenient to define

$$1 - \frac{Qk^0}{\nu} \equiv \rho, \tag{22.10}$$

and thus,

$$\delta\left[(p_f - k + q)^2\right] = \frac{1}{2\nu}\delta(\rho - x).$$

Moreover, we see that when $\cos\theta = 1$,

$$k_{\theta=0,\pi} = (1 - \rho)p_f\,.$$

We can now easily complete the calculation of (22.8):

$$\Phi^{\mu\nu} \overset{\log}{=} -2\pi\int_{-1}^{+1} d\cos\theta \int_0^\infty \frac{dk^0 \cdot k^0}{2} \cdot \frac{1}{\nu}\delta(\rho - x)\frac{1+\rho^2}{1-\rho}$$

$$\times \frac{\operatorname{Tr}\gamma^\mu(\rho \not{p}_f + \not{q})\gamma^\nu \not{p}_f}{2k^0p_f^0\cos\theta - (\mu^2 + 2k^0p_f^0)}$$

$$\overset{\log}{=} \left(\log\frac{Q^2}{\mu^2}\right)\frac{\pi}{2\nu}\int d\rho\,\frac{1+\rho^2}{1-\rho}\operatorname{Tr}\{\gamma^\mu(\rho\not{p}_f + \not{q})\gamma^\nu\not{p}_f\}\,\delta(x - \rho).$$

Therefore, for f_2 and with obvious notation,

$$w_2 = 4C_F\frac{g^2}{16\pi^2}\int d\rho\,\frac{1+\rho^2}{1-\rho}\,\rho\delta(x-\rho)\log\frac{Q^2}{\mu^2}\,. \tag{22.11}$$

Equation (22.11) does not give the full answer; it is, in fact, undefined at $\rho = 1$. This corresponds to a zero-energy gluon, which is a typical infrared singularity. In fact, it may be seen that this singularity is exactly cancelled by the vertex and propagator corrections we have still not taken into account. Since no real gluon is emitted, their contribution to w_2 has to be like that in (22.11), but with $\lambda\delta(\rho - 1)$ instead of the term $(1 + \rho^2)/(1 - \rho)$. With all terms taken into account, we thus write

$$w_2 = \frac{C_2(F)\alpha_g \log Q^2/\mu^2}{\pi}\int d\rho\,\rho\,\delta(x-\rho)\left\{\frac{1+\rho^2}{1-\rho} + \lambda\delta(1-\rho)\right\}. \tag{22.12}$$

We have found the desired correction to (22.7); it is[60]

$$q_f(x,t) = \int_0^1 dy\int_0^1 dz\,\delta(zy - 1)q_f(y,t)\left\{\delta(z-1) + \frac{\alpha_g t}{4\pi}P_{NS}^{(0)}(z)\right\},$$

$$P_{NS}^{(0)}(z) = C_F\left\{3\delta(1-z) - 2\frac{1+z^2}{(1-z)_+}\right\} \tag{22.13a}$$

[60]Equations (22.13a, b) already take into account the correct value of λ.

where, for any function φ, we have defined

$$\int_0^1 dz \frac{1}{(1-z)_+}\varphi(z) \equiv \int_0^1 dz \frac{\varphi(z)-\varphi(1)}{1-z}. \tag{22.13b}$$

Note that if we identify this $P_{NS}^{(0)}$ with the one introduced previously, we may check that Equation (22.4) is indeed satisfied. In fact, it is because of this that we did not bother to calculate the coefficient λ of $\delta(\rho-1)$: the conditions $\gamma_{NS}^{(0)}(1)=0$ (or $\det\gamma^{(0)}(2)=0$ for the singlet case) fix it directly.

The comparison of (22.13) with (22.5) may be carried out at once. In fact, it is sufficient to consider g to be defined at $-\mu^2$ and take $t \to dt$ to be infinitesimal.

There is, however, a more interesting method. We consider that an aribtrary number of gluons can actually be emitted. Thus we must sum all the diagrams where gluons are emitted. Of course, this is an impossible task; but it simplifies enormously if we only consider the *leading logarithms*. In this case, it may be shown [see Gribov and Lipatov (1972)] that only the ladder graphs contribute (Figure 18). It then turns out that we can calculate the diagrams and even sum them. In this way, we recover the results of the standard analysis, with two bonuses. First, we see that the leading order in the running coupling constant is equivalent to summing all the leading

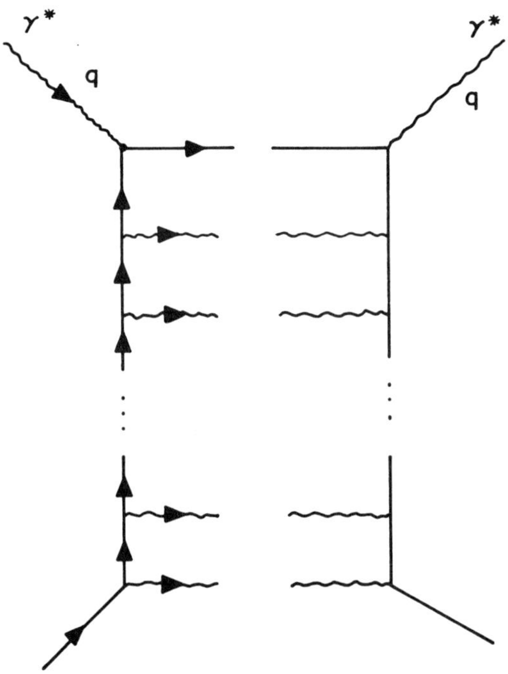

Figure 18. Ladder graph for NS, F scattering.

logarithms in α_g:

$$\alpha_g^n \log^n \frac{Q^2}{\mu^2} .$$

Second, it gives a hint of how to treat processes where the operator product method is not applicable. We will not delve further into this matter, but refer to the lectures of Sachrajda (1979) and work quoted there.

The second-order kernels have been calculated by Curci, Furmanski, and Petronzio (1980) and by Furmanski and Petronzio (1980).

The Altarelli–Parisi methods allow us to give a very simple decomposition of structure functions for various processes into a few "quark densities," $q(x, Q^2)$, for quarks with flavor q. For easy reference, we collect the expressions for a few important processes. We let I be an isoscalar target, p a proton target. Then,

$$\left.\begin{aligned} f_{2ep}^F &= \begin{cases} \frac{2}{9} x(u + \bar{u} + d + \bar{d} + s + \bar{s}), & n_f = 3 \\ \frac{5}{18} x(u + \bar{u} + d + \bar{d} + s + \bar{s} + c + \bar{c}), & n_f = 4; \end{cases} \\ f_{2ep}^{NS} &= \begin{cases} \frac{1}{3} x\left(\frac{2}{3} u - \frac{1}{3} d - \frac{1}{3} s + \frac{2}{3} \bar{u} - \frac{1}{3} \bar{d} - \frac{1}{3} \bar{s}\right), & n_f = 3 \\ \frac{1}{6} x(u - d - s + \bar{u} - \bar{d} - \bar{s} + c + \bar{c}), & n_f = 4 \end{cases} \end{aligned}\right\} \tag{22.14a}$$

$$f_{2eI}^F = f_{2ep}^F; \qquad f_{2eI}^{NS} = \begin{cases} \frac{1}{18} x(u + \bar{u} + d + \bar{d} - 2s - 2\bar{s}), & n_f = 3 \\ \frac{1}{6} x(c - s + \bar{c} - \bar{s}), & n_f = 4. \end{cases} \tag{22.14b}$$

$$f_{2\nu I}^{NS} = 0, \qquad f_{2\nu I} = f_{2\nu I}^F = \begin{cases} \frac{9}{2} f_{2ep}^F, & n_f = 3 \\ \frac{18}{5} f_{2ep}^F, & n_f = 4. \end{cases} \tag{22.14c}$$

$$f_{3\nu I}^F = 0, \qquad f_{3\nu I} = f_{3\nu I}^{NS} = \begin{cases} x(u - \bar{u} + d - \bar{d} + s - \bar{s}), & n_f = 3 \\ x(u - \bar{u} + d - \bar{d} + s - \bar{s} + c - \bar{c}), & n_f = 4. \end{cases} \tag{22.14d}$$

Some of these we had obtained before. Furthermore, one can define the "valence" quarks q_v as the excess of quarks over antiquarks (so a proton has $\int_0^1 dx\, u_v = 2$, $\int_0^1 dx\, d_v = 1$), the "sea" as the rest, etc. A detailed treatment may be found in the excellent reviews of Buras (1980) and Altarelli (1982).

23 General Consequences of QCD for Structure Functions

i Sum Rules

We have stated repeatedly that, in general, the matrix elements A^n cannot be calculated; but there are cases where the corresponding composite operators are related to symmetry generators. In this case, they are observable and, thus, at least in principle, their matrix elements are measurable. As discussed in Section 13, such operators do not require renormalization, and the corresponding anomalous dimensions vanish; therefore, for $Q^2 \to \infty$, the A^n can be calculated with the free quark-parton model.[61]

Such operators are those with $n = 1$ for the nonsinglet, and $n = 2$ for the singlet; no others may exist because it is only for these that γ_{NS} (and an eigenvalue of γ) vanish. This means that the integrals

$$\int_0^1 dx\, x^{-1} f_{NS}(x, Q^2), \tag{23.1a}$$

and a combination of

$$\int_0^1 dx\, \vec{f}(x, Q^2) \tag{23.1b}$$

can, at least in principle, be calculated in absolute value (besides their evolution with Q^2). This is possible in practice only in some favorable cases where the integrals in (23.1) can be related to observables on which information is available: they give rise to sum rules, many of which had already been discovered with the help of the parton model and which become exact theorems in QCD. Here we will discuss a few typical cases.

We begin with the nonsinglet case. For $f_{2,3}^{NS}$ the operators with $n = 1$ are combinations of the

$$N_{NS\mu}^{a\pm} = \tfrac{1}{2} i : \bar{q} \lambda^a \gamma_\mu (1 \pm \gamma_5) q : ,$$

which, indeed, generate chiral symmetry transformations (Section 10). As expected, $\gamma_{NS}^{(0)}(1) = \gamma_{NS}^{(1)-}(1) = 0$. For electroproduction for the three flavors u, d, and s (the decomposition is different for four flavors), we symbolically have [cf., Equation (19.4)]

$$iTJ_{\text{em}}^{\mu}(z) J_{\text{em}}^{\nu}(0)\Big|^{NS}_{p^\mu p^\nu;\, n=1} \underset{z^2 \to 0}{=} \tfrac{1}{3} \bar{C}_{2NS}^{1}(z^2) J_{\text{em}}(0),$$

or, more precisely,

$$\frac{1}{i} \bar{A}_{2NS}^{l} p^\mu = \langle p | J_{\text{em}}^{\mu}(0) | p \rangle = 2(2\pi)^{-3} p^\mu Q_N ,$$

[61] In general, we have to go to $Q^2 \to \infty$ because of the residual dependence on the interaction due to the Wilson coefficients.

where Q_N is the charge of the target (in units of e). Therefore, and taking into account second-order corrections as well,

$$\int_0^1 dx\, x^{-1} f_2^{NS}(x, Q^2) = \tfrac{1}{3} Q_N \left\{1 + \frac{13 + 8\zeta(3) - 2\pi^2}{33 - 2n_f} \cdot \frac{\alpha_s(Q^2)}{3\pi}\right\}. \tag{23.2}$$

Similarly, for neutrino scattering, the Adler sum rule is valid exactly for all Q^2:

$$\int_0^1 dx\, x^{-1}\{f_2^{\bar{\nu}p} - f_2^{\nu p}\} = 2. \tag{23.3}$$

The relevant operator here is the isospin one.

Equation (23.3) has no corrections because it may be related to an equal-time current algebra commutator [cf., Section 10 and Adler (1966)]. For electroproduction, because f_2 is even, the relevant correction involves $\gamma_{NS}^{(1)^+}(1) \neq 0$. For a discussion, see López and Ynduráin (1981).

For f_3, we have the Gross–Llewellyn Smith sum rule [Gross and Llewellyn Smith (1969)]:

$$\int_0^1 dx\, x^{-1} f_3^{\nu I}(x, Q^2) = 3\left\{1 - \frac{\alpha_s(Q^2)}{\pi} + O(\alpha_s^2)\right\}. \tag{23.4}$$

Other nonsinglet sum rules may be found collected in the review of Buras (1980); see also Bardeen *et al.* (1978). We now turn to the singlet.

In this case, the conserved operator corresponds to $n = 2$. This is reflected by $\det \gamma^{(0)}(2) = \det \gamma^{(1)}(2) = 0$. (Because singlet structure functions are always even, it is unnecessary to distinguish $\gamma^{(1)\pm}$; only $\gamma^{(1)+} \equiv \gamma^{(1)}$ enters.) Indeed, [González–Arroyo and López (1980)],

$$\gamma^{(0)}(2) = \tfrac{1}{9}\begin{pmatrix} 64 & -12n_f \\ -64 & 12n_f \end{pmatrix}, \tag{23.5a}$$

$$\gamma^{(1)}(2) = \tfrac{1}{243}\begin{pmatrix} 65[367 - 39n_f] & -3666n_f \\ -64[367 - 39n_f] & 3666n_f \end{pmatrix}. \tag{23.5b}$$

The normalization of the function f^V is a priori arbitrary; we have chosen it so that the eigenvector corresponding to the zero eigenvalue of γ will be precisely the sum. Now the conserved operator is the energy-momentum tensor [recall Equation (10.2)]

$$\Theta^{\mu\nu} = i\sum_f \bar{q}_f \gamma^\mu D^\nu q_f + g_{\alpha\beta} G^{\mu\alpha} G^{\beta\nu} - g^{\mu\nu}\mathcal{L}.$$

The term $g^{\mu\nu}\mathcal{L}$ contributes only $O(M^2/Q^2)$. It can thus be neglected, and we then find

$$\int_0^1 dx\, \{f_2^F(x, Q^2) + f_2^V(x, Q^2)\} = \delta\left\{1 + c_2 \frac{\alpha_s(Q^2)}{\pi} + O(\alpha_s^2)\right\}, \tag{23.6}$$

where δ, c_2 depend on the process. For electroproduction,

$$\delta^{ep} = \langle Q_f^2 \rangle, \qquad c_2 = -5/9,$$

where by $\langle Q_f^2 \rangle$ we mean the average charge of the various quark flavors excited. For νI, νp

$$\delta^{\nu I} = 1, \qquad \delta^{\nu p} = 2/3.$$

In fact, one can calculate the individual integrals of each of the f_2^i, $i = 1,2$ for $Q^2 \to \infty$. This is so because, for $n = 2$,

$$d_+(2) = 0, \qquad d_-(2) = \frac{2}{3} \cdot \frac{16 + 3n_f}{33 - 2n_f} > 0;$$

hence, to leading order in α_s, we can write [with **S** defined in (21.12)]

$$\vec{\mu}(2, Q^2) \underset{Q^2\to\infty}{=} \mathbf{S}(2)\vec{b}(2), \qquad \vec{b}(2) = b\begin{pmatrix}1\\0\end{pmatrix}$$

with some b independent of Q^2. Therefore,

$$\int_0^1 dx\, f_2^F(x, Q^2) \underset{Q^2\to\infty}{=} \delta \frac{3n_f}{16 + 3n_f}, \qquad \int_0^1 dx\, f_2^V(x, Q^2) \underset{Q^2\to\infty}{=} \delta \frac{16}{16 + 3n_f}. \tag{23.7}$$

Unfortunately, the corrections are of the form

$$K\left[\alpha_s(Q^2)\right]^{-d_-(2)}$$

with K a quantity that cannot be calculated at present. [However, the $O(\alpha_s)$ corrections to (23.7) are known. See, for example, López and Ynduráin (1981).] Equations (23.7) are among those that provide the best evidence for the existence of gluons. If they did not exist, one would expect all momentum to be carried by quarks, and hence

$$\int_0^1 dx\, f_2^F(x, Q^2) \approx \delta,$$

which for, say, $n_f = 4$ is twice as great as the experimental value. For example, in νI scattering [De Groot *et al.* (1979)],

$$\int_0^1 dx\, f_2^{\text{exp}}(x, Q^2) \approx 0.43 \pm 0.03, \qquad Q^2 \approx 30 \text{ to } 100 \text{ GeV}^2,$$

and theoretically (23.7) gives[62]

$$\int_0^1 dx\, f_2^{th}(x, Q^2) \approx \frac{12}{28} = 0.43.$$

The leading order analysis of these relations was performed by Gross and

[62]Note that the ν's or e/μ's only probe *quarks*, so the experimentally measured function is precisely f^F. To obtain f^V directly, one requires probes that act on gluons.

Wilzcek (1974), although the momentum-sum rule (only at the quark level) had already been discussed by Llewellyn Smith (1972).

ii Behavior at the Endpoints

Let us begin by considering the limit $x \to 1$ for nonsinglet functions. We will assume that

$$f^{NS}(x, Q^2) \underset{x\to 1}{\approx} A(Q^2)(1-x)^{\nu(\alpha_s)} \tag{23.8}$$

with eventual logarithms (see below). Actually, (23.8) can be proved in QCD, but we will not do so here.[63] On general grounds, we would expect that the $x \to 1$ behavior be tied to the large n behavior of the moments. It is easy to verify that

$$d(n) \underset{n\to\infty}{\approx} \frac{-16}{33-2n_f}\left(\log n - \frac{3}{4} + \gamma_E + O\left(\frac{1}{n}\right)\right). \tag{23.9}$$

Using (23.8), we obtain

$$\mu_{NS}(n, Q^2) \underset{n\to\infty}{\approx} A(Q^2)\frac{\Gamma(n-1)\Gamma[1+\nu(\alpha_s)]}{\Gamma[n+\nu(\alpha_s)]}$$

and, from (23.9) and (20.6),

$$\frac{\mu_{NS}(n, Q^2)}{\mu_{NS}(n, Q_0^2)} \underset{n\to\infty}{\approx} \exp\left\{\left[\log\frac{\alpha_s(Q^2)}{\alpha_s(Q_0^2)}\right]\frac{16}{33-2n_f}\left(\log n - \frac{3}{4} + \gamma_E\right)\right\}.$$

Equating, we find the explicit form of A, ν [Gross (1974)]:

$$f^{NS}(x, Q^2) \underset{x\to 1}{\approx} A_{0NS}[\alpha_s(Q^2)]^{-d_0}\frac{(1-x)^{\nu_{NS}(\alpha_s)}}{\Gamma[1+\nu_{NS}(\alpha_s)]}, \tag{23.10a}$$

$$\begin{aligned} \nu_{NS}(\alpha_s) &= \nu_{NS0} - \frac{16}{33-2n_f}\log\alpha_s(Q^2), \\ d_0 &= \frac{16}{33-2n_f}\left(\frac{3}{4} - \gamma_E\right). \end{aligned} \tag{23.10b}$$

The constants ν_0, A_{0NS} are not given by the theory, although one expects [Feynman and Field (1977)] $\nu_{0NS} \approx 2$ to 3.

For the singlet, the calculations are more complicated because of the matrix character of the equations. One finds that for gluons, (23.8) must be modified, but for quarks, a behavior similar to that of the nonsinglet is obtained [Martin (1979); López and Yndurain (1981), where the extension

[63] Cf., Brodsky and Lepage (1980) and references therein.

to second order may also be found]:

$$f^F(x, Q^2) \underset{x\to 1}{\approx} A_{0S}[\alpha_s(Q^2)]^{-d_0} \frac{(1-x)^{\nu_S(\alpha_s)}}{\Gamma[1+\nu_s(\alpha_s)]}, \tag{23.11}$$

$$f^V(x, Q^2) \underset{x\to 1}{\approx} \frac{2}{5} A_{0S}[\alpha_s(Q^2)]^{-d_0} \frac{(1-x)^{\nu_S(\alpha_s)+1}}{\Gamma(2+\nu_S(\alpha_s))|\log(1-x)|}. \tag{23.12}$$

Here d_0 is as before, and ν_S is given by a formula like that for ν_{NS}:

$$\nu_S(\alpha_s) = \nu_{0S} - \frac{16}{33-2n_f}\log\alpha_s(Q^2).$$

A_{0S}, ν_{0S} are not given by perturbative QCD. Two facts may be stated. First, the gluon structure function approaches zero rapidly compared to the quark singlet. Second, *all* structure functions decrease, for $x \to 1$, when $Q^2 \to \infty$, and this occurs fairly rapidly. This is supported by all experimental evidence.

Second-order corrections modify the functional form of these behaviors. For example, for the nonsinglet, one finds [González–Arroyo, López and Ynduráin (1979)]

$$f^{NS}(x, Q^2) \underset{x\to 1}{\approx} A_{0NS}[\alpha_s(Q^2)]^{-d_0} \frac{e^{a(\alpha_s)\alpha_s(Q^2)}}{\Gamma[1+\nu_{1NS}(\alpha_s)]} \times (1-x)^{\nu_{1NS}(\alpha_s)+2\alpha_s[\log(1-x)]/3\pi}. \tag{23.13}$$

Here

$$\nu_{1NS}(\alpha_s) = \nu_{NS}(\alpha_s) - \psi(\nu_{NS}(\alpha_s)+1)\frac{4\alpha_s(Q^2)}{3\pi} - a_1\alpha_s(Q^2),$$

$$a(\alpha_s) = a_0 + a_1\psi(\nu_{NS}(\alpha_s)+1) + \frac{2}{3\pi}\left\{[\psi(\nu_{NS}(\alpha_s)+1)]^2 - \psi'(\nu(\alpha_s)+1)\right\}.$$

$$a_0 \approx 1.18, \qquad a_1 \approx 0.66.$$

It is interesting to note that because of the term

$$(1-x)^{2\alpha_s[\log(1-x)]/3\pi}, \tag{23.14}$$

we obtain corrections as large as we wish if x is near enough to 1. Of course, this simply indicates that perturbation theory fails for $x = 1$, as we expected: when $x = 1$, we encounter bound states (the elastic contribution to $\gamma^* + N \to$ all, viz., $\gamma^* + N \to N$). Actually, there are also other reasons why the perturbative QCD analysis fails if x is too close to unity. From (23.14), we see that (23.13), say, should only be applied in an intermediate region,

$$1 - x \ll 1 \quad \text{but} \quad \frac{2\alpha_s}{3\pi}|\log(1-x)| \ll 1. \tag{23.15}$$

This leading behavior as $x \to 1$ (or, equivalently, $n \to \infty$) has been recently calculated to all orders for the dominant $(\alpha_s \log n)^N$ contributions [Amati *et al.* (1980); Ciafalloni and Curci (1981)]. The quark singlet term is like (23.13), substituting

$$A_{0NS} \to A_{0S}, \qquad \nu_{1NS} \to \nu_{1S}.$$

Let us now turn to $x \approx 0$. We have to obtain this limit by keeping Q^2 fixed, large enough for perturbation theory to be valid, and letting $\nu \to \infty$. This is the Regge limit,[64] and since the structure functions can be interpreted as cross sections for scattering of virtual γ (or W, Z) with invariant $(\text{mass})^2$ equal to $-Q^2$, we assume [Abarbanel, Goldberger, and Treiman (1969)]

$$f(x, Q^2) \underset{x \to 0}{\approx} b(Q^2)\nu^{\alpha_R(0)}, \qquad x = \frac{Q^2}{2\nu}, \tag{23.16}$$

where R is the relevant Regge trajectory. Unlike the case of the behavior for $x \to 1$, one cannot prove (23.16) from QCD in the present state of the art.

We will rewrite (23.16) in the more convenient form

$$f_{NS}(x, Q^2) \underset{x \to 0}{\approx} B_{NS}(Q^2)x^{\lambda}, \tag{23.17a}$$

$$f_i(x, Q^2) \underset{x \to 0}{\approx} B_i(Q^2)x^{-\lambda_i}, \qquad i = F, V. \tag{23.17b}$$

In principle, one may allow the λ to depend on Q^2; but Regge theory and QCD show that they are actually constant (at least for Q^2 large) to $O(M^2/Q^2)$ terms.

The behavior[65] of f as $x \to 0$ is linked to the singularities of the quantity $\mu(n, Q^2)$. This requires analytical continuation of $\mu(n, Q^2)$ in n. Because $\mu(n, Q^2) = A^n C^n$, the relevant singularity will be that of A^n or C^n, depending on which is further to the right. It may be shown that (23.16) and (23.17) are only possible if the right-most singularity of A^n is to the right of that of C^n. Moreover, if n_0 is this singularity, then

$$n_0 = 1 - \lambda \quad (NS)$$

$$n_0 = 1 + \lambda_s \quad (\text{singlet}),$$

and necessarily $\lambda_F = \lambda_V \equiv \lambda_s$.

Since the singularities of C^n are those of $d(n)$ or $\mathbf{D}(n)$, it follows that

$$\lambda < 1, \qquad \lambda_s > 0.$$

For scattering of off-shell particles, the second inequality implies the existence of a singularity above the Pomeron. There is, in fact, some independent evidence for this from Gribov–Regge calculations [see, e.g., the review of Bartels (1979) and references therein].

[64] For Reggeology, see, for example, Barger and Cline (1969).

[65] The details of the following proofs may be found in Martin (1979) for the nonsinglet and in López and Ynduráin (1981) for both singlet and nonsinglet to first and second order. In these references, behaviors other than Regge-like ones are also discussed.

Let us now consider the singlet case. From the Equations (20.7), it follows that

$$\left[\alpha_s(Q^2)\right]^{\mathbf{D}(n)}\vec{\mu}(n, Q^2)$$

is independent of Q^2. Let $\mathbf{S}(n)$ be the matrix that diagonalizes $\mathbf{D}(n)$. We standardize it by writing it as in Equation (21.12), and let

$$\mathbf{S}^{-1}(n)\mathbf{D}(n)\mathbf{S}(n) = \hat{\mathbf{D}}(n) = \begin{pmatrix} d_+(n) & 0 \\ 0 & d_-(n) \end{pmatrix}. \tag{23.18}$$

Using (23.17) and taking $n = 1 + \lambda_s + \epsilon$, we find

$$\vec{\mu}(1 + \lambda_s + \epsilon) = \frac{\vec{B}(Q^2)}{\epsilon}. \tag{23.19}$$

Therefore,

$$\left[\alpha_s(Q^2)\right]^{\mathbf{D}(1+\lambda_s+\epsilon)}\vec{B}(Q^2) \equiv \vec{b}$$

is independent of Q^2. Applying $\mathbf{S}(1 + \lambda_s + \epsilon)$ and letting $\epsilon \to 0$, we get

$$\vec{B}(Q^2) = \mathbf{S}(1 + \lambda_s)\begin{pmatrix} \alpha_s^{-d_+(1+\lambda_s)} & 0 \\ 0 & \alpha_s^{-d_-(1+\lambda_s)} \end{pmatrix}\vec{b}.$$

We label the eigenvalues so that $d_+ > d_-$; hence, to leading order, we can neglect $\alpha_s^{-d_-}$ compared to $\alpha_s^{-d_+}$ and, thus, finally obtain

$$f_i(x, Q^2) \underset{x\to 0}{\approx} B_{0i}\left[\alpha_s(Q^2)\right]^{-d_+(1+\lambda_s)} x^{-\lambda_s}, \tag{23.20a}$$

$$\frac{B_{0V}}{B_{0F}} = \frac{d_+(1+\lambda_s) - D_{11}(1+\lambda_s)}{D_{12}(1+\lambda_s)}. \tag{23.20b}$$

The constants B_{0F}, λ_s are not given by QCD, although one can expect $\lambda_s \approx 0.1$–0.6.

For the nonsinglet,

$$f_{NS}(x, Q^2) \underset{x\to 0}{\approx} B_{0NS}\left[\alpha_s(Q^2)\right]^{-d(1-\lambda)} x^{\lambda}. \tag{23.21}$$

One does not know B_{0NS}, but since λ is related to the *intercept of* the ρ trajectory we expect

$$\lambda = 1 - \alpha_\rho(0) \approx 0.5.$$

Three comments are in order. First, unlike the case $x \to 1$, higher orders do not disrupt the result; they merely amount to multiplying (23.20) and (23.21) by factors $1 + b_1\alpha_s$ with b_1 known. Second, because with the expected values of λ, λ_s both $d(1-\lambda)$ and $d_+(1+\lambda_s)$ are positive, it follows that *all* structure functions grow with Q^2 for small x. This, too, is confirmed by experiment. Last, and unlike the case $x \to 1$, at $x \approx 0$ the gluon is larger than the quark singlet. In fact, the right-hand side of (23.20b) is ~4–8 for the expected values of λ_s.

24 Comparison with Experiment: Parametrizations Compatible with QCD and Pointlike Evolution of Structure Functions

Since the theoretical predictions are simpler for moments of structure functions, it would seem that we should compare QCD predictions with moments. This, however, is not very convenient for the following reasons. First, to experimentally obtain the moments for a large set of Q values, we would require detailed measurements of the structure functions for a tight mesh of values of x. This is not generally available; but even when good data exist, we have a problem for high moments. In fact, these involve integrals of the f with x^{n-2}, which is strongly peaked at $x \approx 1$. Since it is here that structure functions are smallest, experimental errors become amplified and, even in very favorable cases, get out of hand for $n \gtrsim 6$: we lose an enormous amount of experimental information. For this reason, other methods have been devised.

A possibility is to write reasonable parametrizations for the f, which embody QCD results and can be fitted to experiment. Although not very rigorous, this method presents the advantages of simplicity and of giving an explicit representation of the structure functions which can then be used for other related processes (Drell–Yan, high p_t hadron-hadron scattering, or scattering of off-shell hadrons).

The first such parametrizations were introduced by Feynman and Field (1977); they are of the form

$$f_a(x, Q^2) = C_a x^{\lambda_a}(1-x)^{\nu_a}, \tag{24.1a}$$

or, with two Regge poles,

$$f_a(x, Q^2) = (C_a x^{\lambda_a} + C'_a x^{\mu_a})(1-x)^{\nu_a}. \tag{24.1b}$$

If we take the C, λ, ν to be constant, we obtain exact scaling.

Buras and Gaemers (1978) noted that if we allow the λ, ν to depend on α_s,

$$\lambda = \lambda_0 + \lambda_1 \log \alpha_s\,, \qquad \nu = \nu_0 + \nu_1 \log \alpha_s\,,$$

we can fix C (by using sum rules; see Section 23i) as a known function of λ_0, λ_1, ν_0, ν_1, α_s. We then require simultaneous fits to the QCD equations for the moments *and* experimental values of f, which also determine the λ, ν.

A further step was taken by López and Ynduráin (1981) (to leading order and next-to-leading order), who remarked that one can use the results of Section 23ii to *calculate* λ_1 (it actually vanishes) and ν_1. In this way, one obtains extremely simple parametrizations; they are exactly compatible with QCD only at the endpoints and for the integrals related to known sum rules, but actually they deviate from the exact evolution equations by less than 1%. To leading order, we find

$$f_2^{NS}(x, Q^2) = \Big\{ B_{0NS}\big[\alpha_s(Q^2)\big]^{-d(1-\lambda)}(x^{\lambda} - x^{\mu_{NS}(\alpha_s)})$$

$$+ A_{0NS}\big[\alpha_s(Q^2)\big]^{-d_0} \frac{\Gamma(\nu_{0NS}+1)}{\Gamma(\nu_{NS}(\alpha_s)+1)} x^{\mu_{NS}(\alpha_s)} \Big\} (1-x)^{\nu_{NS}(\alpha_s)}, \tag{24.2a}$$

$$f_2^{F}(x, Q^2) = \Big\{ B_{0F}\big[\alpha_s(Q^2)\big]^{-d_+(1+\lambda_s)}(x^{-\lambda_s} - x^{-\mu_F(\alpha_s)})$$

$$+ A_{0S}\big[\alpha_s(Q^2)\big]^{-d_0} \frac{\Gamma(\nu_{0S}+1)}{\Gamma(\nu_s(\alpha_s)+1)} x^{-\mu_F(\alpha_s)} \Big\} (1-x)^{\nu_S(\alpha_s)}, \tag{24.2b}$$

$$f_2^{V}(x; Q^2) = \Big\{ B_{0F} \frac{d_+(1+\lambda_s) - D_{FF}(1+\lambda_s)}{D_{FV}(1+\lambda_s)} \big[\alpha_s(Q^2)\big]^{-d_+(1+\lambda_s)}$$

$$\times (x^{-\lambda_s} - x^{-\mu_V(\alpha_s)}) + \frac{2}{5} A_{0S}\big[\alpha_s(Q^2)\big]^{-d_0} x^{-\mu_V(\alpha_s)} \frac{\Gamma(\nu_{0S}+1)}{\Gamma(\nu_S(\alpha_s)+2)} \Big\}$$

$$\times \frac{(1-x)^{\nu_S(\alpha_s)+1}}{1 + |\log(1-x)|} \tag{24.2c}$$

where

$$\nu_i(\alpha_s) = \nu_{0i} - \frac{16}{33 - 2n_f} \log \alpha_s(Q^2), \qquad i = S, NS. \tag{24.2d}$$

λ is given by the ρ trajectory, $\lambda \approx 1 - \alpha_\rho(0) \approx 0.5$, and we can find the μ in terms of the other constants by using the sum rules of Section 23. Thus we have a set of simple expressions that parametrize three functions f_2^{NS}, f_2^F, f_2^V in terms of seven parameters: ν_{0NS}, ν_{0S}, A_{0S}, A_{0NS}, B_{0NS}, B_{0F}, λ_s (apart from Λ). They should, of course, be fitted to experiment. Actually, and without increasing the number of parameters, we can also calculate f_L, so the fact that we are able to achieve agreement with data is an important check of QCD.[66] This is shown in Figure 19*a*.

We now turn to the method of exact reconstruction. Consider a nonsinglet case, and change variables, $\log x = -\xi$. Then the evolution equations

[66] Particularly since one can argue that $\nu_{0NS} \approx \nu_{0S} \approx 2$–$2.5$, $0 < \lambda_s < 1$.

may be rewritten as

$$\mu_{NS}(n, Q^2) = \int_0^\infty d\xi\, e^{-(n-1)\xi} f_{NS}(e^\xi, Q^2),$$

$$\mu_{NS}(n, Q^2) = \left[\frac{\alpha_s(Q_0^2)}{\alpha_s(Q^2)}\right]^{d(n)} \mu_{NS}(n, Q_0^2), \tag{24.3}$$

so we can use the well-known convolution theorem for Laplace transforms to invert (24.3):

$$f_{NS}(x, Q^2) = \int_x^1 dy\, b(x, y; Q^2, Q_0^2) f_{NS}(y, Q^2),$$

where the kernel b can be calculated in terms of the γ, C^n. To leading order [Gross (1974)]:

$$b = b^{(0)}(x, y; Q^2, Q_0^2) = \sum_{j=0}^{\infty} G_j(r) b_0(x, y; r + j),$$

$$r = \frac{16}{3\beta_0} \log \frac{\alpha_s(Q_0^2)}{\alpha_s(Q^2)},$$

and

$$G_0(r) = 1, \qquad G_1(r) = -\frac{r}{2}, \qquad G_2(r) = r\frac{3r + 14}{24}, \ldots,$$

$$b_0(x, y; r + j) = \frac{x}{y^2} \frac{1}{\Gamma(r + j)} \left(\log \frac{y}{x}\right)^{r+j} e^{(3/4 - \gamma_E) r}.$$

To second order [González–Arroyo, López, and Ynduráin (1979)],

$$b = b^{(0)} + \frac{\alpha_s(Q^2) - \alpha_s(Q_0^2)}{4\pi} b^{(1)},$$

$$b^{(1)}(x, y; Q^2, Q_0^2) = \sum_{p=0}^{2} \sum_{j=0}^{\infty} a_{pj}(r) b_p(x, y; r + j),$$

and

$$b_1 = \left\{\psi(r + j) - \log\log \frac{y}{x}\right\} b_0,$$

$$b_2 = \left\{\left[\psi(r + j) - \log\log \frac{y}{x}\right]^2 - \psi'(r + j)\right\} b_0.$$

Finally, the a can be expressed in terms of the G:

$$a_{pj} = \sum_{l=0}^{j} H_{pl} G_{j-l}(r);$$

a list of values of the coefficients H may be found in González-Arroyo *et al.* (1979).

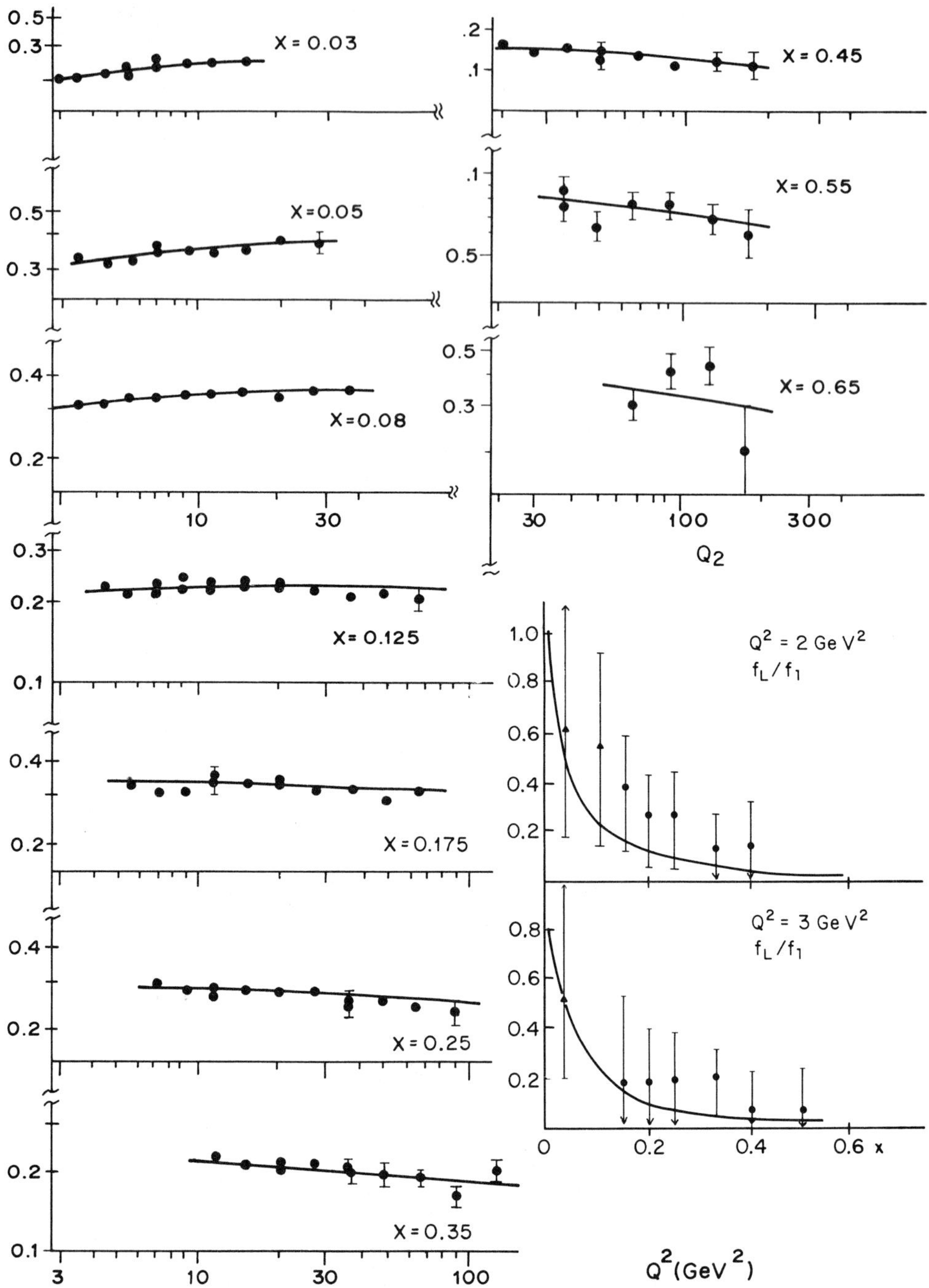

(a)

Figure 19. (*a*) Fits to $f_2(x, Q^2)$ from the μp data of Aubert *et al.* (1982) and to s_l/s_1 with data from Anderson *et al.* (1979) and Bodek *et al.* (1979) with the parametrizations of Equations (24.2) including second order corrections. The corresponding value of Λ is 110 MeV. The same value is obtained in the exact calculation of Aubert *et al.* (1982). [From unpublished work by B. Escoubes, M. J. Herrero, C. López and F. J. Yndurain].

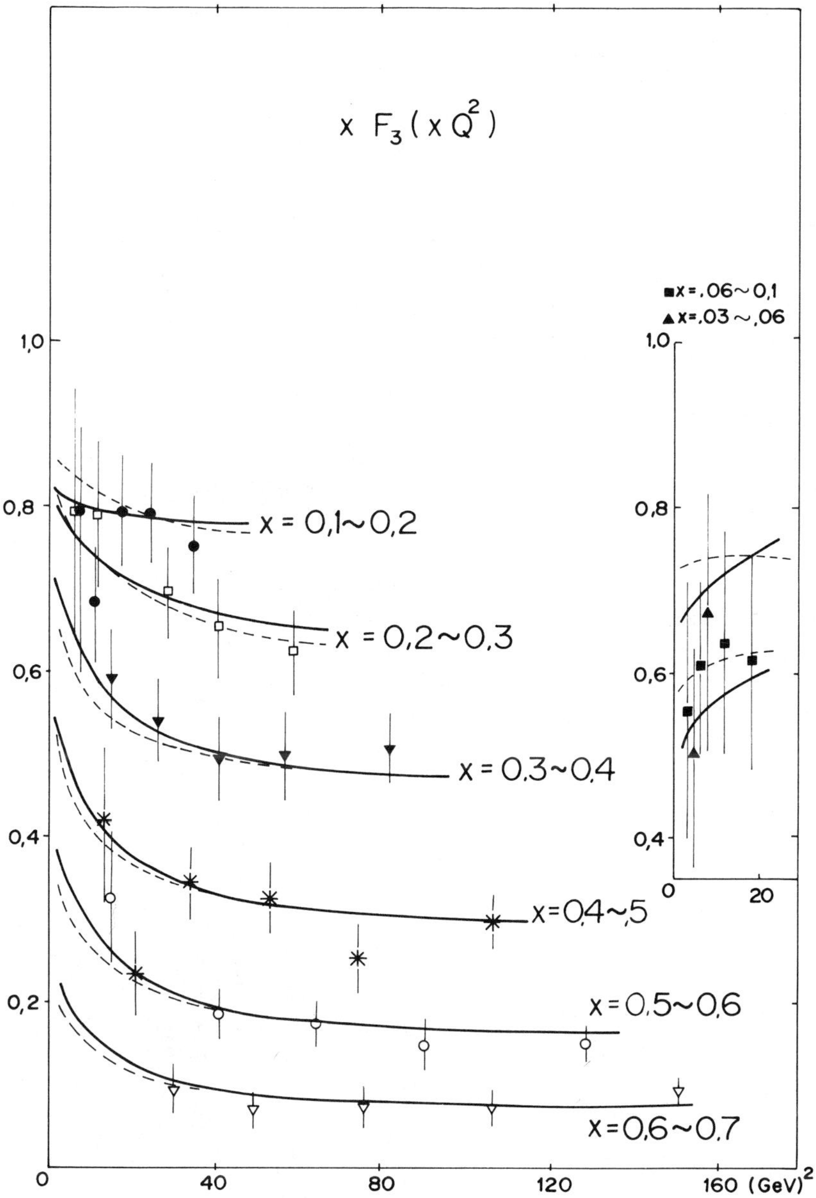

(b)

Figure 19. (*b*) Fit to the ν scattering data of De Groot *et al.* (1979), including second order QCD corrections. The value of Λ (400 ± 250 MeV) decreases to 180 ± 130 MeV if using the re-analyzed data (H. Abramowicz *et al.*, CERN preprint EP/81-168, 1981, to be published in Z. Phys. C.)

The extension to the singlet case is nontrivial. It may be found in the paper of González–Arroyo and López (1980). The comparison with experiment depends only on the input $f(x, Q_0^2)$ for a fixed Q_0^2 (usually taken of 2 to 3 GeV). The result is depicted in Figure 19*b*.

An alternate method uses the Altarelli–Parisi evolution equations directly; it may be found in Abbott, Atwood, and Barnett (1980).

25 Target Mass Corrections

Consider a moment of the nonsinglet part of f. In principle, μ_{NS} depends not only on n, α_s, but also on a set of masses: the masses of the target, m_N, quark masses m_q, and, eventually, nonperturbative masses. Let us neglect the latter for now. Quark masses and m_N will yield corrections of $O(m_q^2/Q^2)$, $O(m_N^2/Q^2)$. As will be argued in Section 32, the u, d, and s have small masses, the largest being $\hat{m}_s \approx 0.3$ GeV. With the values of Λ we have found, perturbative QCD will hardly make sense, except if $Q^2 \gtrsim 1.5$ GeV; thus, even the s quark mass contributes only $\sim 5\%$ at the lower end. Heavy quarks contribute differently: $m_c \approx 1.5$ GeV, $m_b \approx 6.5$ GeV, but we will neglect them for the moment. The target mass gives corrections $\sim m_N^2/Q^2$, which is quite large. In this section, we will show how to take these corrections into account.

The effect of target mass corrections was estimated by Nachtmann (1973); it leads to so-called ξ scaling. Here we will follow the method of Georgi and Politzer (1976). Recall the expansion (19.3) and (19.11). In general, it contains other terms which are of two types; there are terms that correspond to the operators

$$g^{\mu\nu}\bar{q}\not{\partial} D^{\mu_1}\dots q \quad \text{and} \quad g^{\mu\nu}\bar{q}\partial^2\gamma^{\mu_1}D^{\mu_2}\dots q.$$

For the free-field case, $\not{\partial} q = -im_q q$; hence, they will yield terms proportional to m_q^2/Q^2 which we are now neglecting. However, terms

$$\langle p|N_{NS}^{\mu\nu\mu_1\cdots\mu_n}(0)|p\rangle = g^{\mu_i\mu_j}\dots g^{\mu_l\mu_m}p^{\mu_{k_1}}\dots p^{\mu_{k'_{n-m}}}(p^2)^m \bar{A}'^{n}_{NS},$$

give, as will be seen shortly, m_N^2/Q^2 corrections. We neglected these corrections earlier, but we shall focus on them now. Consider $N^{\mu_1\cdots\mu_n}$; later we will replace $n \to n+2$ and identify $\mu_{n+1} \to \mu$, $\mu_{n+2} \to \nu$. Because it is symmetrized, its matrix elements can be written quite generally as (n = even)

$$i\langle p|N_{NS}^{\mu_1\cdots\mu_n}(0)|p\rangle = \sum_{j=0}^{n/2}(-1)^j\frac{(n-j)!}{2^j n!}\left\{\sum_{\text{permutations}} g^{\mu_{i_1}\mu_{i'_1}}\cdots g^{\mu_{i_j}\mu_{i'_j}}\right\}(p^2)^j$$

$$\times \sum_{\text{permutations}} p^{\mu_{k_1}}\dots p^{\mu_{k_{n-2j}}}\bar{A}_{NS,j}^{(\text{TMC})n-2}, \tag{25.1}$$

$$N_{NS}^{\mu_1\cdots\mu_n} = \mathcal{S}\bar{q}\gamma^{\mu_1}D^{\mu_2}\dots D^{\mu_n}q|_{NS}\,.$$

Now, $g^{\mu_i\mu_j}\langle p|N_{NS_{\mu_1\ldots\mu_n}}|p\rangle = 0$: we obtain sufficiently many relations that we can solve for all the A_j^n in terms of A_0^n. Then,

$$T_{2NS}^{(\mathrm{TMC})}(x, Q^2) = \frac{1}{2}\sum_n x^{-n-1}\sum_{j=0}^{\infty}\left(\frac{p^2}{Q^2}\right)^j \frac{(n+j+2)!\,(n+2j)!}{j!\,n!\,(n+2j+2)!}$$

$$\times A_{NS}^{(0)n+2j}C_{NS}^{n+2j}, \tag{25.2}$$

$$A_{NS}^{(0)n} \equiv A_{NS,j=0}^{(\mathrm{TMC})n}.$$

Therefore, we obtain the result

$$\mu_{NS}^{(\mathrm{TMC})}(n, Q^2) = \sum_{j=0}^{\infty}\left(\frac{m_N^2}{Q^2}\right)^j \frac{(n+j)!}{j!\,(n-2)!}\,\frac{C_{NS}^{n+2j}}{(n+2j)(n+2j-1)}A_{NS}^{(0)n+2j}, \tag{25.3a}$$

$$\mu_{NS}^{(\mathrm{TMC})}(n, Q^2) = \int_0^1 dx\, x^{n-2} f_2^{(\mathrm{TMC})}(x, Q^2). \tag{25.3b}$$

It is convenient to define the function f_2 to be the limit $m_N^2 \to 0$ of $f_2^{(\mathrm{TMC})}$, and set

$$\mu_{NS}(n, Q^2) = \int_0^1 dx\, x^{n-2} f_2(x, Q^2). \tag{25.4}$$

It is to the μ, f_2 that the equations derived in the previous sections apply. To obtain the moments *with TMC* taken into account, we use (25.3*a*),

$$\mu_{NS}^{(\mathrm{TMC})}(n, Q^2) = \sum_{j=0}^{\infty}\left(\frac{m_N^2}{Q^2}\right)^j \frac{(n+j)!}{j!\,(n-2)!}$$

$$\times \frac{1}{(n+2j)(n+2j-1)}\,\mu_{NS}(n, Q^2), \tag{25.5}$$

but we do not have to go through the moments. After some simple manipulations, we find that (25.5) is equivalent to (ξ-*scaling*)

$$f_2^{(\mathrm{TMC})}(x, Q^2) = \frac{x^2/\xi^2}{(1+4x^2m_N^2/Q^2)^{3/2}} f_2(\xi, Q^2)$$

$$+ \frac{6m_N^2}{Q^2}\cdot\frac{x^3}{(1+4x^2m_N^2/Q^2)^2}\int_\xi^1 \frac{d\xi'}{\xi'^2} f_2(\xi', Q^2)$$

$$+ \frac{12m_N^4}{Q^4}\cdot\frac{x^4}{(1+4x^2m_N^2/Q^2)^{5/2}}\int_\xi^1 d\xi'\int_{\xi'}^1 \frac{d\xi''}{\xi''^2} f_2(\xi'', Q^2), \tag{25.6}$$

where ξ is Nachtmann's variable,

$$\xi = \frac{2x}{1 + \left(1 + 4x^2 m_N^2 / Q^2\right)^{1/2}}. \tag{25.7}$$

A few features of these formulas are worth noting. First, for small x, and since TMCs behave like $x^2 m_N^2 / Q^2$, we can neglect them completely. TMCs are only relevant for large—but not too large—values of x. Indeed, if applied at $x \to 1$, inconsistencies develop. There are two reasons for this: first, higher twists (to be discussed later) are largest for $x \to 1$. Although one expects higher twist corrections to be of the form $3M^2/Q^2$ with $M \approx \Lambda$, i.e., half an order of magnitude smaller than TMCs, cancellations may (and probably do) occur.[67] Second, as we saw in Section 23, perturbation theory fails for $x \to 1$.

Because of this, it is perhaps more consistent to expand (25.6) in powers of m_N^2/Q^2 and retain only the leading term. The expression for TMCs then simplifies to

$$f^{(\mathrm{TMC})}(x, Q^2) = f(x, Q^2) + \frac{x^2 m_N^2}{Q^2} \left\{ 6x \int_x^1 dy \, \frac{f(y, Q^2)}{y^2} - x \frac{\partial}{\partial x} f(x, Q^2) - 4f(x, Q^2) \right\}, \tag{25.8}$$

and one forbids oneself to apply QCD when the second-order corrections,

$$\sim \left\{ \frac{x^3 \nu(\alpha_s) m_N^2}{(1-x) Q^2} \right\}^2$$

are large or, alternatively, one takes these as a measure of the intrinsic error of the calculation: it is difficult to maintain that m_N^4/Q^4 effects are to be taken into account while neglecting the M^2/Q^2 effects.

26 Nonperturbative Effects in e^+e^- Annihilation and Higher Twists in Deep Inelastic Scattering

We treat both of these effects in the same section because they are, from our point of view, clearly related. Let us begin with the first. As discussed in Section 15, we have to consider the quantity $\Pi^{\mu\nu}$ from Equations (15.4).

Let us look at the product

$$TJ^{\mu}(x)J^{\nu}(0)$$

from the OPE point of view. We can write a short-distance expansion for it;

[67] For a discussion of this, see the papers of De Rújula, Georgi, and Politzer (1977*a*, *b*).

in momentum space and with $Q^2 = -q^2$,

$$\begin{aligned} i\int d^4x\, e^{iq\cdot x} TJ^\mu(x)J^\nu(0) \\ &= (-g^{\mu\nu}q^2 + q^\mu q^\nu) \\ &\quad\times\Big\{C_0\big[Q^2/\nu^2, g(\nu)\big]\cdot 1 + \sum_f C_f\big[Q^2/\nu^2, g(\nu)\big]m_f :\bar{q}_f(0)q_f(0): \\ &\qquad + C_G\big[Q^2/\nu^2, g(\nu)\big]\alpha_s :\sum G_a^{\mu\nu}(0)G_{a\mu\nu}(0): + \cdots\Big\}. \end{aligned} \tag{26.1}$$

In Section 15 we considered only the first term, $C_0 1$. This was done for two reasons. First, on purely dimensional grounds,

$$C_f \approx \frac{(\text{constant})}{Q^4}, \qquad C_G \approx \frac{(\text{constant})}{Q^4}. \tag{26.2}$$

Second, to all orders of perturbation theory,

$$\langle :\bar{q}q: \rangle_0 = 0, \qquad \langle :G^2: \rangle_0 = 0. \tag{26.3}$$

However, it will be argued later (Section 30ff.) that the physical vacuum is not that of perturbation theory, but must incorporate nonperturbative effects. Using "vac" to denote the physical vacuum, it is very likely that, indeed,

$$\langle :\bar{q}q: \rangle_{\text{vac}} \neq 0, \qquad \langle :G^2: \rangle_{\text{vac}} \neq 0.$$

Let us return to (26.1). At $Q^2 \to \infty$, any power $[1/\log Q^2/\Lambda^2]^n$ decreases *less* rapidly than, and hence overwhelms, the terms $(M^2/Q^2)^r$. But it is clear that there may (and likely will) exist intermediate regions where, for example, the terms (26.2) are still important compared to the second-order correction to C_0, which is the purely perturbative term. Thus, for practical applications,[68] it is interesting to look at the entire Equation (26.1).

We already known C_0,

$$\begin{aligned} C_0\big(Q^2/\nu^2; g(\nu), \nu\big) \\ &= 3\sum_f Q_f^2 \frac{-1}{12\pi^2}\left\{\log\frac{-q^2}{\nu^2} + \frac{3}{4}\cdot\frac{4C_F}{\beta_0}\log\log\frac{-q^2}{\nu^2} + \cdots\right\} \\ &\quad + O\left(\frac{m_f^2}{Q^2}\right). \end{aligned} \tag{26.4}$$

It should be noted that the calculation of Section 15 neglected perturbative contributions due to quark masses; these are the $O(m_f^2/Q^2)$ terms in (26.4). It may seem unjustified to take into account the terms in (26.2) while neglecting the m_f^2/Q^2 terms. The m_f^2/Q^2 terms are indeed very important

[68] Some of the applications may be found in the extensive and pioneering work of Shifman, Vainshtein, and Zakharov (1979*a*, *b*).

for heavy quarks c and b; their incorporation does not cause any problem, and may be found in Section 28. For light quarks (u, d, and s), we will argue that $\bar{m}_s \approx 200$ MeV at $Q^2 \gtrsim 2$ GeV^2. The fact that they can be neglected is then a matter of numerology; m_f^2/Q^2 is, for the relevant values of Q^2, much smaller than the other terms.

The coefficients C_f, C_G can be calculated using standard techniques [the details in a typical case may be found in Section 36, Equations (36.4)–(36.8)]. One finds (Shifman, Vainshtein, and Zakharov, 1979*a*, *b*)

$$C_f = \frac{2}{3} Q_f^2 \frac{1}{Q^4}, \qquad C_G = \left(3 \sum Q_f^2\right) \frac{1}{36\pi Q^4}. \tag{26.5}$$

It is important to realize that the anomalous dimensions of the combinations $m:\bar{q}q:$ and $\alpha_s:G^2:$ vanish to lowest order, which is why the coefficients C_f and C_G do not depend on ν. This is proved for $m:\bar{q}q:$ by combining our calculation of Z_m (Section 14) with that of Z_M (Section 13). For $\alpha_s:G^2:$, see Kluberg-Stern and Zuber (1975) and Tarrach (1982). With all this, we find

$$\Pi^{\mu\nu} = \left(3 \sum_f Q_f^2\right)\left(-q^2 g^{\mu\nu} + q^\mu q^\nu\right)$$
$$\times \left\{ -\frac{1}{12\pi^2}\left[\log\frac{Q^2}{\nu^2} + \frac{3C_F}{\beta_0}\log\log\frac{Q^2}{\nu^2} + \cdots + O\left(\frac{m_f^2}{Q^2}\right)\right.\right.$$
$$\left. + \frac{2}{3}\cdot\frac{m_f\langle:\bar{q}_f(0)q_f(0):\rangle_{\text{vac}}}{Q^4} + \frac{1}{36\pi}\cdot\frac{\alpha_s\langle:G^2(0):\rangle_{\text{vac}}}{Q^4} + O\left(\frac{M^2}{Q^6}\right)\right\}. \tag{26.6}$$

Let us now turn to deep inelastic scattering. In the operator product expansion (Section 19), we considered only the leading twist operators. As for e^+e^- annihilation, there are likely regions of Q^2 where higher twists compete with, say, second-order perturbative contributions. Some higher twist contributions are those that give TMC; others are quark mass corrections.[69] Yet others are genuinely new dynamical effects, related to the "primordial" transverse momentum of quarks inside a nucleon, or to the fact that the last has a finite radius. Higher twist operators are much more complicated to handle than the leading ones; for example, the potential mixing of ghosts with the gluon operators (19.2), which can be proved not to occur for leading twists, does occur for higher ones. Moreover, such higher twist terms introduce new unknown matrix elements like the A of (19.11), only more, due to mixing. These are the reasons why the treatment of higher twist effects is in its infancy and likely to remain so for quite some time. All we have at present are partial theoretical calculations [e.g.,

[69] See Nachtmann (1973), Georgi and Politzer (1976), and Barbieri *et al.* (1976).

Gottlieb (1978)] and heuristic arguments [De Rújula, Georgi, and Politzer (1977)]. The latter indicate that the contributions of higher twist operators is probably of the form

$$f^{(HT)}(x, Q^2) \approx \frac{k_1}{Q^2} \cdot \frac{x}{1-x} f^{(2)}(x, Q^2) + \frac{k_2}{Q^2} f^{(2)}(x, Q^2) \quad (26.7)$$

where $f^{(2)}$ is the structure function with only twist two operators taken into account. The terms k_1, k_2 are phenomenological parameters, expected to be $|k_i| \approx p_t^2$, R_N^{-2} with R_N the nucleon radius.[70] We will not delve further into this matter.

27 Other Processes

i Inclusive Processes: Timelike Deep Inelastic Scattering; OZI Forbidden Decays; Drell–Yan Processes; Large p_t Scattering of Hadrons

If we allow for timelike momentum transfer, there are other deep inelastic processes that are observable. The more important ones are $\gamma^* + (\pi, K, \gamma) \to$ all, observable in $e^+e^- \to (\pi, K, \gamma)$ + all. Beyond the features they share with ordinary, spacelike deep inelastic scattering on nucleons, these processes present some specific features. First, we have the matter of continuation to timelike momentum (negative Q^2) common to all these processes. In addition, the scattering $\gamma^*\gamma \to$ all presents the important peculiarity of being *calculable* for $x \gg 0$. That is to say, not only the evolution, but also the absolute normalization is known in this case (except in a region $x \approx 0$). The reason is that the dominating part is given by a new set of operators built solely from the photon field whose matrix elements are known. The first-order calculation was performed by Witten (1977),[71] and, to second order, by Bardeen and Buras (1979). We will not say more about this subject which is reviewed in Buras (1981). We continue with OZI forbidden decays.

Many people will agree that the first spectacular success of asymptotic freedom was the explanation of the narrowness of the $\psi(J)$ resonances [Appelquist and Politzer (1975); De Rújula and Glashow (1975)]. More generally, this was an instance of the application of QCD to justify the small widths of so-called OZI forbidden decays.

[70] The same order of magnitude for the k's ($k^{1/2} \approx 0.1$ to 0.3 GeV) was obtained in a recent calculation in the bag model by Jaffe and Soldate (1981). Very recently, a detailed study of higher twist effects has been carried out by R. K. Ellis, W. Furmanski and R. Petronzio, CERN preprint TH. 3301, 1982, to appear in Nuclear Physics B.

[71] See, also, Kingsley (1973) and Walsh and Zerwas (1973).

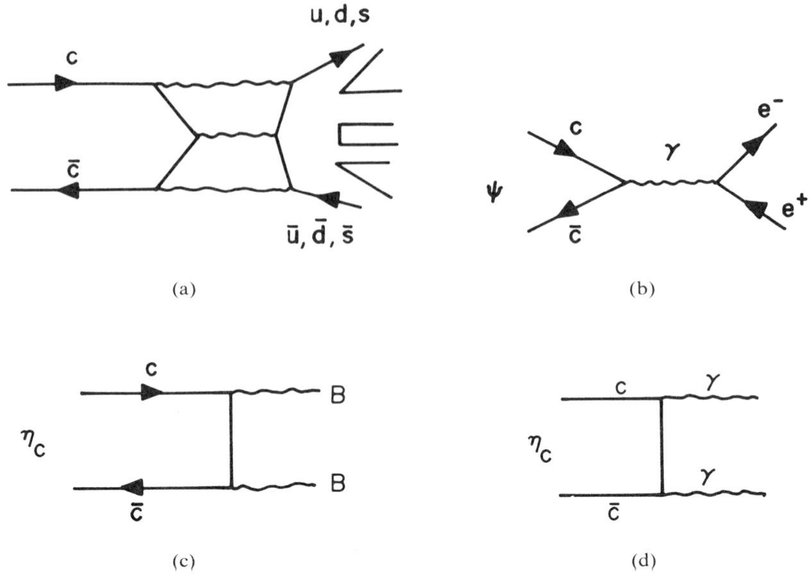

Figure 20. ψ, η_c decays

The Zweig (1964) or OZI[72] rule states that decays of heavy resonances which involve disconnected quark graphs (i.e., graphs that can be connected only via gluons) are suppressed. The rule works well for resonances like the ϕ or f', and very well for the ψ or Υ: in fact, the heavier the quarks and the resonance, the better the rule works. Consider, for example, the ψ (3.1) made up of $\bar{c}c$ quarks. Because the lightest charmed particles (D) are too heavy for the ψ to decay into them, the process $\psi \to$ hadrons necessarily has to occur via gluons. Due to the quantum numbers of the ψ, we require at least three gluons (Figure 20a); so the width is $\Gamma(\psi \to \text{hadrons}) \approx \alpha_s^3 m_\psi$. It may be argued that the relevant constant is the running coupling constant evaluated at $Q^2 = -m_\psi^2$, so we write, by analogy with three-photon decay of positronium to which $\Gamma \to 3$ gluons equals up to the color factor C_D and the replacement $\alpha \to \alpha_s$,

$$\Gamma(\psi \to \text{hadrons}) = \frac{64 C_D}{9}(\pi^2 - 9)\frac{|^3S_1(0)|^2}{m_\psi^2}\left[\alpha_s(-m_\psi^2)\right]^3,$$
$$C_D = \frac{1}{16 n_c}\sum_{abc} d_{abc}^2 = \frac{5}{18}. \tag{27.1}$$

Here $^3S_1(0)$ is the $\bar{c}c$ wave function at the origin. This formula may be shown to be correct, to leading and next-to-leading order in QCD, with calculable corrections (see below). One can obtain $|^3S_1(0)|$ from models, but a model-independent prediction is obtained if we normalize to the leptonic

[72]Okubo (1963); Iizuka, Okada, and Shito (1966).

width $\psi \to e^+e^-$ (Figure 20*b*), for then $|^3S_1(0)|$ is dropped and we obtain, to leading order,

$$B^{\psi}_{h/l} \equiv \frac{\Gamma^0(\psi \to \text{hadrons})}{\Gamma^0(\psi \to e^+e^-)} = \frac{10(\pi^2 - 9)\alpha_s^3(-m_\psi^2)}{81\pi\alpha Q_c^2}. \tag{27.2}$$

Most important are second-order corrections recently calculated. The corrections stem from two sources: corrections to the leptonic width Γ_l [Barbieri *et al.* (1975)] which yield

$$\Gamma_l = \Gamma_l^0 \left\{ 1 - \frac{16}{3} \cdot \frac{\alpha_s(m_\psi^2)}{\pi} \right\},$$

and corrections to the hadronic width [(Mackenzie and Lepage (1981)],

$$\Gamma_h = \Gamma_h^0 \left\{ 1 + (3.8 \pm 0.5) \frac{\alpha_s(m_\psi^2)}{\pi} \right\}.$$

The error is due to the fact that the calculations are made numerically. In addition, we have finite mass corrections (phase-space, velocity corrections, etc.). These are large for the ϕ ($\sim$70%), less for the ψ (20%), and small ($\sim$16%) for the Υ. We may then write, for $V = \psi$ or Υ,

$$B^V_{h/l} = \frac{10(\pi^2 - 9)\alpha_s^3(m_V^2)}{81\pi\alpha Q_q^2} \left\{ 1 + (9.1 \pm 0.5) \frac{\alpha_s(m_V^2)}{\pi} - \frac{M_V^2}{m_V^2} \right\}.$$

Perhaps the best procedure for comparison with experiment is to treat M_V^2/m_V^2 as an *error*, requiring it to be of the order of magnitude discussed above. In this way, we find $\Lambda = 120^{+60}_{-30}$ for Υ decay, and, for the ψ, $\Lambda = 60^{+100}_{-10}$. The agreement between these two determinations is a nontrivial test of QCD, as is the fact that the values we found are compatible with the deep inelastic data of Section 24. (To obtain these values for Λ, corrections $O(\alpha)$ were also taken into account.)

The decays of pseudoscalar resonances (like η_c) are similar: the decay now proceeds via two gluons (Figure 20*c*) and is normalized to $\eta_c \to \gamma\gamma$ (Figure 20*d*):

$$\frac{\Gamma(\eta_c \to \text{hadrons})}{\Gamma(\eta_c \to \gamma\gamma)} = \frac{2}{9Q_c^4} \left(\frac{\alpha_s(m_{\eta_c}^2)}{\alpha} \right)^2. \tag{27.3}$$

The second-order corrections here have been calculated by Barbieri *et al.* (1979); they are also fairly large. For heavy enough quarks, one can obtain rigorous results not only on ratios like (27.2, 3), but even on exclusive decays [Duncan and Muller (1980*a*)].

Next we turn to the Drell–Yan (1971) mechanism. Here a quark from a hadron, and an antiquark from another hadron, annihilate to give a virtual photon with large invariant mass $-Q^2$, which subsequently materializes into a e^+e^- or $\mu^+\mu^-$ pair (Figure 21). Applying the Altarelli–Parisi

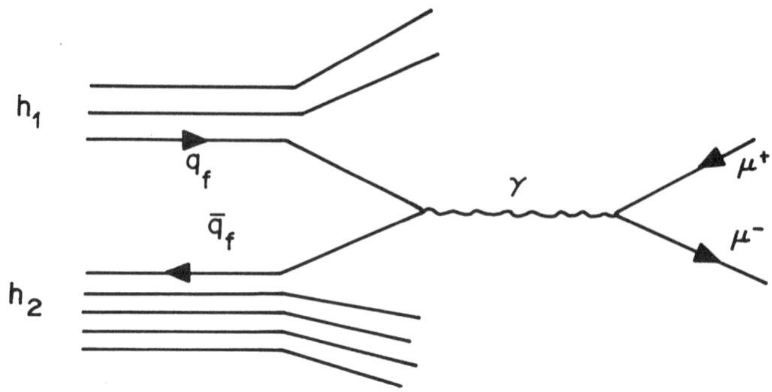

Figure 21. Drell–Yan mechanism.

formalism, it can be shown that, at least in the leading logarithmic approximation, one can write the cross-section as[73]

$$\frac{d\sigma}{dQ^2} = \frac{4\pi\alpha^2}{9Q^2}\sum_f Q_f^2 \int_0^1 dx_1 \int_0^1 dx_2\, x_1 x_2 \delta(x_1 x_2 - Q^2/s)$$
$$\times \left\{ q_{f,h_1}(x_1, Q^2)\bar{q}_{f,h_2}(x_2, Q^2) + \bar{q}_{f,h_1}(x_1, Q^2) q_{f,h_2}(x_2, Q^2) \right\}, \tag{27.4}$$

where the q are the distribution functions of Section 22 and $s = (p_{h_1} + p_{h_2})^2$. The second-order corrections are very important here: they embody the effect of the continuation to timelike photon momentum. They have been calculated recently.[74] The calculation is very complicated due to the interplay of mass singularities. They modify Equation (27.4), in particular, by the inclusion of a factor

$$1 + \frac{\alpha_s(Q^2)}{4\pi}\cdot\frac{8}{3}\left(1 + \frac{4\pi^2}{3}\right), \tag{27.5}$$

where the π^2 on the right-hand side comes from the analytical continuation. The corrections are therefore very large, of order unity, so it would seem that, at present energies, QCD would only give qualitative estimates. This may not be so if, as conjectured, the π^2 terms exponentiate and one could replace (27.5) by

$$e^{8\pi\alpha_s(Q^2)/3}\left\{1 + \frac{8}{3}\cdot\frac{\alpha_s(Q^2)}{4\pi}\right\}, \tag{27.6}$$

with the exponential being exact to all orders. If this is the case, one finds good quantitative agreement with experiment.

[73] See Ellis *et al*., (1979); Sterman and Libby (1978).

[74] Kubar-André and Paige (1979); Altarelli, Ellis, and Martinelli (1978), (1979); Floratos, Kounnas, and Lacaze (1981); Humpert and van Neerven (1981). The calculation has been completed by Harada and Muta (1980) and Ellis, Martinelli, and Petronzio CERN preprint TH.3186, 1982 (to be published).

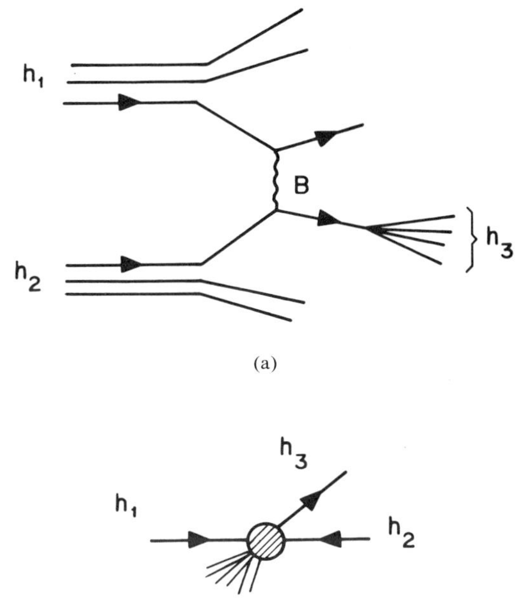

(b)

Figure 22. Large p_t scattering of hadrons.

Still farther removed from the direct application of the QCD methods is the scattering of hadrons at large p_t, Figure 22. The experimental set-up is that of Figure 22*b*: one scatters hadrons h_1, h_2 and triggers hadron h_3 with large momentum, transverse to the collision. It may be argued that the mechanism is that of Figure 22*a*: the cross section would then be, to lowest order,

$$\frac{d\sigma(h_1 + h_2 \to h_3 + \text{all})}{d^3_{p_{h_3}}}$$
$$= \frac{1}{\pi} E_{h_3} \sum \int_0^1 dx_a \int_0^1 dx_b \int_0^1 dx_{b'} q_{a,h_1}(x_a) q_{b,h_2}(x_b) q_{b'_1 h_3}(x_{b'})$$
$$\times \frac{s'\delta(s' + t' + u')}{x_{b'}^2} \cdot \frac{d\sigma(a + b \to a' + b')}{dt'} . \tag{27.7}$$

Here

$$s' = x_a x_b s, \qquad t' = x_a t / x_b', \qquad u' = x_b u / x_{b'},$$
$$s = (p_{h_1} + p_{h_2})^2, \qquad t = (p_{h_1} - p_{h_3})^2, \qquad u = (p_{h_1} + p_{h_3})^2.$$

The elementary cross section $d\sigma / dt'$ is to be calculated to lowest order in perturbation theory. We have not written $q(x, Q^2)$ because it is unclear, at least to us, what Q^2 to take, as well as the region of applicability of (27.7). For a discussion of these processes, see, for example, Sachrajda (1979),

Gribov and Lipatov (1972), and the reviews of Ellis and Sachrajda (1980), Jacob and Landshoff (1978).

ii Jets

We now turn to jets. Jets are a subject in themselves; here we will only give the briefest of sketches. The basic remark is that, for example, in e^+e^- annihilations, the leading diagram is the absorptive part of that of Figure 23*a*, namely, the square of that of Figure 23*b*. If quarks were real particles, this would tell us that the cross section would be

$$\frac{d\sigma(e^+e^- \to \bar{q}q)}{d\Omega} \approx (1+\cos^2\theta)\{1+O(\alpha_s)\}.$$

However, this cannot be true, for we have seen that processes with colinear particles (Figure 23*c*) give rise to divergences. But it is likely that inclusive cross sections are finite,[75] even in QCD. Then the trick is not to consider

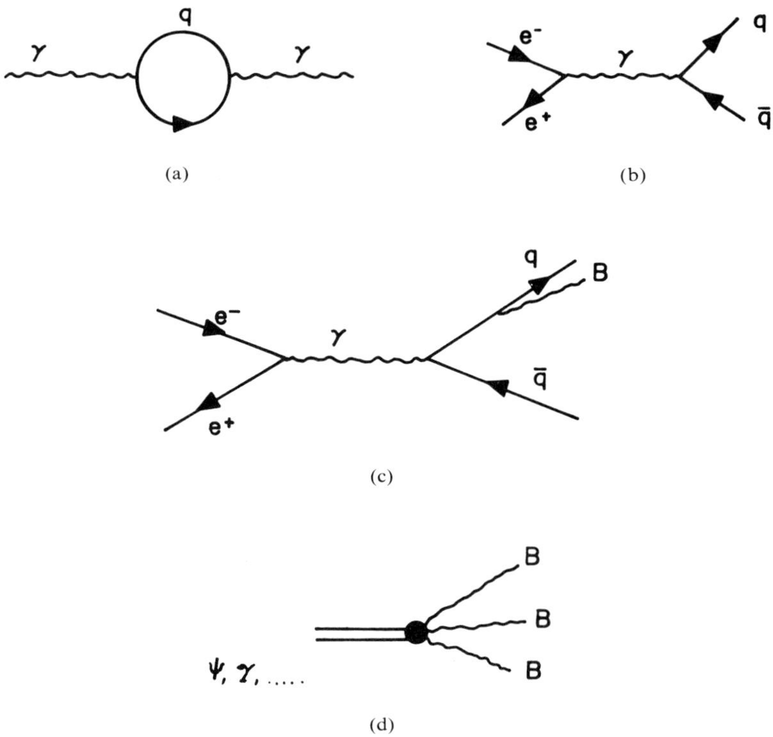

Figure 23. Jets.

[75] For QED, this is known as the Bloch–Nordseik (1937) theorem. In QCD, similar results follow from generalizations [Lee and Nauenberg (1964)] of Kinoshita's (1962) theorem.

processes where the quarks and/or gluons have well-defined momenta, $\vec{p}_1, \ldots, \vec{p}_n$ generally divergent, but to smear the cross sections with smooth functions $\phi(\vec{p}_1, \ldots, \vec{p}_n)$, i.e., consider cross sections over bunches of final states. Typically,

$$\sigma(\langle \vec{p}_1 \rangle, \ldots, \langle \vec{p}_n \rangle) = \int \frac{d\vec{p}_1}{2p_1^0} \cdots \frac{d\vec{p}_n}{2p_n^0} \phi(\vec{p}_1, \ldots, \vec{p}_n)\sigma(i \to \vec{p}_1, \ldots, \vec{p}_n),$$

with the $\phi(\vec{p})$ peaked around the mean values $\langle \vec{p} \rangle$.

Since, of course, quarks or gluons are not detected directly, one has to devise a method by which the jetlike character of the cross sections are made apparent. Basically, the method consists of measuring observables which are infrared finite [Sterman and Weinberg (1977)] and indicate the deviations from sphericity in the momentum distributions of the final states. For example, the thrust [Farhi (1977)],

$$T = \max_{\vec{v}} \frac{\sum |\vec{p}_i \cdot \vec{v}|}{\sum |\vec{p}_i|} \; ;$$

for a pure two jet event, $T = 1$; for a spherical event, $T = 1/2$. One then expects that, in e^+e^- annihilations, $T \simeq 1 - O(\alpha_s)$.

We shall not delve further into jets, but refer to De Rújula, Ellis, Floratos, and Gaillard (1978) for a comprehensive study of two and, mainly, three jet events (as in the decays of Υ; see Figure 23*d*), Mendez (1978) for jets in deep inelastic scattering, or the reviews of Ellis and Sachrajda (1981). We only add that two and three jet events have been clearly seen; the latter provide direct evidence for gluons and for the quark-quark-gluon coupling. The value of this coupling corresponds to [Ali (1982)] $\alpha_s(Q^2 \simeq (35\ \text{GeV})^2) \simeq 0.125 \pm 0.01$, or $\Lambda = 110^{+70}_{-50}$ MeV, in impressive agreement with previous determinations.

iii Exclusive Processes

We will give a simplified discussion of the pion form factor; this will, we hope, pave the way for the extension to other processes for which we will give only the results.

One can define the *pion form factor* F_π via

$$\begin{aligned} V^\mu(p_1, p_2) &= (2\pi)^3 \langle \pi(p_2) | J^\mu_{\text{em}}(0) | \pi(p_1) \rangle \\ &= (p_1^\mu + p_2^\mu) F_\pi(q^2), \qquad q = p_2 - p_1, \end{aligned} \tag{27.8}$$

and F_π is normalized to $F_\pi(0) = 1$. Suppressing the index em in J^μ, we write

$$V^\mu(p_1, p_2) = (2\pi)^3 \langle \pi(p_2) | T J_0^\mu(0) e^{i\int d^4x \mathscr{L}^0_{\text{int}}(x)} | \pi(p_1) \rangle.$$

To second order, this yields (as usual, the $q_u^0 \equiv q_0$, $B_u^0 \equiv B_0, \ldots$ are free fields)

$$V^{\mu}(p_1, p_2) = -(2\pi)^3 \frac{g^2}{2!} \sum_{f=u,d} Q_f \int d^4x\, d^4y \langle \pi(p_2)| T\bar{q}_{f0}(0)\gamma^{\mu} q_{f0}(0)$$

$$\times \sum_{a,b} \left\{ \bar{u}_0(x)\gamma^{\rho} t^a u_0(x) \bar{d}_0(y)\gamma^{\sigma} t^b d_0(y) + (x \leftrightarrow y) \right\} B^a_{0\rho}(x)$$

$$\times B^b_{0\sigma}(y)|\pi(p_1)\rangle. \tag{27.9}$$

The various combinations give rise to the diagrams of Figure 24*a* and *b*. We have dropped the terms corresponding to the diagrams of Figure 24*a* because they do not contribute to the final result. The contribution of Figure 24*b* is, with the color indices i, j, and k, and the Dirac indices α, β, and δ, and dropping the free-field 0 indices,

$$V^{\mu}(p_1, p_2) = -(2\pi)^3 g^2 \sum \int d^4x\, d^4y \langle \pi(p_2)| \bar{u}^i_{\alpha}(0) d^{k'}_{\delta'}(y) \gamma^{\mu}_{\alpha\alpha'} S_{\alpha'\beta}(-x) t^a_{ii'} t^b_{kk'}$$

$$\times \gamma^{\rho}_{\beta\beta'} \gamma^{\sigma}_{\delta\delta'} D_{\rho\sigma}(x-y) \delta_{ab} u^{i'}_{\beta'}(x) \bar{d}^k_{\delta}(y)|\pi(p_1)\rangle + \text{"cross,"}$$

where "cross" corresponds to the other contraction. We can perform a space-time shift by y. We then are left with the expression ($z = x - y$),

$$V^{\mu}(p_1, p_2) = (2\pi)^3 g^2 \sum \int \frac{d^4z}{(2\pi)^4} \int \frac{d^4y}{(2\pi)^4} \int d^4k \int d^4p\, e^{iz\cdot(p-k)} e^{iy\cdot(p+p_2-p_1)}$$

$$\times \langle \pi(p_2)| \bar{u}^i_{\alpha}(-y) d^{k'}_{\delta}(0) \sum_F |F\rangle\langle F| u^i_{\beta'}(z) \bar{d}^k_{\delta}(0)|\pi(p_n)\rangle \gamma^{\mu}_{\alpha\alpha'}$$

$$\times \frac{-\not{p}_{\alpha'\beta}}{p^2k^2} \gamma^{\rho}_{\beta\beta'} \gamma^{\sigma}_{\delta\delta'} g_{\rho\sigma} t^c_{ii'} t^c_{kk'} + (p_1 \leftrightarrow p_2).$$

The $(p_1 \leftrightarrow p_2)$ term comes from the "cross" term. We have not explicitly written the contribution of the gauge terms since they yield zero, to leading order. We have inserted a complete set of states; to leading order, only the vacuum contributes:

$$\sum_F |F\rangle\langle F| \simeq |0\rangle\langle 0| + O(\alpha_s).$$

We have evaluated $D_{\rho\sigma}$ in the Fermi–Feynman gauge, but the result (after adding the $p_1 \leftrightarrow p_2$ term) is, of course, gauge invariant. Next, with n_c = number of colors = 3,

$$u^i_{\beta'}(z)\bar{d}^k_{\delta}(0) = \frac{\delta_{ik}}{4n_c} (\gamma^{\lambda}\gamma_5)_{\beta'\delta} \bar{d}(0)\gamma_{\lambda}\gamma_5 u(z)$$

$$- \frac{\delta_{ik}}{4n_c} (\gamma_5)_{\beta'\delta} \bar{d}(0)\gamma_5 u(z) + \cdots; \tag{27.10}$$

other terms will not contribute because of the pseudoscalar and colour singlet nature of the pion. In fact, $\bar{d}\gamma_5 u$ is a twist-three operator and, hence,

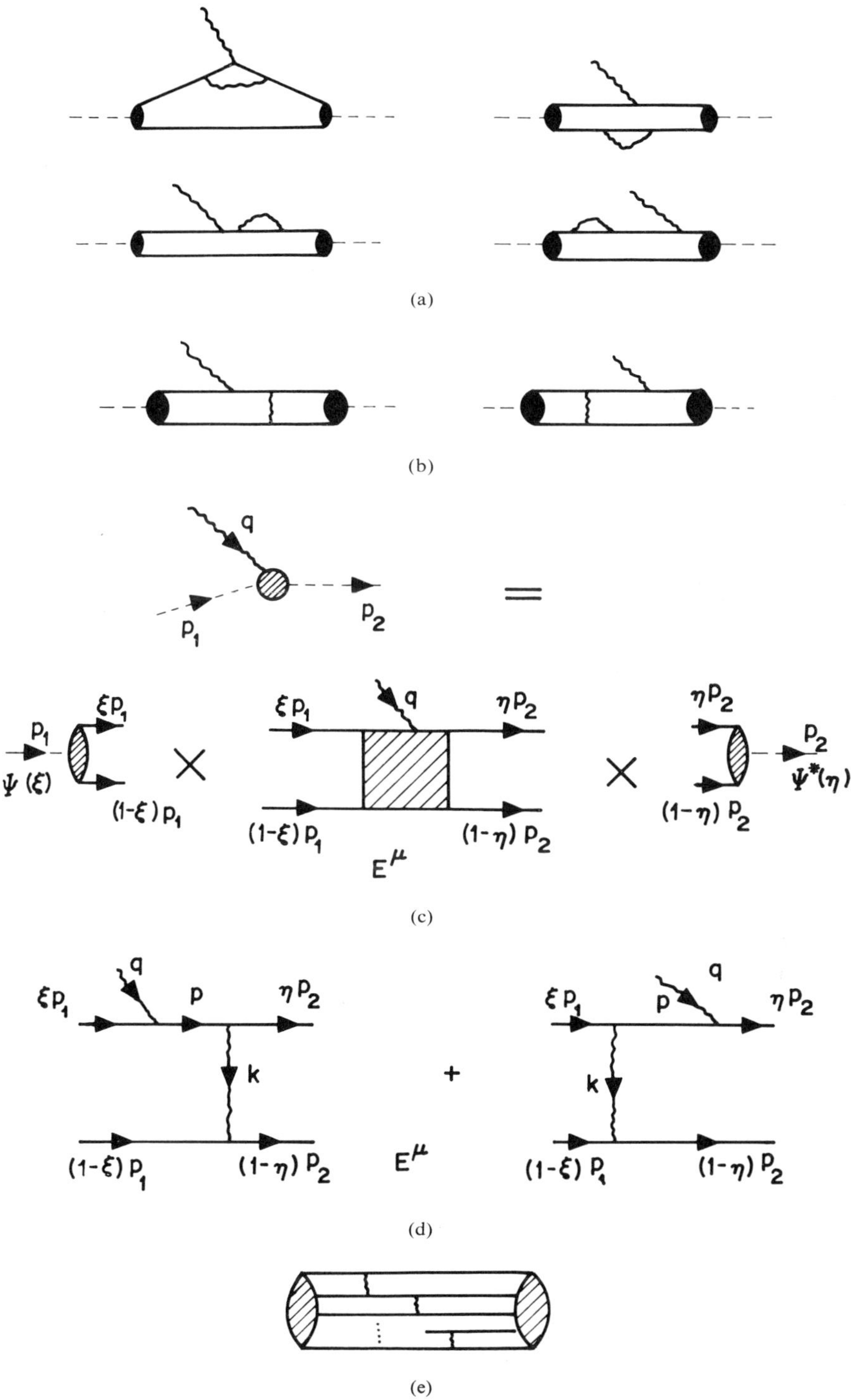

Figure 24. Diagrams relevant for exclusive processes. (*a–d*), pion form factor.

to leading order, can be dropped. We thus find

$$V^\mu(p_1, p_2) = (2\pi)^3 \frac{C_F g^2}{48} \int \frac{d^4z}{(2\pi)^4} \int \frac{d^4y}{(2\pi)^4} \int d^4k \int d^4p e^{iz\cdot(p-k)} e^{-iy\cdot(p+p_2-p_1)}$$
$$\times \frac{\operatorname{Tr} \gamma^\mu \not{p} \gamma^\rho \gamma^\lambda \gamma_5 \gamma_\rho \gamma^\tau \gamma_5}{p^2k^2} \langle 0|\bar{d}(0)\gamma_\lambda\gamma_5 u(z)|\pi(p_1)\rangle$$
$$\times \langle \pi(p_2)|\bar{u}(y)\gamma_\tau\gamma_5 d(0)|0\rangle + (p_1 \leftrightarrow p_2). \tag{27.11}$$

Let us concentrate on the $\langle 0| \dots |\pi\rangle$ terms. We can expand them in powers of z and y: for example,

$$\langle 0|\bar{d}(0)\gamma_\lambda\gamma_5 u(z)|\pi(p_1)\rangle$$
$$= \sum_n \frac{z^{\mu_1} \dots z^{\mu_n}}{n!} \mathscr{S} \langle 0|\bar{d}(0)\gamma_\lambda\gamma_5 D_{\mu_1} \dots D_{\mu_n} u(0)|\pi(p_1)\rangle \tag{27.12a}$$

and, neglecting terms proportional to the pion mass,

$$(2\pi)^{3/2}\langle 0|\bar{d}(0)\gamma_\lambda\gamma_5 D_{\mu_1} \dots D_{\mu_n} u(0)|\pi(p_1)\rangle \equiv i^{n+1} p_{1\lambda} p_{1\mu_1} \dots p_{1\mu_n} A_n. \tag{27.12b}$$

All of this has been accomplished formally. When renormalizing, we will have to replace $g \to g(\nu^2)$, and realize that $A_n = A_n(\nu^2)$. To avoid $\log Q^2/\nu^2$ terms, we will take $\nu^2 = Q^2 = -(p_2 - p_1)^2$. Now (27.11) may be cast in a physically very transparent form if we define the "parton wave function" Ψ such that

$$\int_0^1 d\xi \xi^n \Psi = A_n, \tag{27.12c}$$

for then

$$(2\pi)^{3/2}\langle 0|\bar{d}(0)\gamma_\lambda\gamma_5 u(z)|\pi(p_1)\rangle = ip_{1\lambda} \int_0^1 d\xi \Psi(\xi, \nu^2) e^{i\xi p_1\cdot z}, \tag{27.13}$$

and thus we may carry over the z, y'; k, p integrations in (27.11), obtaining

$$V^\mu(p_1, p_2) = \frac{C_F g^2(\nu)}{48} \int_0^1 d\xi \Psi(\xi, \nu^2) \int_0^1 d\eta\, \Psi^*(\eta, \nu^2) \frac{\operatorname{Tr} \gamma^\mu \not{p} \gamma^\rho \not{p}_1 \gamma_5 \gamma_\rho \not{p}_2 \gamma_5}{p^2k^2}$$
$$+ (p_1 \leftrightarrow p_2), \tag{27.14a}$$

$$p = p_1 - (1-\eta)p_2,$$
$$k = (1-\eta)p_2 - (1-\xi)p_1. \tag{27.14b}$$

We have succeeded in splitting the vertex into a "soft part," buried in the wave functions Ψ, Ψ^*, an a "hard part," E^μ (See Figures 24c,d). The variables ξ, η describe the fraction of momentum in each quark. Evaluating the trace in (27.14), we finally find

$$F_\pi(q^2) = \frac{4\pi C_F \alpha_s(Q^2)}{6Q^2} \left| \int_0^1 d\xi \frac{\Psi(\xi, Q^2)}{1-\xi} \right|^2 + O\left(\frac{M_\pi^2}{Q^2}\right) + O(\alpha_s^2), \quad (27.15)$$

$$Q^2 \equiv -q^2.$$

The last task is the evaluation of the Q^2 evolution of the Ψ. The operators that defined Ψ via Equations (27.12) are the same as those for the nonsinglet part of deep inelastic scattering (cf., Sections 19 and 20). We have, however, an extra complication: because the matrix elements are nondiagonal, total derivatives yield a nonzero contribution. The operators $N_{A,n,k}^{\lambda\mu_1 \ldots \mu_n}$, $k = 0, \ldots, n$,

$$N_{A,n,k}^{\lambda\mu_1 \ldots \mu_n} = \partial^{\mu_{k+1}} \ldots \partial^{\mu_n} \bar{d}(0)\gamma^\lambda\gamma_5 D^{\mu_1} \ldots D^{\mu_k} u(0) \quad (27.16)$$

mix under renormalization. They are thus renormalized by a matrix:

$$N_{A,n,k} \to \sum_{k'} Z_{n+1,k'} N_{A,n,k'} \quad (27.17a)$$

For $k = n$, the $Z_{n+1,n}$ values coincide with those calculated in Section 20:

$$Z_{n+1,n} = 1 + \frac{g^2 N_\epsilon}{16\pi^2} C_F \left\{ 4S_1(n+1) - 3 - \frac{2}{(n+1)(n+2)} \right\}; \quad (27.17b)$$

for $k \leqslant n-1$, we get

$$Z_{n+1,k} = \frac{g^2 N_\epsilon}{16\pi^2} C_F \left\{ \frac{2}{n+2} - \frac{2}{n-k} \right\}. \quad (27.17c)$$

To obtain the operators with a definite behavior[76] as $Q^2 \to \infty$, we have to diagonalize Z. Let S be the matrix that accomplishes this; we then have, with the $\hat{A}_k$ diagonal,

$$A_n(Q^2) = \sum_{k=0}^{n} S_{nk} \hat{A}_k(Q^2). \quad (27.18)$$

The anomalous dimensions of $\hat{A}_k$ are the eigenvalues of Z. But because Z is triangular, it follows that its eigenvalues are simply its diagonal elements. Therefore,

$$\hat{A}_k(Q^2) \underset{Q^2 \to \infty}{\approx} \left[\alpha_s(Q^2)\right]^{d_{NS}(k+1)} \hat{A}_{k0}.$$

To leading order, and since $d_{NS}(k+1) > d_{NS}(1) = 0$, we need to retain only one term in (27.18) so that

$$A_n(Q^2) \underset{Q^2 \to \infty}{\to} S_{n0} \hat{A}_{00},$$

[76] An alternate elegant method entails properties of conformal invariance [Ferrara, Gatto, and Grillo (1972)].

and we get

$$\int_0^1 d\xi \frac{\Psi(\xi, Q^2)}{1-\xi} \underset{Q^2\to\infty}{\to} \hat{A}_{00} \sum_{n=0}^{\infty} S_{n0}.$$

The values of the S_{n0} are easily verified to be

$$S_{n0} = \frac{1}{n+2} - \frac{1}{n+3}.$$

In addition, the value of $\hat{A}_{00}$ is also known. Because of the PCAC equation [cf., Section 31, especially Equation (31.2*b*) later],

$$(2\pi)^{3/2}\langle 0|\bar{d}(0)\gamma^\lambda\gamma_5 u(0)|\pi(p)\rangle = ip^\lambda\sqrt{2}\, f_\pi, \qquad f_\pi \approx 93\ MeV;$$

hence,

$$A_0 = \int_0^1 d\xi \Psi(\xi, Q^2) = \sqrt{2}\, f_\pi \qquad \text{independent of } Q^2.$$

From this,

$$\hat{A}_{00} = 6\sqrt{2}\, f_\pi,$$

so we finally obtain the result[77]

$$F_\pi(t) \underset{Q^2\to\infty}{\approx} \frac{12\pi C_F f_\pi^2 \alpha_s(-t)}{-t}. \tag{27.19}$$

The corrections are $O(\alpha_s^{d_{NS}(3)} = \alpha_s^{0.6})$; even values of n actually yield zero due to charge conjugation invariance.

The example of the pion form factor is instructive in several respects. If we compare (27.19) with experimental results, it appears that the theoretical expression is much too small. One may of course attribute this to a large next-to-leading perturbative contribution; but what probably occurs is that the higher twist corrections become especially important due to the smallness of the u and d quark masses. In fact, consider the contribution of the pseudoscalar term in (27.10). The mixed term gives contributions suppressed by m_π^2/t with respect to (27.19) and may be safely ignored; but the pseudoscalar-pseudoscalar contribution contains the square of $\langle O|\bar{d}\gamma_5 u|\pi\rangle$, which, using the equations of motion, is easily seen to be proportional to $f_\pi m_\pi^2/(m_u + m_d)$. Repeating the above analysis, we find

$$F_\pi(t) = \frac{12\pi C_F f_\pi^2 \alpha_s(-t)}{-t}\left\{1 + \frac{4m_\pi^4 \log(-t/m_\pi^2)}{-(\bar{m}_u + \bar{m}_d)^2 t}\right\}, \tag{27.20}$$

a result that is exact in the chiral limit (cf., Sections 29ff.) in leading and next-to-leading twist. Although (27.20) can be made to agree with experiment for reasonable choices of the quark masses, it is difficult to take the

[77] Farrar and Jackson (1979); Brodsky, Frishman, Lepage, and Sachrajda (1980); Efremov and Radyushin (1980*a*), which we have followed. The same result may be obtained using so-called light cone perturbation theory [Brodsky and Lepage (1980)].

result very seriously, because the higher twist contribution is infrared divergent (actually, what we have here is a mass singularity); the simple factorization used to obtain (27.20) is not valid. Thus the coefficient of the correction is not yet known.

This brings us to the second point. The "hard" part of the pion form factor appears to be infrared divergent (the term $1/(1-\xi)$ in Equation [27.15]). However, for the leading order, we are lucky: we found that

$$\int_0^1 d\xi\, \Psi(\xi, Q^2)\xi^n \underset{Q^2\to\infty}{\to} S_{n0}\hat{A}_{00},$$

which implies that

$$\Psi(\xi, Q^2) \underset{Q^2\to\infty}{\to} \xi(1-\xi)\hat{A}_{00},$$

thus cancelling the potential divergence. This is the reason why some exclusive processes are, after all, amenable to a simple perturbation treatment. This property is not generally true, however, as is seen for the subleading twist term in (27.20). In fact, for some processes, this infrared divergence already appears at the leading twist level. For example, we may consider the scalar form factor

$$D(t) = (2\pi)^{-3}\langle \pi|\sigma_{us}(0)|K^0\rangle, \qquad \sigma_{us}(x) = i\partial_\mu \bar{u}(x)\gamma^\mu s(x). \quad (27.21a)$$

The calculation again follows that of the pion form factor. The only novelty is that here the mixed pseudoscalar-axial vector contribution (nominally higher twist) is in fact leading. We find, after using the PCAC relations of Section 31,

$$D(t) \approx \frac{12\pi C_F \alpha_s(-t) f_\pi f_K}{-t}\left\{\left(\overline{m}_s^2 - \overline{m}_u^2\right) + \left(m_K^2 - m_\pi^2\right)\right\}\log\frac{t}{M^2} \quad (27.21b)$$

The first term is the axial-axial, the second is the mixed contribution; but both are infrared divergent (if we use naive factorization). These two examples [subleading twist in the pion form factor and the scalar form factor (27.21)] show that unlike inclusive ones, exclusive processes are very sensitive to infrared divergences and that one should check each individual process to determine whether perturbative QCD is directly applicable. With this in mind, we end this section with a very brief resume of results for other exclusive processes.

From the example of the pion form factor, we may infer the general rule: The amplitude for an exclusive process is of the form (see Figure 24e)

$$\mathscr{A} = \int \phi^+ K \phi,$$

where ϕ is the corresponding wave function for the bound state B,

$$\phi \approx \langle 0|Tq_1(x_1)\ldots q_n(x_n)|B\rangle,$$

and T is the kernel,

$$K \approx \left[\frac{\alpha_s(Q^2)}{Q^2}\right]^{n-1}.$$

This yields the counting rules of Brodsky and Farrar (1973); for example, for nucleon form factors, one gets the celebrated dipole behavior

$$F_N \approx \left[\frac{\alpha_s(-t)}{-t} \right]^2$$

for the deuteron,

$$F_d \approx \left[\frac{\alpha_s(-t)}{-t} \right]^5,$$

a behavior that is, in fact, seen experimentally. For fixed-angle scattering,

$$\left. \frac{d\sigma(A+B \to C+D)}{dt} \right|_{\theta \text{ fixed}} \approx \frac{\alpha_s^2(t)}{-t} F_A(t) F_B(t) F_C(t) F_D(t) f(\theta).$$

Further details and references may be found in the paper of Brodsky and Lepage (1980). Many of these results have been rigorized and reformulated in terms of renormalization-group analyses by Duncan and Muller (1980*b*) [see also the review of Duncan (1981) and work cited therein].

CHAPTER IV

Quark Masses, PCAC, Chiral Dynamics, and the QCD Vacuum

If the Lord Almighty had consulted me before embarking upon creation, I should have recommended something simpler.

ALPHONSE X "The Wise", (1221–1284), King of Castille and León, on having the Ptolemaic system of epicycles explained to him.

28 Heavy and Light Quarks: The Symanzik–Appelquist–Carrazzone Theorem

The $\overline{MS}$ scheme is independent of quark masses; therefore, when calculating the β_n or $\gamma^{(n)}$, one has to take into account all existing quark flavors. For simplicity, let us concentrate on the β function and work with an axial gauge so that the entire Q^2 evolution may be obtained with just the gluon propagator. Furthermore, we will simplify the discussion by considering only two quark flavors: one essentially massless, $\hat{m}_l = 0$, and a heavy one, $\hat{m}_h \gg \Lambda$. In the $\overline{MS}$ scheme, we have

$$\alpha_s(Q^2, \Lambda^2) = \frac{12\pi}{(33 - 2n_f)\log Q^2/\Lambda^2}\{1 - \ldots\}, \tag{28.1}$$

with $n_f = 2$. However, it stands to reason that while $n_f = 2$ will be a good expression for $Q \gg m_h$, there will be a region $m_h \gg Q \gg \Lambda$ where one would expect that (28.1) with $n_f = 1$ would be a more reasonable choice. This is made more dramatic if we set m_h extremely large, for instance, 1 gram. Clearly microscopic physics can hardly depend upon the existence or nonexistence of this particle.

This is basically the content of a theorem proved by Symanzik (1973) and rediscovered by Appelquist and Carrazzone (1975)[78] which states that

[78] Actually, the result is essentially contained in the basic paper of Kinoshita (1962). For a discussion using functional methods, see Weinberg (1980).

when $Q \ll m_h$, we can neglect the existence of this quark up to terms Q^2/m_h^2. Equation (28.1) is valid as it stands only if $Q^2 \ll m^2$, where m is any relevant mass, in particular, m_h. If we want to keep the functional form (28.1), we must allow for a Q^2 dependence other than the merely logarithmic.

Since the problem arises because we neglected masses, we must re-derive (28.1), but now with masses taken into account. We recall that the running coupling constant was defined by $\alpha_s = \bar{g}^2/4\pi$, with $\bar{g}$ a solution of Equations (12.6),

$$\frac{d\bar{g}}{d\log Q/\nu} = \bar{g}\beta(\bar{g}), \qquad \bar{g}|_{Q=\nu} = g(\nu), \tag{28.2a}$$

with

$$\frac{\nu d}{d\nu} g(\nu) = g(\nu)\beta(g(\nu)), \qquad \beta = -Z_g^{-1}\frac{\nu d}{d\nu}Z_g. \tag{28.2b}$$

We now consider the behavior of the transverse part of the gluon propagator, which we write as in Equation (6.9). From (12.1), (12.7), and with $Q/\nu = \lambda$,

$$D_{\mathrm{tr}}\big(q^2;\, g(\nu), m(\nu);\, \nu^2\big)$$
$$= D_{\mathrm{tr}}\big(\nu^2;\, \bar{g}(\lambda), \bar{m}(\lambda);\, \nu^2\big)\exp\Big\{-\int_0^{\log\lambda} d\log\lambda'\,\gamma_D[\bar{g}(\lambda')]\Big\}. \tag{28.3}$$

In the physical gauge we are using [recall Equation (9.18)], $\gamma_D = 2\beta_0 g^2/16\pi^2$, and hence

$$D_{\mathrm{tr}}\big(q^2;\, g(\nu), m(\nu);\, \nu^2\big) = \frac{2}{\log Q^2/\nu^2} D_{\mathrm{tr}}\big(\nu^2;\, \bar{g}(\lambda), \bar{m}(\lambda);\, \nu^2\big). \tag{28.4}$$

Next we require $D_{\mathrm{tr}}(\nu^2;\, \bar{g}(\lambda),\, \bar{m}(\lambda);\, \nu^2)$ *exactly* in $\bar{m}$. We have

$$D_{\mathrm{tr}}\big(\nu^2;\, \bar{g}(\lambda), \bar{m}(\lambda);\, \nu^2\big) = K_\nu + \frac{2\alpha_s(Q^2)T_F}{\pi}$$
$$\times \int_0^1 dx\, x(1-x)\log\frac{\bar{m}^2 + x(1-x)\nu^2}{\nu^2},$$

where K_ν is a constant. To begin, we choose $\nu = \Lambda$:

$$D_{\mathrm{tr}}\big(q^2;\, g(\nu), m(\nu);\, \nu^2\big) = \frac{2}{\log Q^2/\Lambda^2}\Bigg\{K + \frac{2\alpha_s(Q^2)T_F}{\pi}$$
$$\times \int_0^1 dx\, x(1-x)\log\Big[x(1-x) + \frac{\bar{m}^2(Q^2)}{\Lambda^2}\Big]\Bigg\}. \tag{28.5}$$

If $m \gg \Lambda$, this becomes

$$D_{\mathrm{tr}}\big(q^2;\, g(\nu), m(\nu);\, \nu^2\big) \approx \Big\{K + \frac{\alpha_s(Q^2)T_F}{\pi}\log\frac{\bar{m}^2(Q^2)}{\Lambda^2}\Big\}\frac{2}{\log Q^2/\Lambda^2}. \tag{28.6}$$

Now, if $m^2 \gg Q^2$, the correction to K in (28.6) is large, so that the approach is not very useful. This was to be expected; the $\overline{MS}$ scheme, or any mass-independent scheme [like that of Weinberg (1973b)], must necessarily destroy the convergence when there is a mass larger than the momentum scale. The solution to this problem is to take n_f to be dependent on the momentum scale so that, for example,[79]

$$n_f(Q^2) = \sum_{f=1}^{n_f} \left(1 - \frac{4\hat{m}_f^2}{Q^2}\right)^{1/2} \left(1 + \frac{2\hat{m}_f^2}{Q^2}\right) \theta(Q^2 - 4\hat{m}_f^2). \tag{28.7}$$

We still have to prove that such a procedure is consistent. That (28.7) is right above all quark masses we already know; the corrections are $O(\hat{m}_q^2/Q^2)$. We will complete the proof by showing that such is also the case for $Q^2 \ll m^2$, for the gluon propagator. This will indicate how to extend the proof to the general case.

Since the contribution of quarks and gluons to D_{tr} is additive, we need only consider the former. To leading order, then,

$$D_{\text{tr}}^{(\text{quark})} = 1 - \frac{\alpha_g}{\pi} \int_0^1 dx\, x(1-x) \log \frac{x(1-x)Q^2 + \hat{m}^2}{\nu^2}, \tag{28.8}$$

to this order, we need not consider the renormalization of α_g or m. Now, for $Q^2 \ll \hat{m}^2$, we obtain

$$D_{\text{tr}}^{(\text{quark})} = 1 - \frac{\alpha_g}{6\pi} \log \frac{\hat{m}^2}{\nu^2} - \frac{\alpha_g}{30\pi} \cdot \frac{Q^2}{\hat{m}^2}, \tag{28.9}$$

i.e., constant up to terms $O(Q^2/\hat{m}^2)$. Therefore, it coincides (up to these terms) with the gluon propagator calculated with *zero* flavors but for a different value of ν'^2, namely, $\nu^2 = \nu^2\{1 + \log \hat{m}^2/\nu^2\}$. Because physical observables are independent of ν, one can drop the heavy quark, which only contributes terms $O(Q^2/\hat{m}^2)$.

The case of the gluon propagator is particularly simple; in general, the order of the corrections can be $\log(\hat{m}^2/Q^2)(Q^2/\hat{m}^2)$.

The decoupling theorem is particularly transparent in the μ scheme of renormalization. Consider again the quark contribution to the gluon propagator. We work to second order and then, recalling (9.21),

$$\begin{aligned} D_{u\text{tr}}^{(\text{quark})}(q^2) = {} & \frac{i}{q^2} \\ & + T_F \frac{q^2}{16\pi^2} \left\{ \frac{2}{3} N_\epsilon n_f - 4 \int_0^1 dx\, x(1-x) \right. \\ & \qquad \left. \times \sum_{f=1}^{n_f} \log \frac{m_f^2 - x(1-x)q^2}{\nu_0^2} \right\} + \cdots \end{aligned}$$

[79] Other interpolation formulas or procedures are possible [cf., the paper of Coquereaux (1981*b*) and especially Weinberg (1973*b*) where a complete discussion, including the calculation of the dependence on the effective value of n_f, may be found.] Which scheme one uses is largely irrelevant because the dependence of QCD on n_f is very slight between three light quarks and six.

It will be recalled that the μ scheme is obtained by requiring $D_{R\mathrm{tr}}^{(\mathrm{quark})}(q^2 = -\mu^2) = D_{\mathrm{free\ tr}}(-\mu^2)$; therefore,

$$D_{R\mathrm{tr}}^{(\mathrm{quark})}(q^2) = \frac{i}{q^2}$$

$$+ T_F \frac{q^2}{16\pi^2} \left\{ -4 \int_0^1 dx\, x(1-x) \sum_f \log \frac{m_f^2 - x(1-x)q^2}{m_f^2 + x(1-x)\mu^2} \right\}$$

Take $Q^2 = -q^2$. For $Q^2, \mu^2 \gg m_f^2$,

$$\int_0^1 dx\, x(1-x) \log \frac{m_f^2 + x(1-x)Q^2}{m_f^2 + x(1-x)\mu^2} \simeq \frac{1}{6} \log \frac{Q^2}{\mu^2} + O\left(\frac{m_f^2}{\mu^2}, \frac{m_f^2}{Q^2} \right);$$

for $m_f^2 \gg \mu^2, Q^2$,

$$\int_0^1 dx\, x(1-x) \log \frac{m_f^2 + x(1-x)Q^2}{m_f^2 + x(1-x)\mu^2} \simeq O\left(\frac{\mu^2}{m_f^2}, \frac{Q^2}{m_f^2} \right).$$

29 Mass Terms and Invariances: Chiral Invariance

In the previous section, we have seen that at energies $Q \gg \Lambda$ where perturbation theory in the running coupling constant may be meaningful, one can neglect quarks with masses $m \gg Q$. In this section, we will consider the opposite situation, viz., quarks with masses $m \ll \Lambda$. Because the only intrinsic dimension parameter in QCD is, we believe, Λ,[80] we may expect that to some approximation we may neglect the masses of such quarks which will only yield contributions of order m^2/Λ^2 or m^2/Q^2.

Let us return to the discussion of Section 10. Consider the QCD Lagrangian,

$$\mathscr{L} = -\sum_{l=1}^{n} m_l \bar{q}_l q_l + i \sum_{l=1}^{n} \bar{q}_l \not{D} q_l - \frac{1}{4}(D \times B)^2 + \text{gauge fixing} + \text{ghost terms}. \tag{29.1}$$

The sum runs only over *light* quarks, i.e., quarks for which $\hat{m}^2 \ll \Lambda^2$; the eventual existence of heavy quarks will have no effect on the following. We consider the set of transformations $W^{\pm}$ in $U_L(n) \times U_R(n)$ (left-handed times right-handed)

$$\frac{1 \pm \gamma_5}{2} q_l \to \sum_{l'} W_{ll'}^{\pm} \frac{1 \pm \gamma_5}{2} q_{l'}; \qquad W^{\pm} \text{ unitary.} \tag{29.2}$$

[80] It is, of course, unclear whether the meaningful parameter is Λ or Λ_0 defined by $\alpha_s(\Lambda_0^2) \approx 1$. Likewise, the meaning of the expression $m \ll \Lambda$ is ambiguous. Clearly, $\Lambda \approx \Lambda_0$, so there is really no guide but heuristics to help us decide which quarks are light in borderline cases. There is little doubt that u and d should be classed as "light," but the situation is less clear for s.

Clearly, the only term that is not invariant under all of (29.2) is the mass term of $\mathscr{L}$,

$$\mathscr{M} = \sum_{l=1}^{n} m_l \bar{q}_l q_l . \tag{29.3}$$

When written in this form, the mass term is invariant under the set of transformations $[U(1)]^n$,

$$q_l \to e^{i\theta_l} q_l , \tag{29.4}$$

but this would not be the case if we had allowed for nondiagonal terms in the mass matrix. To resolve the question of which are the general invariance properties of a general mass term, we will prove two theorems.

Theorem 1. *Any general mass matrix can be written in the form* (29.3) *by appropriate redefinition of the quark fields.*

Moreover, we may assume that $m \geqslant 0$. So (29.3) is actually the most general possible mass term.

PROOF. Let

$$q_L = \frac{1}{2}(1-\gamma_5)q, \qquad q_R = \frac{1}{2}(1+\gamma_5)q.$$

The most general mass term compatible with hermiticity is

$$\mathscr{M}' = \sum_{ll'} \{ \bar{q}_{lL} M_{ll'} q_{l'R} + \bar{q}_{l'R} M^*_{ll'} q_{lL} \}. \tag{29.5}$$

Let $\mathbf{M}$ be the matrix with components $M_{ll'}$. The well-known polar decomposition of any matrix allows us to write

$$\mathbf{M} = \mathbf{mU},$$

where $\mathbf{m}$ is positive-definite, so all its eigenvalues are $\geqslant 0$, and $\mathbf{U}$ is unitary. Equation (29.5) then becomes

$$\mathscr{M}' = \sum \{ \bar{q}_{lL} m_{ll'} q'_{l'R} + \bar{q}'_{lR} m_{ll'} q_{l'L} \}, \qquad q'_{lR} = \sum_{l'} U_{ll'} q_{lR} , \tag{29.6}$$

and we have used the fact that $\mathbf{m}$ is self-adjoint. Define $q' = q'_R + q_L$; then (29.6) in terms of q' is

$$\mathscr{M}' = \sum \bar{q}'_l m_{ll'} q'_{l'} ,$$

where we have profited from the fact that $\bar{q}_R q_R = \bar{q}_L q_L = 0$. It then suffices to transform the q' by the matrix $\mathbf{V}$ that diagonalizes $\mathbf{m}$ to obtain (29.3). The fact that the m_l are positive follows because they are the eigenvalues of $\mathbf{m}$. (Note that the term $\bar{q}\not{D}q$ in the Lagrangian is invariant under these transformations.)

Theorem 2. *If all the m_l are nonzero and different, then the only invariance left is the $[U(1)]^n$ of* (29.4).

Let us assume that $W_+ = W_- = W$; to show that this must actually be the case is left as an exercise. The condition of invariance of $\mathscr{M}$ yields the relation

$$\mathbf{W}^{+}\mathbf{mW}=\mathbf{m}, \quad \text{i.e.,} \quad \mathbf{mW}=\mathbf{Wm}. \tag{29.7}$$

It is known that any diagonal matrix can be written as $\sum_{k=0}^{n-1}c_k\mathbf{m}^k$, if, as occurs in our case, all the eigenvalues of $\mathbf{m}$ are different and nonzero. Because of (29.7), it then follows that $\mathbf{W}$ commutes with all diagonal matrices; hence, it must be diagonal. Because it is also unitary, it may be written as the product of transformations (29.4) as was to be proved. We leave it to the reader to check that the conserved quantity corresponding to $U(1)$ that acts on flavor q_f is the corresponding flavor number.

In the preceding theorems, we have not worried about whether or not the m were bare, running, or invariant masses. This is because, in the $\overline{MS}$ scheme, the mass matrix becomes renormalized as a whole,

$$\mathbf{M}=Z_m^{-1}\mathbf{M}_u,$$

where Z_m is a *number*. The proof is very easy. In fact, all we have to do is repeat the analysis of Sections 7, 8, 9, and 14, allowing for the matrix character of M, Z_m. We find, for the divergent part, and in an arbitrary covariant gauge,

$$\begin{aligned}\mathbf{S}_{\mathrm{R}}^{\xi}(p)=&\frac{i}{\not p-\mathbf{M}}\\&+\frac{1}{\not p-\mathbf{M}}\Big\{-\left[\boldsymbol{\Delta}_F(\not p-\mathbf{M})+(\not p-\mathbf{M})\boldsymbol{\Delta}_F^{+}\right]-\delta\mathbf{M}\\&\qquad-(1-\xi)(\not p-\mathbf{M})N_\epsilon C_F\frac{g^2}{16\pi^2}+3N_\epsilon C_F\mathbf{M}\frac{g^2}{16\pi^2}\Big\}\frac{i}{\not p-\mathbf{M}},\end{aligned}$$

and we have defined

$$\mathbf{M}=\mathbf{M}_u+\delta\mathbf{M}, \qquad \mathbf{Z}_F=1+\boldsymbol{\Delta}_F.$$

The renormalization conditions then yield

$$\boldsymbol{\Delta}_F^{+}+\boldsymbol{\Delta}_F=-(1-\xi)N_\epsilon C_F\frac{g^2}{16\pi^2}=\text{diagonal},$$

$$\boldsymbol{\Delta}_F\mathbf{M}=\mathbf{M}\boldsymbol{\Delta}_F, \qquad \mathbf{M}\delta\mathbf{M}=(\delta\mathbf{M})M,$$

$$\delta\mathbf{M}=3N_\epsilon C_F\frac{g^2}{16\pi^2}\mathbf{M}.$$

Thus, the set of fermion fields and the mass matrix are renormalized as a whole:

$$\mathbf{Z}_F^{-1}=1+N_\epsilon C_F\frac{g^2}{16\pi^2}, \qquad \mathbf{Z}_m=1-3N_\epsilon C_F\frac{g^2}{16\pi^2}, \tag{29.8a}$$

i.e.,

$$\mathbf{Z}_F=Z_F\cdot\mathbf{1}, \qquad \mathbf{Z}_m=Z_m\cdot\mathbf{1}. \tag{29.8b}$$

We have proved this only to the lowest order, but the renormalization group equations guarantee the result to leading order in α_s.

This result can be understood in yet another way. The invariance of $\mathscr{L}$ under the transformations (29.4) implies that we may always choose the counterterms to satisfy the same invariance, so the mass matrix will remain diagonal after renormalization. In fact, this proof shows that in mass independent renormalization schemes (like the $\overline{MS}$), Equations (29.8*b*) actually hold to all orders.

The results we have derived show that if all $\hat{m}_l$ are different and nonvanishing[81], the only global symmetries of the Lagrangian are those associated with flavor conservation, (29.4). As stated above, however, it may be a good approximation to neglect the m_l. Under this circumstance, all the transformations of Equation (29.2) become symmetries of the Lagrangian. The measure of the accuracy of the symmetry is given by, for example, the divergences of the corresponding generators. Although these have already been considered in Section 10, we will now present some extra details.

Let us parametrize the W as $\exp\{(i/2)\Sigma\theta_a\lambda^a\}$, where the λ are Gell–Mann's matrices.[82] We may denote by $U_\pm(\theta)$ the operators that implement (29.2):

$$U_\pm(\theta)\frac{1\pm\gamma_5}{2}q_l U_\pm^{-1}(\theta)=\sum_{l'}\left(e^{(i/2)\Sigma\theta_a\lambda^a}\right)_{ll'}\frac{1\pm\gamma_5}{2}q_{l'}. \tag{29.9}$$

For infinitesimal values of θ we write

$$U_\pm(\theta)\approx 1-\frac{i}{2}\sum L_\pm^a\theta_a,\qquad (L_\pm^a)^+=L_\pm^a,$$

so that (29.9) yields

$$\left[L_\pm^a, q_{l\pm}(x)\right]=-\sum_{l'}\lambda_{ll'}^a q_{l'\pm}(x),\qquad q_{l\pm}\equiv\frac{1\pm\gamma_5}{2}q_l. \tag{29.10}$$

Because the U leave the interaction part of the Lagrangian invariant, we may solve (29.10) for free fields. The result is[83]

$$L_\pm^a(t)=:\int d\vec{x}\sum_{ll'}\bar{q}_{l\pm}(x)\gamma^0\lambda_{ll'}^a q_{l'_\pm}(x):,\qquad t=x^0. \tag{29.11}$$

These will be recognized as the charges corresponding to the currents

$$J_\pm^{a\mu}(x)=:\sum_{ll'}\bar{q}_l(x)\lambda_{ll'}^a\gamma^\mu\frac{1\pm\gamma_5}{2}q_{l'}(x):. \tag{29.12}$$

If the symmetry is exact, $\partial_m J_\pm^{a\mu}=0$, and then one can easily see that the $L_\pm^a(t)$ are actually independent of t. Otherwise, we have to define *equal time* transformations and modify (29.9) and (29.10), writing, for example,

$$\left[L_\pm^a(t), q_{l\pm}(x)\right]=-\sum_{l'}\lambda_{ll'}^a q_{l'\pm}(x),\qquad t=x^0. \tag{29.13}$$

[81] This seems to be the case in nature. As we will see (Section 31), one expects $\hat{m}_d/\hat{m}_u\approx 2$, $\hat{m}_s/\hat{m}_d\approx 20$, $\hat{m}_u\approx 6$ MeV.

[82] We consider the case $n=3$. For $n=2$, replace the λ by the σ of Pauli.

[83] To verify (29.11), we may use the free-field commutation relations (Appendix F), which is justified because QCD is free at distance 0.

The set of transformations

$$U_{\pm}(\theta,t) = \exp\left\{ \frac{-i}{2} \sum L_{\pm}^{a}(t)\theta_{a} \right\},$$

builds the group of *chiral transformations* generated by the currents (29.12). In our case, $n = 3$, we find the chiral $SU_F^+(3) \times SU_F^-(3)$ group. Its generators may be rearranged in terms of the set of vector and axial currents[84] $V_{ll'}^{\mu}(x)$, $A_{ll'}^{\mu}(x)$ introduced in Section 10. An important subgroup of $SU_F^+(3) \times SU_F^-(3)$ is that generated by the vector currents, which is simply the flavor $SU_F(3)$ of Gell–Mann and Ne'eman.

The exactness of the symmetry is related to the time independence of the charges $L_{\pm}$, which, in turn, is linked to the divergence of the currents. Except for the diagonal axial currents, these divergences are proportional to differences $m_l - m_{l'}$ for the vector and sums $m_l + m_{l'}$ for the axial currents [see (10.5)]. Thus, we conjecture that $SU_F(3)$ will be good to the extent that $|m_l - m_{l'}|^2 \ll \Lambda^2$ and chiral $SU_F^+(3) \times SU_F^-(3)$ to the extent that $m_l^2 \ll \Lambda^2$. In reality, it appears that mass differences are of the order of the masses themselves, so we expect chiral symmetries to be almost as good as flavor symmetries. This seems to be the case.[85]

30 Wigner–Weyl and Nambu–Goldstone Realizations of Symmetries

The fact that flavor and chiral $SU(3)$ are expected to be valid to a similar order of approximation does not mean that these symmetries are realized in the same manner. In fact, we will see that there are good theoretical and "experimental" reasons why they are very different.

Let us begin by introducing the charges with definite parity,

$$Q^a = L_+^a + L_-^a, \qquad Q_5^a = L_+^a - L_-^a. \tag{30.1}$$

Their *equal time commutation relations* are

$$\begin{aligned} \left[Q^a(t), Q^b(t) \right] &= 2i \sum f^{abc} Q^c(t), \\ \left[Q^a(t), Q_5^b(t) \right] &= 2i \sum f^{abc} Q_5^c(t), \\ \left[Q_5^a(t), Q_5^b(t) \right] &= 2i \sum f^{abc} Q^c(t). \end{aligned} \tag{30.2}$$

The set Q^a builds $SU_F(3)$. In the limit $m_l \to 0$, all Q, Q_5 are t independent, and

$$[Q^a, \mathscr{L}] = [Q_5^a, \mathscr{L}] = 0. \tag{30.3}$$

[84]Not all the diagonal elements are in $SU_F(3) \times SU_F(3)$, but they are in the group $U_F(3) \times U_F(3)$.

[85]Chiral symmetry and chiral dynamics is a subject in itself. Here we only touch upon some of its aspects that are related to QCD. This omits many important applications. The interested reader may consult the reviews of Pagels (1975) and Scadron (1981) and work quoted therein.

The difference, however, lies in the vacuum. In general, given a set of generators L^j of symmetry transformations of $\mathscr{L}$, we have two possibilities:

$$L^j\,|0\rangle = 0, \tag{30.4}$$

which is called *Wigner–Weyl* symmetry, or

$$L^j|0\rangle \neq 0, \tag{30.5}$$

or *Goldstone–Nambu* symmetry. Of course, in general, we will have a mixture of the two, with some L^i, $i = 1, \ldots, r$ verifying (30.4) and the rest, L^k, $k = r + 1, \ldots, n$, satisfying (30.5). It is clear that if L^1, L^2 verify (30.4), so does its commutator. Hence, the set of Wigner–Weyl symmetries is a subgroup.

Two theorems are especially relevant with respect to these questions. The first, due to Coleman (1966) asserts that "the invariance of the vacuum is the invariance of the world," or, in more straightforward terms, the physical states (including bound states) are invariant under the transformations of a group of Wigner–Weyl symmetries. If we assumed that chiral symmetry was all realized in the Wigner–Weyl mode, we would conclude that the masses of all mesons would be degenerated, except for corrections of order m^2/m_h, where h are the hadron masses. This is true of the ω, ρ, K^*, ϕ or f', A_2, f^0, but if we include parity doublets, this is certainly not the case. Thus it is strongly suggested that $SU_F(3)$ is a Wigner–Weyl symmetry, but chiral $SU_F^+(3) \times SU_F^-(3)$ contains Goldstone–Nambu type generators. We assume, therefore,

$$Q^a(t)|0\rangle = 0, \qquad Q_5^a(t)|0\rangle \neq 0. \tag{30.6}$$

The second relevant theorem is Goldstone's (1961). It states that for each generator that fails to annihilate the vacuum, there must exist a massless boson with the quantum numbers of that generator. Therefore, we "understand" the smallness of the π, K masses[86] because, in the limit $m_u, m_d, m_s \to 0$, we would have $m_\pi \to 0, m_K \to 0$. Indeed, we will later show that

$$m_\pi^2 \approx m_u + m_d\,, \qquad m_K^2 \approx m_s + m_{u,d}\,. \tag{30.7}$$

We will not prove either theorem here, but we note that (30.7) affords a more quantitative criterion for the validity of the chiral or flavor symmetries; they hold to corrections of order m_π^2/m_ρ^2 for $SU_F(2)$, $m_K^2/m_{K^*}^2$ for $SU_F(3)$.

We also note that a Nambu–Goldstone realization [Nambu (1960); Nambu and Jona–Lasinio (1961*a,b*)] is never possible in perturbation theory; to all orders of perturbation theory, $Q_5^a(t)|0\rangle = 0$. This means that the physical vacuum is different from the vacuum of perturbation theory in the limit $m \to 0$. We will emphasize this by writing $|0\rangle$ for the perturbation-

[86]The particles with zero flavor quantum numbers present problems of their own [the so-called $U(1)$ problem] that will be discussed later.

theoretic vacuum and $|\text{vac}\rangle$ for the physical one when there is danger of confusion. So we write (30.6) as

$$Q^a(t)|\text{vac}\rangle = 0, \qquad Q_5^a(t)|\text{vac}\rangle \neq 0. \tag{30.8}$$

How this comes about is not difficult to see. Let $a_{m_G}^+(\vec{p})$ be the creation operator for a particle whose mass is allowed to vanish. The states

$$a_{m_G}^+(\vec{0})^{(n)} \ldots a_{m_G}^+(\vec{0})|0\rangle = |n\rangle$$

are all degenerate in the limit $m_G \to 0$. Therefore, the physical vacuum is, in this limit,

$$|\text{vac}\rangle = \sum C_n|n\rangle.$$

We thus expect this to occur in QCD, particularly in the limit $m_q \to 0$.

31 PCAC and Quark Mass Ratios

We are now in a position to obtain quantitative results on the masses of the light quarks. To do so, consider the current

$$A_{ud}^\mu(x) = \bar{u}(x)\gamma^\mu\gamma_5 d(x),$$

and its divergence,

$$\partial_\mu A_{ud}^\mu(x) = i(m_u + m_d)\bar{u}(x)\gamma_5 d(x).$$

The latter has the quantum number of the π^+, so we can use it as a (composite) pion field. We thus write

$$\partial_\mu A_{ud}^\mu(x) = \sqrt{2}\, f_\pi m_\pi^2 \phi_\pi(x). \tag{31.1}$$

The factors in (31.1) are chosen for historical reasons. $\phi_\pi(x)$ is the pion field normalized to

$$\langle 0|\phi_\pi(x)|\pi(p)\rangle = \frac{1}{(2\pi)^{3/2}} e^{-ipx}, \tag{31.2a}$$

where $|\pi(p)\rangle$ is the state of one pion with momentum p. The constant f_π may be obtained experimentally. In fact, consider the weak decay $\pi \to \mu\nu$. With the effective Fermi Lagrangian for weak interactions, $\mathscr{L}_{\text{int}}^F = (G_F/\sqrt{2})\bar{\mu}\gamma_\lambda(1-\gamma_5)\nu_\mu \bar{u}\gamma^\lambda(1-\gamma_5)d + \ldots$, we find

$$F(\pi \to \mu\nu) = \frac{2\pi G_F}{\sqrt{2}} \bar{u}_{(\nu)}(p_2)\gamma_\lambda(1-\gamma_5)v_{(\mu)}(p_1, \sigma)\langle 0|A_{ud}^\lambda(0)|\pi(p)\rangle.$$

Now, on invariance grounds,

$$\langle 0|A_{ud}^\lambda(0)|\pi(p)\rangle = ip^\lambda C_\pi\,; \tag{31.2b}$$

contracting with p_μ, we find $C_\pi = f_\pi\sqrt{2}/(2\pi)^{3/2}$:

$$m_\pi^2 C_\pi = \langle 0|\partial_\lambda A_{ud}^\lambda(0)|\pi(p)\rangle = \sqrt{2}\, f_\pi m_\pi^2 \frac{1}{(2\pi)^{3/2}}; \tag{31.2c}$$

hence,

$$\tau(\pi\to\mu\nu) = \frac{4\pi}{\left(1 - m_\mu^2/m_\pi^2\right)^2 G_F^2 f_\pi^2 m_\pi m_\mu^2}.$$

Therefore, f_π is directly related to the $\pi\to\mu\nu$ decay rate. Experimentally, one has $f_\pi \approx 93.3$ MeV. A remarkable fact is that if we repeat the analysis for kaons,

$$\partial_\mu A_{us}^\mu(x) = \sqrt{2}\, f_K m_K^2 \phi_K(x), \tag{31.3}$$

we find that, experimentally, $f_K \approx 110$ MeV, i.e., it agrees with f_π to 20%. Actually, this is to be expected because in the limit as $m_{s,d,u}\to 0$, there is no difference between pions and kaons, and we would find strict equality. That f_π, f_K are so similar in the real world is a good point in favor of $SU_F(3)$ chiral ideas.

The relations (31.1) and (31.3) are at times called PCAC[87] but this is not very meaningful, for these equations are really *identities*. One may use any pion field one wishes, in particular (31.1), provided it has the right quantum numbers and its vacuum-one pion matrix element is not zero. The nontrivial part of PCAC will be described below.

The next step is to consider the two-point function (we drop the *ud* index from A_{ud})

$$F^{\mu\nu}(q) = i\int d^4x\, e^{iq\cdot x}\langle TA^\mu(x)A^\nu(0)^+\rangle_{\text{vac}},$$

and contract with q_μ, q_ν,

$$\begin{aligned} q_\nu q_\mu F^{\mu\nu}(q) &= -q_\nu\int d^4x\, e^{iq\cdot x}\partial_\mu\langle TA^\mu(x)A^\nu(0)^+\rangle_{\text{vac}}, \\ &= -q_\nu\int d^4x\, e^{iq\cdot x}\delta(x^0)\langle\left[A^0(x), A^\nu(0)^+\right]\rangle_{\text{vac}} \\ &\quad - q_\nu\int d^4x\, e^{iq\cdot x}\langle T\partial A(x)A^\nu(0)^+\rangle_{\text{vac}}, \\ &= 2i\int d^4x\, e^{iq\cdot x}\delta(x^0)\langle\left[A^0(x)\partial A(0)^+\right]\rangle_{\text{vac}} \\ &\quad + i\int d^4x\, e^{iq\cdot x}\langle T\partial A(x)\partial A(0)^+\rangle_{\text{vac}}. \end{aligned}$$

Using Equations (31.1) and evaluating the commutator, we find

$$\begin{aligned} q_\mu q_\nu F^{\mu\nu}(q) &= 2(m_u + m_d)\int d^4x\, e^{iq\cdot x}\delta(x)\langle\bar{u}(x)u(x) + \bar{d}(x)d(x)\rangle_{\text{vac}} \\ &\quad + 2if_\pi^2 m_\pi^4\int d^4x\, e^{iq\cdot x}\langle T\phi_\pi(x)\phi_\pi(0)^+\rangle_{\text{vac}}, \end{aligned}$$

[87] Partially conserved axial current. In fact, in the limit $m_\pi^2\to 0$, the right-hand side of (31.1) vanishes.

or, in the limit $q \to 0$,

$$2(m_u + m_d)\langle \bar{u}(0)u(0) + \bar{d}(0)d(0)\rangle_{\text{vac}} = -2if_\pi^2 m_\pi^4 \int d^4x\, e^{iq\cdot x}\langle T\phi_\pi(x)\phi_\pi(0)^+\rangle_{\text{vac}}\big|_{q\to 0}.$$

The right-hand side of this inequality has a contribution from the pion pole and the continuum; they can be written as

$$i\int d^4x\, e^{iq\cdot x}\langle T\phi_\pi(x)\phi_\pi(0)^+\rangle_{\text{vac}}\big|_{q\to 0} = \left\{ \frac{1}{m_\pi^2 - q^2} + \frac{1}{\pi}\int dt' \frac{\operatorname{Im}\Pi}{t' - q^2} \right\}_{q\to 0}$$

$$= \frac{1}{m_\pi^2} + \frac{1}{\pi}\int dt' \frac{\operatorname{Im}\Pi}{t'};$$

$$\Pi = i\int d^4x\, e^{iq\cdot x}\langle T\phi_n(x)\phi_\pi(0)^+\rangle_{\text{vac}}.$$

The order of the limits is essential; we must first take $q \to 0$ and the chiral limit afterwards. In the limit[88] $m_\pi^2 \to 0$, the first term in the right hand side above diverges and the second remains finite; therefore, we obtain the final result

$$(m_u + m_d)\langle \bar{u}u + \bar{d}d\rangle_{\text{vac}} = -f_\pi^2 m_\pi^2\{1 + O(m_\pi^2)\}. \tag{31.4}$$

This is an indication that $\langle \bar{q}q\rangle_{\text{vac}} \neq 0$ because, in order to ensure that it vanishes, we would require $f_\pi = 0$. We also note that we have not distinguished between bare or renormalized masses and operators; the distinction is not necessary because we know that m and $\bar{q}q$ acquire opposite renormalization so that $m_R(\bar{q}q)_R = m_u(\bar{q}q)_u$.

We may repeat the derivation of (31.4) for kaons. Neglecting the $O(m_\pi^2)$ or $O(m_K^2)$ terms, we find

$$\begin{aligned}(m_u + m_s)\langle \bar{u}u + \bar{s}s\rangle_{\text{vac}} &= -f_K^2 m_{K^+}^2, \\ (m_d + m_s)\langle \bar{d}d + \bar{s}s\rangle_{\text{vac}} &= -f_K^2 m_{K^0}^2.\end{aligned} \tag{31.5}$$

If we assume flavor invariance of the VEV's $\langle \bar{q}q\rangle$, we may eliminate them and obtain

$$\frac{m_s + m_u}{m_d + m_u} \approx \frac{f_K^2 m_{K^+}^2}{f_\pi^2 m_\pi^2}, \qquad \frac{m_d - m_u}{m_d + m_u} \approx \frac{f_K^2}{f_\pi^2}\cdot\frac{m_{K^0}^2 - m_{K^+}^2}{m_\pi^2}.$$

A more careful evaluation requires consideration of the electromagnetic contributions to the observed π, K masses. In this way, one obtains[89]

$$\frac{m_s}{m_d} = 18 \pm 4, \qquad \frac{m_d}{m_u} = 2.0 \pm 0.3 \tag{31.6}$$

[88] This is properly the PCAC limt, for in this limit, the axial current is conserved: $\partial_\mu A^\mu = 0$. (The P stands for partially).

[89] See Weinberg (1978*a*); Dominguez (1978), and Zepeda (1978). The method originates in the work of Glashow and Weinberg (1968), Gell-Mann, Oakes, and Renner (1968), and Leutwyler (1974).

If we couple this with the phenomenological estimate (from meson and baryon spectroscopy) $m_s - m_d \approx 100$ to 200 MeV, $m_d - m_u \approx 4$ MeV, we obtain the masses (in MeV),

$$\bar{m}_u(q \sim m_\rho) \approx 6, \qquad \bar{m}_d(Q \sim m_\rho) \approx 10, \qquad \bar{m}_s(Q \sim m_\rho) \approx 200, \quad (31.7)$$

where the symbol $\approx$ here means that a factor of 2 error would not be very surprising.

This method for obtaining quark masses is admittedly very rough; in the next section, we will describe more sophisticated ones.

32 Bounds and Estimates of Light Quark Masses

In this section we describe a method of obtaining bounds and estimates on quark masses. The method was first used by Vainshtein *et al.* (1978) and further refined by Becchi *et al.* (1981). We start with

$$\Psi^5_{ij}(q^2) = i(m_i + m_j)^2 \int d^4x\, e^{iq\cdot x} \langle TJ^5_{ij}(x) J^5_{ij}(0)^+ \rangle_{\text{vac}}, \qquad (32.1)$$

where

$$J^5_{ij} = \bar{q}_i \gamma_5 q_j \,.$$

To all orders in perturbation theory, the function

$$F_{ij}(Q^2) = \frac{\partial^2}{\partial (q^2)^2} \Psi^5_{ij}(q^2), \qquad Q^2 = -q^2,$$

vanishes as $Q^2 \to \infty$. Hence, we may write a dispersive representation without subtractions,

$$F_{ij}(Q^2) = \frac{2}{\pi} \int_0^\infty dt \frac{\operatorname{Im} \Psi^5(t)}{(t + Q^2)^3} \,. \qquad (32.2)$$

For large values of Q^2, we can calculate the left-hand side from QCD. However, we have to be careful; it is not sufficient to keep the leading term in the OPE for TJ^5J^{5+}; the contribution of the operators $\bar{q}q, x^\alpha \bar{q}\partial_\alpha q$ and $G^2 = \sum_a G_{a\mu\nu} G_a^{\mu\nu}$ is also relevant. One finds, by calculating to two loops and remembering that the combinations $\alpha_s G^2, m\bar{q}q$ are not renormalized to the order we are working,

$$F_{ij}(Q^2) = \frac{3}{8\pi^2} \cdot \frac{\left[\bar{m}_i(Q^2) + \bar{m}_j(Q^2)\right]^2}{Q^2}$$
$$\times \left\{ 1 + O\left(\frac{\bar{m}^2}{Q^2}\right) + \frac{11}{3} \cdot \frac{\alpha_s(Q^2)}{\pi} + \frac{2\pi}{3} \cdot \frac{\alpha_s \langle G^2 \rangle}{Q^4} \right.$$
$$\left. - \frac{16\pi^2}{3Q^4} \left[\left(m_j - \frac{m_i}{2}\right) \langle \bar{q}_i q_i \rangle + \left(m_i - \frac{m_j}{2}\right) \langle \bar{q}_j q_j \rangle \right] \right\}.$$

The contributions of $\langle\bar{q}q\rangle, \langle G^2\rangle$ are evaluated taking into account the nonperturbative parts of quark and gluon propagators (Sec. 35; see Sec. 36 for a detailed sample calculation). The contributions $m\langle\bar{q}q\rangle$ can be estimated with the help of (31.4) and (31.5); they are seen to be negligible, and so are those of $O(\overline{m}^2/Q^2)$. We thus take

$$F_{ij}(Q^2) = \frac{3}{8\pi^2} \cdot \frac{\left[\overline{m}_i(Q^2) + \overline{m}_j(Q^2)\right]^2}{Q^2} \times \left\{1 + \frac{11}{3} \cdot \frac{\alpha_s(Q^2)}{\pi} + \frac{2\pi}{3Q^4}\alpha_s\langle G^2\rangle\right\}. \tag{32.3}$$

We now turn to the right-hand side of (32.2). The contribution of the pion (for $ij = ud$) or kaon (for $ij = us, ud$) can be extracted explicitly, yielding

$$\frac{2}{\pi}\int_0^\infty dt \frac{\operatorname{Im}\Psi^5(t)}{(t+Q^2)^3} = 4f_\pi^2 m_\pi^4 \frac{1}{(m_\pi^2+Q^2)^3} + \frac{2}{\pi}\int_{9m_\pi^2} dt \frac{\operatorname{Im}\Psi^5(t)}{(t+Q^2)^3}, \tag{32.4}$$

for the pion case. Now, the key remark is that $\operatorname{Im}\Psi^5(t) \geqslant 0$; hence, we immediately obtain an inequality relating $m_u + m_d$ to m_π, f_π, $\langle\alpha_s G^2\rangle$:

$$\left[\overline{m}_u(Q^2) + \overline{m}_d(Q^2)\right]^2 \geqslant \frac{32\pi^2 f_\pi^2 m_\pi^4}{3(m_\pi^2+Q^2)^3} \times \left\{1 + \frac{11}{3} \cdot \frac{\alpha_s(Q^2)}{\pi} + \frac{2\pi}{3Q^4}\alpha_s\langle G^2\rangle\right\}^{-1}. \tag{32.5}$$

This bound is not very good because we are losing a great deal. One can improve it by considering N derivatives of $F(Q^2)$ and optimizing with respect to N, Q^2. The details may be found in the paper of Becchi *et al.* (1981). One finds

$$\hat{m}_u + \hat{m}_d \geqslant \sqrt{\frac{2\pi}{3}} \cdot \frac{8m_\pi^2 f_\pi^2}{3\langle\alpha_s G^2\rangle^{1/2}}\{1 \pm \delta\}, \tag{32.6}$$

where δ is a correction, estimated as ~25%. If we use the value of $\langle\alpha_s G^2\rangle_0$ obtained from charmonium spectroscopy by Shifman, Vainshtein, and Zakharov (1979) or in the lattice calculation of Di Giacomo and Rossi (1981), we obtain

$$\hat{m}_u + \hat{m}_d \geqslant (23 \pm 8)\ \text{MeV}, \qquad \langle\alpha_s G^2\rangle \approx 0.044^{+0.014}_{-0.006}\ \text{GeV}^4. \tag{32.7}$$

The bound on the masses does not take into account the errors in $\langle\alpha_s G^2\rangle$. If we add them, we obtain a minimum bound

$$\hat{m}_u + \hat{m}_d \geqslant 13\ \text{MeV}. \tag{32.8}$$

At any rate, this is compatible with (31.7), within errors, although it tends to favor rather large quark masses.

It is possible to employ the method to obtain an *estimate*, rather than a bound, for quark masses. For this, one uses a model for $\operatorname{Im}\Psi^5_{ij}(t)$, taking its QCD expression for large t and parametrizing its lower energy part with a few (often, one) resonances. In this way one obtains [Vainshtein *et al.* (1978); Hubschmidt and Malik (1981); J. Gasser and H. Leutwyler, BUTP-6 preprint, 1982 (to be published in Physics Reports C)]

$$\hat{m}_u + \hat{m}_d \approx (20 \pm 6)\ \text{MeV}. \tag{32.9}$$

An alternate method has recently been devised by Narison and de Rafael (1981); it may be thought of as a QCD refinement of the classical estimate of Leutwyler (1974); they find $\hat{m}_u + \hat{m}_d \approx 27 \pm 8$ MeV for $\Lambda = 130 \pm 50$ MeV. We see that, as stated before, we obtain masses in the range (31.7), with values on the heavy side favored. Incidentally, the estimates show that the bound (32.6) is very tight and that perhaps the relation

$$\hat{m}_u + \hat{m}_d \approx \sqrt{\frac{2\pi}{3}} \cdot \frac{8m_\pi^2 f_\pi^2}{3\langle \alpha_s G^2\rangle^{1/2}}$$

is an equality, at least in some limit.

33 The Decay $\pi^0 \to \gamma\gamma$: The Axial Anomaly

Historically, one of the first motivations for the color degree of freedom came from the decay $\pi^0 \to \gamma\gamma$, which we now consider in some detail.

The amplitude for the process $\pi^0 \to \gamma\gamma$ may be written, using the reduction formulas, as

$$\begin{aligned} &\langle \gamma(k_1,\lambda_1), \gamma(k_2,\lambda_2)|S|\pi^0(q)\rangle \\ &\quad = \frac{-ie^2}{(2\pi)^{9/2}} \epsilon^*_\mu(k_1,\lambda_1)\epsilon^*_\nu(k_2,\lambda_2) \int d^4x_1\, d^4x_2\, d^4z\, e^{i(x_1\cdot k_1 + x_2\cdot k_2 - z\cdot q)} \\ &\quad\quad \times (\partial_z^2 + m_\pi^2)\langle TJ^\mu_{\text{em}}(x_1)J^\nu_{\text{em}}(x_2)\phi_{\pi^0}(z)\rangle_0, \end{aligned} \tag{33.1}$$

and we have assumed that

$$\partial^2 A^\mu(x) = J^\mu_{\text{em}}(x),$$

where A is the photon field.[90] Separating a $\delta(k_1 + k_2 - q)$, we find

$$F(\pi^0 \to \gamma(k_1,\lambda_1), \gamma(k_2,\lambda_2)) = \frac{e^2(q^2 - m_\pi^2)}{\sqrt{2\pi}} \epsilon^*_\mu(k_1,\lambda_1)\epsilon^*_\nu(k_2,\lambda_2) F^{\mu\nu}(k_1,k_2), \tag{33.2a}$$

[90] We leave it as an exercise for the reader to verify this as well as that in this case,

$$\partial^2_{x_1}\partial^2_{x_2} TA^\mu(x_1)A^\nu(x_2)\phi(z) = T(\partial^2 A^\mu(x_1)\partial^2 A^\nu(x_2))\phi(z),$$

i.e., that potential terms where the derivatives act on the $\theta(x_1^0 - z^0)\ldots$ implicit in the T product make no contribution.

where we have defined the VEV

$$F^{\mu\nu}(k_1,k_2)=\int d^4x\,d^4y\,e^{i(x\cdot k_1+y\cdot k_2)}\langle TJ^\mu(x)J^\nu(y)\phi_{\pi^0}(0)\rangle_0,\quad q=k_1+k_2. \tag{33.2b}$$

The electromagnetic index (em) in the J will be understood henceforth. Next, we may use Equation (31.1), generalized to include the π^0:

$$\begin{aligned}\partial_\mu A_3^\mu(x)&=2f_\pi m_\pi^2\phi_{\pi^0}(x),\\ A_3^\mu(x)&=\bar{u}(x)\gamma^\mu\gamma_5u(x)-\bar{d}(x)\gamma^\mu\gamma_5d(x),\end{aligned} \tag{33.3}$$

to write

$$\begin{aligned}F^{\mu\nu}(k_1,k_2)&=\frac{1}{f_\pi m_\pi^2}T^{\mu\nu}(k_1,k_2),\\ T^{\mu\nu}(k_1,k_2)&=\frac{1}{2}\int d^4x\,d^4y\,e^{i(x\cdot k_1+y\cdot k_2)}\langle TJ^\mu(x)J^\nu(0)\partial A_3(0)\rangle_0.\end{aligned} \tag{33.4}$$

To this point, everything has been exact. The next step involves using the PCAC hypothesis in the following form: we assume that $F(\pi\to\gamma\gamma)$ can be approximated by its leading term in the limit $q^2\to 0$. On purely kinematic grounds, this is seen to imply that $q,k_1,k_2\to 0$ also. One may write

$$T^{\mu\nu}(k_1,k_2)=\epsilon^{\mu\nu\alpha\beta}k_{1\alpha}k_{2\beta}\Phi+O(k^3), \tag{33.5}$$

The PCAC hypothesis means that we retain only the first term in Equation (33.5). As will be seen presently, this will lead us to a contradiction, the resolution of which will involve introducing the so-called *axial anomaly* and will allow us to actually calculate $T^{\mu\nu}$ exactly to all orders of perturbation theory (in the PCAC approximation).

The first step is to consider the quantity

$$R^{\mu\nu\lambda}(k_1,k_2)=i\int d^4x\,d^4y\,e^{i(x\cdot k_1+y\cdot k_2)}\langle TJ^\mu(x)J^\nu(y)A_3^\lambda(0)\rangle_0. \tag{33.6}$$

On purely Lorentz invariance grounds, we may write the general decomposition,

$$R^{\mu\nu\lambda}(k_1,k_2)=\epsilon^{\mu\nu\lambda\alpha}k_{1\alpha}\Phi_1+\epsilon^{\mu\nu\lambda\alpha}k_{2\alpha}\Phi_2+O(k^3), \tag{33.7}$$

where the $O(k^3)$ terms are of the general form $\epsilon^{\mu\lambda\alpha\beta}k_{i\alpha}k_{j\beta}k_l^\lambda\Phi_{lij}$ + three permutations, and, for $m\neq 0$, the Φ are regular as $k_i\to 0$. The conservation of the e.m. current, $\partial J=0$, yields two equations:

$$k_{1\mu}R^{\mu\nu\lambda}=k_{2\nu}R^{\mu\nu\lambda}=0; \tag{33.8}$$

the first implies

$$\Phi_2=O(k^2), \tag{33.9a}$$

the second

$$\Phi_1=O(k^2). \tag{33.9b}$$

Now, we have, from (33.4) and (33.6),

$$q_\lambda R^{\mu\nu\lambda}(k_1,k_2) = T^{\mu\nu}(k_1,k_2), \quad \text{i.e.,} \quad \Phi = \Phi_2 - \Phi_1, \tag{33.10}$$

and, hence, from (33.9), we find the result of Veltman (1967) and Sutherland (1967),

$$\Phi = O(k^2). \tag{33.11}$$

Because the scale for k is m_π, this means that $\Phi \approx m_\pi^2$. Now, this is in disagreement with experiment; but worse still, (33.11) contradicts a direct calculation. In fact, we may use the equations of motion to write

$$\partial_\mu A_3^\mu(x) = 2i\left\{ m_u \bar{u}(x)\gamma_5 u(x) - m_d \bar{d}(x)\gamma_5 d(x)\right\}. \tag{33.12}$$

We will calculate to zero order in α_s: clearly, (33.11) should be valid to this order. This involves the diagram of Figure 25a. The result, as first obtained by Steinberger (1949) is, in the limit $k_1, k_2 \to 0$ and with $\delta_u = 1$, $\delta_d = -1$,

$$\begin{aligned} T^{\mu\nu}(k_1,k_2) &= 3\times 2\times \sum_{f=u,d} \delta_f Q_f^2 m_f \\ &\quad \times \int \frac{d^4p}{(2\pi)^4}\cdot \frac{\operatorname{Tr}\gamma_5(\not{p}+\not{k}_1+m_f)\gamma^\mu(\not{p}+m_f)\gamma^\nu(\not{p}-\not{k}_2+m_f)}{\left[(p+k_1)^2-m_f^2\right](p^2-m_f^2)\left[(p-k_2)^2-m_f^2\right]} \\ &= \frac{-1}{4\pi^2}\epsilon^{\mu\nu\alpha\beta}k_{1\alpha}k_{2\beta}\left\{3(Q_u^2-Q_d^2)\right\} + O(k^4) \\ &= \frac{-1}{4\pi^2}\epsilon^{\mu\nu\alpha\beta}k_{1\alpha}k_{2\beta} + O(k^4). \end{aligned}$$

The factor of 2 in the first expression comes from the fact that we have the crossed diagram contribution as well; the factor of 3 comes from the sum over color. Thus, we see that

$$\Phi = \frac{-1}{4\pi^2}, \tag{33.13}$$

which contradicts (33.11). This is the triangle anomaly [Bell and Jackiw (1969); Adler (1969)].

What is wrong here? Clearly, we cannot maintain (33.12), which was obtained by using the free-field equations, $i\not{\partial}q = mq$; we must admit that in the presence of vector fields (the photon field in our case), Equation (33.12) is no longer valid. To obtain agreement with (33.13), we have to write [Adler (1969)]

$$\begin{aligned} \partial_\mu A_3^\mu(x) &= 2i\left\{ m_u \bar{u}(x)\gamma_5 u(x) - m_d \bar{d}(x)\gamma_5 d(x)\right\} \\ &\quad + 3(Q_u^2 - Q_d^2)\frac{e^2}{16\pi^2}F_{\mu\nu}(x)\tilde{F}^{\mu\nu}(x), \end{aligned} \tag{33.14}$$

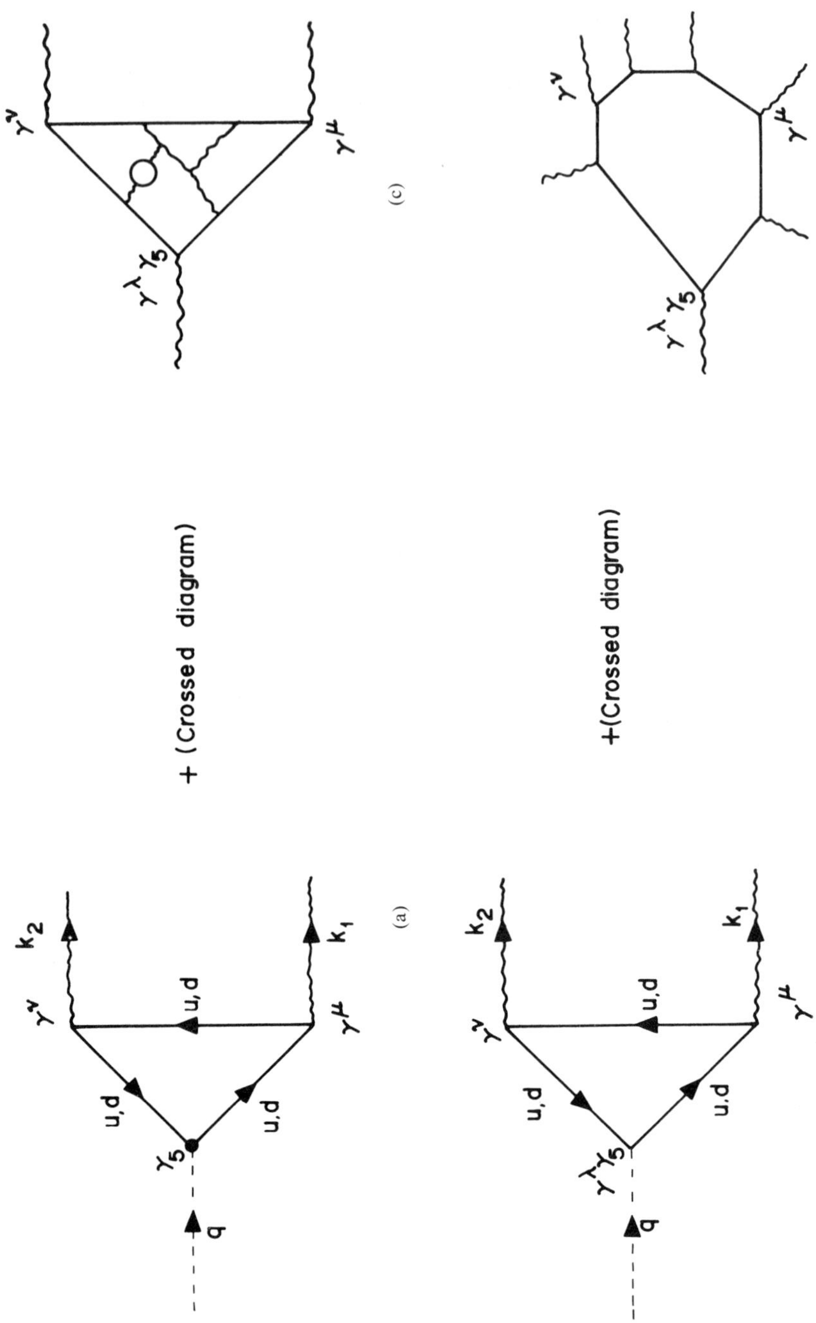

Figure 25. (a, b), diagrams with anomaly; (c, d) nonanomalous diagrams.

where the *dual* $\tilde{F}$ has been defined by

$$\tilde{F}^{\mu\nu} = \frac{1}{2}\epsilon^{\mu\nu\alpha\beta}F_{\alpha\beta}, \qquad F_{\alpha\beta} = \partial_\alpha A_\beta - \partial_\beta A_\alpha,$$

where A is the photon field. More generally, for fermion fields f interacting with vector fields with strength h_f, we find

$$\partial_\mu \bar{f}\gamma^\mu\gamma_5 f = 2im_f\bar{f}\gamma_5 f + \frac{T_F h^2}{8\pi^2}H_{\mu\nu}\tilde{H}^{\mu\nu}; \tag{33.15}$$

$H^{\mu\nu}$ is the vector field strength tensor. Let us return to the decay $\pi^0 \to \gamma\gamma$. From (33.13), we calculate the amplitude, in the PCAC limit $m_\pi \sim 0$,

$$F(\pi^0 \to 2\gamma) = \frac{\alpha}{\pi}\cdot\frac{\epsilon^{\mu\nu\alpha\beta}k_{1\alpha}k_{2\beta}\epsilon_\mu(k_1,\lambda_1)\epsilon_\nu(k_2,\lambda_2)}{\sqrt{2}\pi}, \tag{33.16}$$

and the decay rate

$$\Gamma(\pi^0 \to \gamma\gamma) = \left(\frac{\alpha}{\pi}\right)^2\frac{1}{64\pi}\cdot\frac{m_\pi^3}{f_\pi^2} \approx 7.25 \times 10^{-6}\ \text{MeV},$$

to be compared with the experimental figure

$$\Gamma_{\text{exp}}(\pi^0 \to \gamma\gamma) = 7.95 \times 10^{-6}\ \text{MeV}.$$

In fact, the sign of the decay amplitude can also be measured (from the Primakoff effect), and it agrees with the theory. It is important to note that if we had no color, our result would have been $1/3^2$ of this, and thus off by a full order of magnitude.

One may wonder what credibility to attach to this calculation. After all, it was made to zero order in α_s. In fact, the calculation is exact to all orders in QCD;[91] the only approximation is the PCAC one, $m_\pi \approx 0$. To show this, we will give an alternate derivation of the basic result, Equation (33.13). To accomplish this, let us return to (33.6). To zero order in α_s,

$$R^{\mu\nu\lambda} = \sum \delta_f Q_f^2 \int \frac{d^4p}{(2\pi)^4}\cdot\frac{\operatorname{Tr}\gamma^\lambda\gamma_5(\not{p}+\not{k}_1+m_f)\gamma^\mu(\not{p}+m_f)\gamma^\nu(\not{p}-\not{k}_2+m_f)}{((p+k_1)^2-m_f^2)(p^2-m_f^2)((p-k_2)^2-m_f^2)}$$

$$+ \text{crossed term}$$

(see Figure 25b). More generally, we may consider an arbitrary axial triangle with

$$R_{ijl}^{\mu\nu\lambda} = 2\int\frac{d^Dp}{(2\pi)^D}\cdot\frac{\operatorname{Tr}\gamma^\lambda\gamma_5(\not{p}+\not{k}_1+m_i)\gamma^\mu(\not{p}+m_j)\gamma^\nu(\not{p}-\not{k}_2+m_l)}{\left[(p+k_1)^2-m_i^2\right](p^2-m_j^2)\left[(p-k_2)^2-m_l^2\right]}. \tag{33.17}$$

[91] Actually, to all orders in any vectorlike interactions. The proof is essentially contained in the original paper of Adler and Bardeen (1969). See also Bardeen (1974), Crewther (1972), and Wilson (1969).

We would like to calculate $q_\lambda R^{\lambda\mu\nu}$. Writing $(\not{k}_1+\not{k}_2)\gamma_5 = -(\not{p}-\not{k}_2-m_l)\gamma_5 + (\not{p}+\not{k}_1-m_i)\gamma_5 - (m_i+m_l)\gamma_5$, we find

$$q_\lambda R_{ijl}^{\lambda\mu\nu} = -2(m_i+m_l)$$

$$\times \int \frac{d^4p}{(2\pi)^4} \operatorname{Tr} \frac{\gamma_5(\not{p}+\not{k}_1+m_i)\gamma^\mu(\not{p}+m_j)\gamma^\nu(\not{p}-\not{k}_2+m_l)}{\left((p+k_1)^2-m_i^2\right)\left(p^2-m_j^2\right)\left((p-k_2)^2-m^2\right)} + a_{ijl}^{\mu\nu}, \tag{33.18a}$$

$$a_{ijl}^{\mu\nu} = 2\int d^D\hat{p} \operatorname{Tr}\{(\not{p}-\not{k}_2-m_l)\gamma_5 - (\not{p}+\not{k}_1-m_i)\gamma_5\}$$

$$\times \frac{1}{\not{p}+\not{k}_1-m_i}\gamma^\mu \frac{1}{\not{p}-m_j}\gamma^\nu \frac{1}{\not{p}-\not{k}_2-m_l}. \tag{33.18b}$$

The first term in (33.18a) is what we would have obtained by naive use of the equations of motion, $\partial_\mu \bar{q}_i \gamma^\mu \gamma_5 q_l = i(m_i+m_l)\bar{q}_i\gamma_5 q_l$; the second is the anomaly. If we accepted the commutation relations $\{\gamma^\mu,\gamma_5\}=0$ for dimension D, we could rewrite it as

$$a_{ijl}^{\mu\nu} = -2\int d^D\hat{p}\left\{\operatorname{Tr}\gamma_5 \frac{1}{\not{p}+\not{k}_1-m_i}\gamma^\mu \frac{1}{\not{p}-m_j}\gamma^\nu\right.$$

$$\left. + \operatorname{Tr}\gamma_5\gamma^\mu \frac{1}{\not{p}-m_j}\gamma^\mu \frac{1}{\not{p}-\not{k}_2-m_l}\right\}. \tag{33.18c}$$

Then we would conclude that $a^{\mu\nu}$ vanishes because each of the terms in (33.18c) consists of an antisymmetric tensor that depends on a single vector (k_1 for the first, k_2 for the second), and this should vanish. Incidentally, this shows that a is actually independent of the masses because $(\partial/\partial m)a^{\mu\nu}$ is convergent and thus the argument applies. We may therefore write $a_{ijl}^{\mu\nu} \equiv a^{\mu\nu}$ where $a^{\mu\nu}$ is obtained by setting all masses equal to zero. A similar argument shows that $a^{\mu\nu}$ has to be of the form

$$a^{\mu\nu} = a\epsilon^{\mu\nu\alpha\beta}k_{1\alpha}k_{2\beta}, \qquad a = \text{constant}, \tag{33.19a}$$

so that we may obtain a as

$$a = \left.\frac{\partial^2}{\partial k_{1\alpha}\partial k_{2\beta}} a^{\mu\nu}\right|_{k_i=0} \tag{33.19b}$$

Now this argument shows that, from (33.18c), $a\equiv 0$, in contradiction with the Veltman–Sutherland theorem.

What occurs is that the conclusion $a\equiv 0$ is in fact an illusion. If we had shifted integration variables, say, $p\to p+\xi k_2$ in (33.18c), we would have found a finite but nonzero value, actually ξ dependent, $a=-\xi/2\pi^2$. This shows that the commutation relations[92] $\{\gamma^\mu, \gamma_5\} = 0$ lead to an undefined value for the anomaly. If, however, we start from (33.18*b*) and refrain from commuting γ_μ's and γ_5.

$$a\epsilon^{\mu\nu\alpha\beta} = -2\int d^D\hat{p}\,\mathrm{Tr}\,\gamma_5\left\{\frac{1}{\not p}\gamma^\alpha\frac{1}{\not p}\gamma^\mu\frac{1}{\not p}\gamma^\nu\frac{1}{\not p}\gamma^\beta - \frac{1}{\not p}\gamma^\mu\frac{1}{\not p}\gamma^\nu\frac{1}{\not p}\gamma^\beta\frac{1}{\not p}\gamma^\alpha\right\}.$$

Performing symmetric integration (Appendix B) and using only the rules of Appendix A for $D \neq 4$, we obtain an unambiguous result:

$$a\epsilon^{\mu\nu\alpha\beta} = \frac{8(D-1)(4-D)}{D(D+2)}\cdot\frac{i}{16\pi^2}\cdot\frac{2}{4-D}\cdot\mathrm{Tr}\,\gamma_5\gamma^\mu\gamma^\nu\gamma^\alpha\gamma^\beta$$

$$+\,O(4-D)\xrightarrow[D\to 4]{}\frac{-1}{2\pi^2}\,.$$

This is one of the peculiarities of the anomaly: a *finite* Feynman integral whose value depends on the regularization prescription. Fortunately, we may eschew the problem by using the Veltman–Sutherland theorem to conclude that, at any rate, there is a *unique*[93] value of a compatible with gauge invariance, viz.,

$$a_{ijl}^{\mu\nu} = a^{\mu\nu} = -\frac{1}{2\pi^2}\epsilon^{\mu\nu\alpha\beta}k_{1\alpha}k_{2\beta}\,. \tag{33.20}$$

We have explicitly checked that our regularization leads to precisely this value; to verify that it also respects gauge invariance is left to the reader as a simple exercise.

Before continuing, a few words about the Veltman–Sutherland theorem for zero quark mass are necessary. In this case, the first term in the right-hand side of (33.18*a*) is absent: it would appear that we could not maintain our result for $a^{\mu\nu}$, Equation (33.20), because this would result in

$$q_\lambda R^{\lambda\mu\nu} = -\frac{1}{2\pi^2}\epsilon^{\mu\nu\alpha\beta}k_{1\alpha}k_{2\beta} \neq 0,$$

thus contradicting the V–S conclusion $q_\lambda R^{\lambda\mu\nu} = 0$. This is not so; the relation $q_\lambda R^{\lambda\mu\nu} = a^{\mu\nu}$ and the value of $a^{\mu\nu}$ are correct. What occurs is that, for $m = 0$, the functions Φ_i in (33.7) possess singularities of the type $1/k_1\cdot k_2$. Therefore, the V–S theorem is not applicable in this case. This is

[92] These commutation relations are self-contradictory. For example, using only the relations of Appendix A for $D \neq 4$, we have $\mathrm{Tr}\,\gamma_5\gamma^\alpha\gamma^\mu\gamma^\nu\gamma^\rho\gamma_\alpha\gamma^\sigma = (6-D)\mathrm{Tr}\,\gamma_5\gamma^\mu\gamma^\nu\gamma^\rho\gamma^\sigma$, while if we allow commutation, we can obtain $\mathrm{Tr}\,\gamma_5\gamma^\alpha\gamma^\mu\gamma^\nu\gamma^\rho\gamma_\alpha\gamma^\sigma = -\mathrm{Tr}\,\gamma_5\gamma^\mu\gamma^\nu\gamma^\rho\gamma_\alpha\gamma^\sigma\gamma^\alpha = (D-2)$ $\mathrm{Tr}\,\gamma_5\gamma^\mu\gamma^\nu\gamma^\rho\gamma^\sigma$, which differs from the former by terms $O(4-D)$. These problems, however, only arise with at least four γ_μ's.

[93] Actually, the situation is more complicated; the values of the higher-order corrections to $a^{\mu\nu}$ appear to vary from one regularization to another, even when both respect gauge invariance. A discussion with references may be found in the recent paper by D. R. Jones and J. P. Leveille (The Two Loop Axial Anomaly in $N = 1$ Supersymmetric Yang–Mills Theory, Univ. of Michigan preprint UM. HE. 81–67 (1981), unpublished).

yet another peculiarity of the anomalous triangle: $\lim_{m\to 0} q_\lambda R^{\lambda\mu\nu} = 0$, but, if we begin with $m \equiv 0$,

$$q_\lambda R^{\lambda\mu\nu}_{m\equiv 0} = a^{\mu\nu} \neq 0.$$

Let us return to our original discussion, in particular, to $m \neq 0$. The present method shows how one can prove that the result does not become renormalized. The Veltman–Sutherland theorem is exact; so we have actually shown that it is sufficient to prove that (33.20) is not altered by higher orders. Now, consider a typical higher-order contribution (Fig. 25c). It may be written as an integral over the gluon momenta and an integral over the quark momenta. But there the triangle has become a heptagon (Fig. 25d) for which the quark integral is convergent and thus, the limit $D \to 4$ may be taken; it vanishes identically. In addition, the arguments above have shown that the anomaly is in fact related to the large momentum behavior of the theory and thus we expect that the exactness of (33.13) will not be spoiled by nonperturbative effects.

We will not make the proof more precise, but refer to the literature.[94] However, we will present an alternate derivation [Wilson (1969)] that will clearly reveal the short-distance character of the anomaly. The axial current involves products of two fields at the same point, so it should properly be defined as

$$A_q^\mu(x) = \lim_{\xi\to 0} A^\mu_{\text{gn}}(x,\xi),$$
$$A^\mu_{gn}(x,\xi) \equiv \bar{q}\left(x + \frac{\xi}{2}\right)\gamma^\mu\gamma_5 q\left(x - \frac{\xi}{2}\right). \tag{33.21}$$

For $\xi \neq 0$, however, this is not gauge invariant. To restore gauge invariance, we have to replace (33.21) by (cf., Appendix I)

$$A^\mu_{\text{gi}}(x,\xi) \equiv \bar{q}(x + \xi/2)\gamma^\mu\gamma_5 e^{ie\int_{x-\xi/2}^{x+\xi/2} dy_\mu A^\mu(y)} q(x - \xi/2).$$

Thus,

$$\partial_\mu A^\mu_{\text{gi}}(x,\xi) = \lim_{\xi\to 0}\left\{2im_q\bar{q}(x)\gamma_5 q(x) + igA^\mu_{\text{gi}}(x,\xi)F_{\mu\lambda}\xi^\lambda + O(\xi^2)\right\}.$$

Because $A_f^\mu(x,\xi)$ diverges as $1/\xi$ for $\xi \to 0$, the second term in the right-hand side does not vanish in this limit. The explicit calculation [Wilson (1969) and Crewther (1972)] shows that, as expected, Equation (33.14) is reproduced.

An elegant discussion of the currents with anomalies for arbitrary interactions may be found in Wess and Zumino (1971).

The axial current is not the only one that possesses anomalies. The trace of the energy momentum tensor Θ^μ_μ is also anomalous, due to the fact that renormalization breaks scale invariance. This is discussed in some detail by

[94] For a detailed discussion, see the reviews of Adler (1971) and Ellis (1976). The triangle graph is the only one that has *primitive* anomalies; it does, however, induce secondary anomalies in square and pentagon graphs. The three-axial triangle has an anomaly closely related to the axial-vector one.

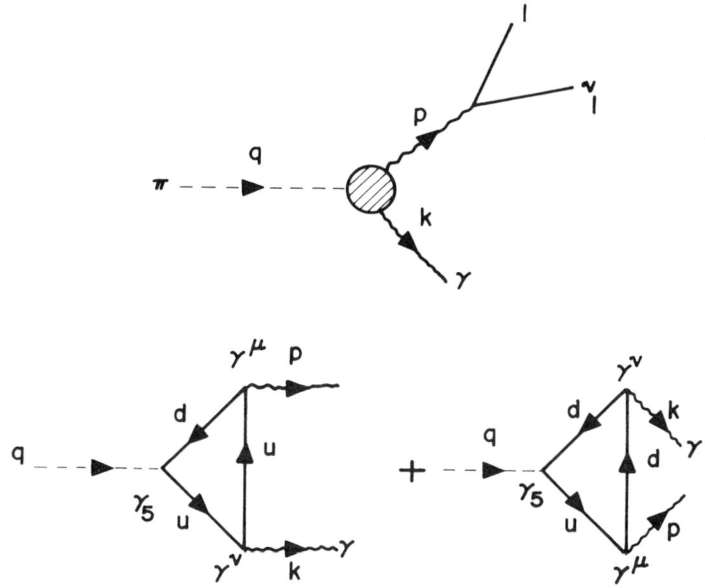

Figure 26. The process $\pi \to (l\nu_l)_{\text{vector}} + \gamma$.

Callan, Coleman, and Jackiw (1970) and, in the context of QCD, by Collins, Duncan, and Joglekar (1977). However, this anomaly is rather harmless; indeed, its analysis is closely related to that of the renormalization group.

34 Quark Mass Effects in Meson Decays

i Light Quarks and Radiative Decays

Our next subject is the radiative decays of mesons,[95]

$$\pi^+ \to l^+ \nu_l \gamma, \qquad K^+ \to l^+ \nu_l \gamma, \qquad l = e, \mu,$$

closely related to $\pi^0 \to \gamma\gamma$. We consider the former; the latter can be treated with obvious changes (s replaced by d, etc.).

The decay consists of two parts: one in which the lepton current is axial, which will not be considered here, and the second in which it is a vector. This last is related to the VEV (the kinematics are those of Figure 26)

$$T_W^{\mu\nu}(p,k) = \int d^4x\, d^4y\, e^{i(x\cdot p + y\cdot k)} \langle TV^\mu(x) J^\nu(y) \partial A(0) \rangle_0, \qquad (34.1)$$

[95] Details of the general features of these decays may be found in the book of Marshak, Riazzuddin, and Ryan (1969).

where, as before, J is the electromagnetic current and

$$V^\mu = \bar{u}\gamma^\mu d, \qquad A^\lambda = \bar{d}\gamma^\lambda\gamma_5 u.$$

Use of the equations of motion yields

$$\partial_\mu V^\mu = i(m_u - m_d)\bar{u}d, \tag{34.2}$$

and the naive relation

$$\partial_\lambda A^\lambda = i(m_u + m_d)\bar{d}\gamma_5 u. \tag{34.3}$$

Of course, we still have $\partial_\nu J^\nu = 0$. As for $\pi^0 \to \gamma\gamma$, we consider

$$R_W^{\mu\nu\lambda}(p,k) = i\int d^4x\, d^4y\, e^{i(x\cdot p + y\cdot k)}\langle TV^\mu(x)J^\nu(y)A^\lambda(0)\rangle_0, \tag{34.4}$$

and expand

$$R_W^{\mu\nu\lambda}(p,k) = \epsilon^{\mu\nu\lambda\rho}p_\rho\Phi_1 + \epsilon^{\mu\nu\lambda\rho}k_\rho\Phi_2 + O(p^3,k^3). \tag{34.5}$$

The contraction with k_ν yields $\Phi_1 = 0$, but if the masses m_u, m_d are different, we cannot require $p_\mu R^{\mu\nu\lambda} = 0$. What we have, instead, is

$$p_\mu R^{\mu\nu\lambda} = \epsilon^{\mu\nu\lambda\rho}p_\mu k_\rho \Phi_2,$$

$$q_\lambda R^{\mu\nu\lambda} = \epsilon^{\mu\nu\lambda\rho}p_\lambda k_\rho \Phi_2.$$

Thus, we find a relation between the two,

$$p_\mu R^{\mu\alpha\beta} = q_\lambda R^{\alpha\beta\lambda}. \tag{34.6}$$

If we substitute (34.2), we obtain

$$p_\mu R^{\mu\alpha\beta} = i(m_d - m_u)\int d^4x\, d^4y\, e^{i(x\cdot p + y\cdot k)}\langle TS(x)J^\alpha(y)A^\beta(0)\rangle_0, \tag{34.7}$$

$$q_\lambda R^{\alpha\beta\lambda} = i(m_d + m_u)\int d^4x\, d^4y\, e^{i(x\cdot p + y\cdot k)}\langle TP(0)J^\beta(y)V^\alpha(x)\rangle_0 + a^{\alpha\beta}, \tag{34.8}$$

where

$$S(x) \equiv \bar{u}(x)d(x), \qquad P(x) \equiv \bar{d}(x)\gamma_5 u(x),$$

and $a^{\alpha\beta}$ is the anomaly. If we define $S^{\alpha\beta} = p_\mu R^{\mu\alpha\beta}$, Equations (34.6) and (34.8) give

$$T_W^{\alpha\beta} + S^{\alpha\beta} = a^{\alpha\beta}.$$

As shown in the previous section, $a^{\alpha\beta}$ is independent of quark masses, so we finally obtain

$$T_W^{\alpha\beta} + S^{\alpha\beta} = \frac{-1}{2\pi^2}\epsilon^{\alpha\beta\rho\sigma}p_\rho k_\sigma. \tag{34.9}$$

If m_u were equal to m_d, we would have $S^{\alpha\beta} = 0$, and thus

$$T_W^{\alpha\beta} = T^{\alpha\beta} = \frac{-1}{2\pi^2}\epsilon^{\alpha\beta\rho\sigma}p_\rho k_\sigma, \tag{34.10}$$

i.e., the vector part of $\pi^+ \to e^+ \nu\gamma$ would equal, up to known factors, the amplitude for $\pi^0 \to \gamma\gamma$ [see, e.g., Adler (1971)]. Since $m_u \neq m_d$, we expect corrections to this naive relation. In general, these corrections cannot be calculated, but there is one case in which we can prove an exact result to all orders of perturbation theory in QCD.[96] If $m_u = 0$, the transformation

$$u \to \gamma_5 u, \qquad d \to d,$$

is a symmetry of the QCD Lagrangian. Under it,

$$S^+ \to P, \qquad A \to V,$$

and thus $T_W^{\alpha\beta} = S^{\alpha\beta}$: we modify (34.10) to read

$$T_W^{\alpha\beta} = \frac{-1}{4\pi} \epsilon^{\alpha\beta\rho\sigma} p_\rho k_\sigma, \tag{34.11}$$

i.e., a variation by a factor of 1/2 in the amplitude, and of 1/4 in the rate. Although there are some experimental indications that the effect is there, the situation is not sufficiently clear, and we will leave the subject here.

ii Heavy Quarks and the GIM Mechanism

In the preceding sections we mainly considered light quarks, i.e., quarks whose masses were small compared to Λ. We will now consider heavy quarks for which $m \gg \Lambda$. This includes the c and b quarks.

Unlike the case of light quarks, the smoothness assumptions somewhat loosely denoted by PCAC are hardly expected to work; thus, we have to look elsewhere to gain an understanding of heavy quark masses.

The first remark is that the property

$$\langle \bar{u}u \rangle \approx \langle \bar{d}d \rangle \approx \langle \bar{s}s \rangle$$

is very unlikely to be extended to $\langle \bar{c}c \rangle, \langle \bar{b}b \rangle$. We expect, however, that

$$\langle \alpha_s G^2 \rangle^{1/4}, |\langle \bar{q}_h q_h \rangle|^{1/3} \ll m_h, \qquad h = c, b.$$

If we accept this we see that most of the mass of a heavy hadron can be attributed to the perturbative mass of its constituent quarks, and thus

$$\hat{m}_c \approx \frac{m_\psi}{2} \approx 1.6 \text{ GeV}, \qquad \hat{m}_b \approx \frac{m_\Upsilon}{2} \approx 6 \text{ GeV}.$$

The best estimates of heavy quark masses are still obtained with sum rules like those of Section 32, and 36; we refer to the review of Narison (1982) and work quoted there, and we turn to another important effect involving quark masses, viz., the GIM mechanism [Glashow, Iliopoulos, and Maiani (1970)]. As a matter of fact, the value $m_c \sim 1.6$ GeV was *predicted* before the ψ particles were discovered in the work by Glashow *et al.* cited above, and,

[96] See Bernabeu, Tarrach, and Ynduráin (1979). Unlike the case of anomaly, one does not know whether this result is affected by nonperturbative corrections.

particularly, in its refinements due to Gaillard and Lee (1974). We refer to this work, as well as to the review of Gaillard (1978), for the detailed treatment of a variety of cases and will consider here a typical example, viz., the decay $K^0 \to \mu^+\mu^-$. To the lowest order in weak interactions, and zero order in α_s, it is given by diagrams like that of Figure 27, which is the only one we will calculate explicitly. The corresponding amplitude will then be[97]

$$\mathscr{A} = g_W^4 \sum_{f=u,c} \delta_f \int \frac{d^4k}{(2\pi)^4}$$
$$\times \left\{ \frac{\left(\gamma_\mu \frac{1-\gamma_5}{2} \not{k} \gamma_\nu \frac{1-\gamma_5}{2}\right)\left(\gamma^\nu \frac{1-\gamma_5}{2}(\not{k} - \not{p}_1' + \not{p}_1 + m_f)\gamma^\mu \frac{1-\gamma_5}{2}\right)}{k^2\left[(k-p_1')^2 - M_W^2\right]\left[(k-p_1'+p_1)^2 - m_f^2\right]\left[(k-p_2')^2 - M_W^2\right]} \right.$$
$$\left. + \frac{\left(\gamma_\mu \frac{1-\gamma_5}{2} \not{k} \gamma_\nu \frac{1-\gamma_5}{2}\right)\left(\gamma^\mu \frac{1-\gamma_5}{2}(\not{k} - \not{p}_1' + \not{p}_1 + m_f)\gamma^\nu \frac{1-\gamma_5}{2}\right)}{k^2\left[(k-p_1')^2 - M_W^2\right]\left[(k-p_1'+p_1)^2 - m_f^2\right]\left[(k-p_2')^2 - M_W^2\right]} \right\}$$
$$= g_W^2 \sum_f \delta_f \int d^D\hat{k} \frac{1}{k^2\left[(k-p_1')^2 - M_W^2\right]\left[(k-p_1'+p_1)^2 - m_f^2\right]\left[(k-p_2')^2 - M_W^2\right]}$$
$$\times \left\{ \left(\gamma_\mu \not{k} \gamma_\nu \frac{1-\gamma_5}{2}\right)\left[\gamma^\nu(\not{k} - \not{p}_1' + \not{p}_2)\gamma^\mu \frac{1-\gamma_5}{2} + (\mu \Leftrightarrow \nu)\right]\right\}$$
$$+ O(m_f^2/M_W^4). \tag{34.12}$$

Here $\delta_u = \cos\theta_C \sin\theta_C$, $\delta_c = -\cos\,\theta_C \sin\theta_C$, θ_C being the Cabibbo angle. Although the integrals are convergent, we have defined them for arbitrary D for reasons that will become apparent. It should be clear that if $m_c = m_u$, (34.12) vanishes: hence, the rate $K^0 \to \mu^+\mu$ must be proportional to $(m_c^2 - m_u^2)$. We will work in the approximation $m_u \approx 0$; then (34.12) may be rewritten as

$$\mathscr{A} = -g_W^4(\cos\theta_C \sin\theta_C)$$
$$\times \int d^D\hat{k} \frac{m_c^2}{k^2\left[(k-p_1')^2 - M_W^2\right]\left[(k-p_1'+p_1)^2 - m_c^2\right](k-p_1'+p_2)^2}$$
$$\times \frac{(\gamma_\mu \not{k} \gamma_\nu (1-\gamma_5)/2)\left[\gamma^\nu(\not{k} - \not{p}_1 + \not{p}_2)\gamma^\nu(1-\gamma_5)/2 + (\mu \Leftrightarrow \nu)\right]}{(k-p_2')^2 - M_W^2}. \tag{34.13}$$

We now have 10 powers of k in the denominator and two in the numerator; therefore, we can work in the limit $M_W^2 \to \infty$ and pick no more than a

[97] The first bracket is to be sandwiched between the spinors of the leptons, the second between those of the quarks.

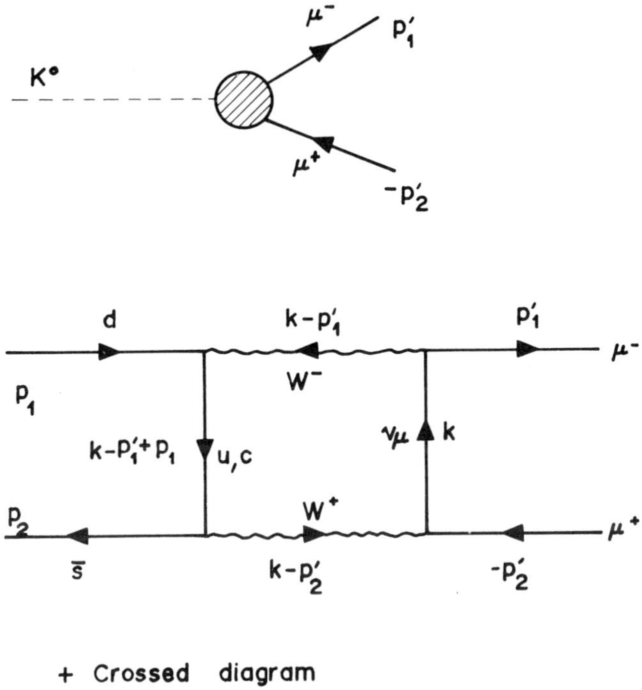

Figure 27. The process $K^0 \to \mu^+\mu^-$ and a typical diagram contributing to it.

logarithmic singularity. In fact, this singularity is cancelled by the contribution of other diagrams (notably, the γ, Z mediated decay). We have, neglecting terms of relative order m_K^2/M_W^2,

$$\mathscr{A} = -g_W^4(\cos\theta_C\sin\theta_C)\frac{m_c^2}{M_W^4}\cdot\frac{1}{4}$$

$$\times\int d^D\hat{k}\,\frac{(\gamma_\mu k\!\!\!/\gamma_\nu(1-\gamma_5))(\gamma^\nu k\!\!\!/\gamma^\mu(1-\gamma_5)+(\mu\Leftrightarrow\nu))}{k^4(k^2-m_c^2)}$$

$$= -g_W^4(\cos\theta_C\sin\theta_C)\frac{m_c^2}{4M_W^4}\left[\gamma_\mu\gamma_\alpha\gamma_\nu(1-\gamma_5)\right]\left[\gamma^\nu\gamma^\alpha\gamma^\mu(1-\gamma_5)\right.$$

$$\left.+\gamma^\mu\gamma^\alpha\gamma^\nu(1-\gamma_5)\right]\frac{i}{16\pi^2}\left(N_\epsilon-\log\frac{m_c^2}{\nu_0^2}-1/2\right).$$

As explained, the $N_\epsilon - \log m_C^2/\nu_0^2 - 1/2$ term is replaced by -2 when adding the other diagrams. (Because of this cancellation the process is actually dominated by the mechanism $K\to 2\gamma\to\mu^+\mu^-$). The final result only involves constant terms, and is sensitive to the ratio m_c^2/M_W^4. We have neglected strong interactions here; a detailed analysis would incorporate them. We send the interested reader to the quoted literature.

35 Perturbative and Spontaneous Effects in Quark and Gluon Propagators

In Chapter II we calculated the perturbative contributions to the quark and gluon propagators. These were expressed in terms of the masses that appear in the Lagrangian, usually called perturbative, mechanical (or, for the quarks, "current algebra") masses. As is known, however, reasonable success has been gained in so-called constituent models in which "constituent" masses, for the u, d, and s quarks of some 400 MeV and the gluons ($\sim$800 MeV) are used. In this and the following sections, we will see that nonperturbative contributions to S and D simulate masses. While these masses cannot be identified with constituent ones, we shall show that the effects are, in fact, responsible for the masses of hadrons like the ρ.

We begin with the quark propagator,

$$S_{\xi}^{ij}(p) = \int d^D x\, e^{ip\cdot x}\langle Tq^i(x)\bar{q}^j(0)\rangle_{\text{vac}}, \tag{35.1}$$

which will be evaluated for large p. We write an OPE for it, neglecting terms that yield zero when sandwiched with $|\text{vac}\rangle$:

$$Tq^i(x)\bar{q}^j(0) = \delta_{ij}\left\{C_0(x)\cdot 1 - C_1(x)\sum_l :\bar{q}^l(0)q^l(x): + \cdots\right\}; \tag{35.2}$$

only the perturbative coefficient, $C_0(x)$, was considered in Sections 7 and 9. To zero order in α_g, we have, with α, β Dirac indices,

$$:\bar{q}_\beta(0)q_\alpha(x): \underset{x\to 0}{\approx} \frac{1}{4}\left\{\delta_{\alpha\beta} - \frac{im_q x_\mu}{D}\gamma^\mu_{\alpha\beta}\right\} :\bar{q}(0)q(0):.$$

If we let S_p, resp., S_{NP}, be the perturbative and nonperturbative contributions of (35.2) to (35.1), respectively, we find (cf. Figure 28)

$$S = S_P + S_{NP},$$

$$S_{NP}^{(0)ij}(p) = -(2\pi)^D \frac{\delta_{ij}\langle\bar{q}q\rangle_{\text{vac}}}{4n_c}\left\{1 - \frac{m_q}{D}\gamma^\mu\frac{\partial}{\partial p^\mu}\right\}\delta(p), \qquad n_c = 3. \tag{35.3}$$

This is identically vanishing for $p \neq 0$: although we will see that terms like (35.3) are important for the masses of observable particles ($\rho, \phi, \ldots$). The second-order correction to S_{NP} is most easily evaluated by writing

$$S_{NP}^{(2)ij}(p) = \sum \frac{1}{\not{p} - m_q} g^2 \int d^D\hat{k}\, i\gamma_\mu t^a_{ik} S^{kk'}(p+k) i\gamma_\nu t^b_{k'j}\delta_{ab}$$

$$\times \frac{-g^{\mu\nu} + \xi k^\mu k^\nu/k^2}{k^2}\cdot\frac{i}{\not{p} - m_q},$$

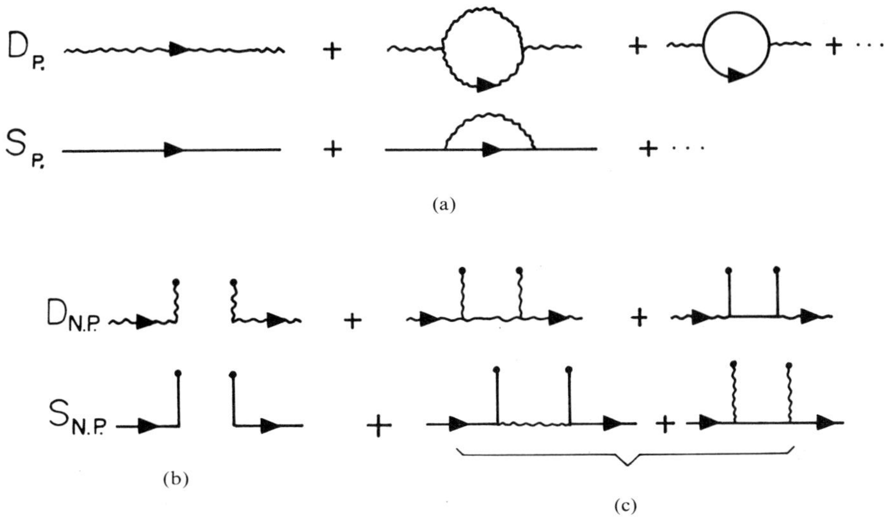

Figure 28. The gluon and quark propagators (*a*), perturbative contributions; (*b*), first nonperturbative contributions; (*c*), first connected nonperturbative contributions.

and replacing in the right-hand side $S^{kk'}$ by $S_{NP}^{(0)kk'}$. We thus find,

$$S_{NP} = S_{NP}^{(0)} + S_{NP}^{(2)} + \cdots,$$

$$S_{\xi NP}^{(2)ij}(p) = -i\delta_{ij}\alpha_g \frac{\pi C_F \langle \bar{q}q \rangle_{\text{vac}}}{3p^4} \left\{ D - \xi - \frac{2(D-2)}{D}(1-\xi)\frac{m_q \not{p}}{p^2} \right\}$$

$$+ O\left(\frac{m_q}{p^6}\right) + O\left(\frac{m_q^4}{p^2}\right). \tag{35.4}$$

Note that this is gauge dependent, so one cannot interpret the expression

$$M_\xi(p) = \frac{-\pi\alpha_g C_F \langle \bar{q}q \rangle_{\text{vac}}}{3p^2}(4-\xi)$$

as a physical mass.

A similar calculation can be performed for the gluon propagator (Figure 28),

$$D_{\xi ab}^{\mu\nu}(k) = \int d^4x\, d^{ik\cdot x} \langle TB_a^\mu(x) B_b^\nu(0) \rangle_{\text{vac}},$$

$$TB_a^\mu(x)B_b^\nu(0) = \delta_{ab}\left\{ C_0^{\mu\nu}(x)\cdot 1 + C_1^{\mu\nu}(x)\sum_c :G_c^{\alpha\beta}(0)G_{c\alpha\beta}(0): + \cdots \right\}, \tag{35.5}$$

obtaining,

$$D = D_P + D_{NP},$$

$$D_{NPab}^{(0)\mu\nu}(k) = (2\pi)^D \delta_{ab} \frac{\langle G^2 \rangle_{\text{vac}}}{4(n_c^2 - 1)D(D-1)(D+2)}$$

$$\times \{(D+1)g^{\mu\nu}\partial^2 - 2\partial^\mu\partial^\nu\}\delta(k). \tag{35.6}$$

It is to be noted that $D_{NP}^{(0)}$ is transverse. This term also contributes to $S_{NP}^{(2)}$: it adds to (34.4) the expression

$$S_{G^2NP}^{(2)}(p) = \frac{2C_F}{3(n_c^2-1)} \cdot \frac{\pi\langle \alpha_s G^2 \rangle_{\text{vac}}}{p^4} \cdot \frac{i}{\not p}. \tag{35.7}$$

We can also evaluate the contributions of $\langle \bar{q}q \rangle, \langle G^2 \rangle$ to D. These would give a masslike term, unfortunately gauge dependent. In fact, as we will see in the next section, the masses of physical particles are *not* related to terms like M_ξ or the equivalent for gluons; these contribute only at a nonleading order. The main contribution comes from Equations (34.3) and (34.6). A thorough discussion of this in connection with the $\langle \bar{q}q \rangle$ term may be found in Pascual and de Rafael (1982).

36 Hadron Masses

Instead of considering the general analysis,[98] we will discuss a typical example, viz., the mass of the ϕ resonance. For this, consider the two-point function

$$\Pi_\phi^{\mu\nu}(q) = i\int d^4x\, e^{iq\cdot x} \langle T\phi^\mu(x)\phi^\nu(0)\rangle_{\text{vac}}$$

$$\equiv (-g^{\mu\nu}q^2 + q^\mu q^\nu)\Pi_\phi(q^2), \tag{36.1}$$

where ϕ^μ is an operator with the quantum numbers of the ϕ. Specifically,

$$\phi^\mu(x) = C_\phi \bar{s}(x)\gamma^\mu s(x).$$

The constant C_ϕ may be obtained from the process $\phi \to e^+e^-$, but we will not discuss this here. The function $\Pi(q^2)$ behaves like $\log q^2$; hence, any derivative

$$\frac{d^N \Pi_\phi(q^2)}{(dq^2)^N} \equiv \Pi_\phi^{(N)}(q^2)$$

[98] The method we will follow was developed originally by Shifman, Vainshtein, and Zakharov (1979*a*, *b*). It has been further developed by these same authors and by Reinders, *et al.* (1980), and work quoted there. Recently, it has been extended to baryons by Ioffe (1981) and by J. Chung, H. Dosch, M. Kremer, and D. Schall (Heidelberg, preprint, 1981). For more details on sum rules like the ones described here, see the excellent review by Narison (1982).

with $N \geqslant 1$ will satisfy an unsubtracted dispersion relation. For $|q^2|$ near m_ϕ^2, we may approximate $\Pi^{(N)}(q^2)$ by the contribution of the ϕ. Hence, we write

$$\Pi_\phi^{(N)}(q^2) \approx \frac{N!\,a}{\left(m_\phi^2 - q^2\right)^{N+1}}.$$

By taking the ratio of two consecutive derivatives, we then find

$$r_\phi(q^2) \equiv \frac{\Pi_\phi^{(N)}(q^2)}{\Pi_\phi^{(N+1)}(q^2)} \approx \frac{1}{N+1}\left(m_\phi^2 - q^2\right). \tag{36.2}$$

If we calculated with QCD, using only the perturbative masses, we would have obtained

$$\Pi_\phi^{(N)}(q^2) \approx \frac{3C_\phi^2}{12\pi^2}(N-1)!\,\frac{1}{\left(-q^2\right)^N}\left\{1 + \frac{\hat{m}_s^2}{q^2} + O\left[\alpha_s(-q^2)\right]\right\}. \tag{36.3}$$

Now, with the values we have obtained for $\hat{m}_s$, it is impossible to fit (36.2) with the physical mass of the ϕ. This indicates that the nonperturbative contributions are essential. These can be implemented most easily by using the calculations of the previous section for the nonperturbative parts of S and D. To lowest order in α_s, we only need (35.3) and (35.6); the result is that (36.3) is modified to

$$\Pi_\phi^{(N)}(q^2) \approx \frac{3C_\phi^2}{12\pi^2}(N-1)!\,\frac{1}{\left(-q^2\right)^N} \times\left\{1 + \frac{\hat{m}_s^2}{(-q^2)} - \frac{4\pi^2 N(N+1)}{q^4}m_s\langle\bar{s}s\rangle_{\text{vac}} - \frac{3\pi N(N+1)}{8q^4}\langle\alpha_s G^2\rangle_{\text{vac}} + O(\alpha_s) + O(q^{-6})\right\}. \tag{36.4}$$

We see that, in the limit $-q^2/N \to \infty$, the mass m_ϕ receives an important contribution (indeed, most of it) from $\langle\alpha_s G^2\rangle$. Thus, in a sense, it is possible to simulate the masses of the $\rho, \omega, \phi \ldots$ by using "constituent" masses of the order of $\langle\alpha_s G^2\rangle^{1/4}$. We will not pursue the matter further, but will make two comments. First, the use of "constituent" masses can be, at best, a rough approximation. This is because the contribution of $\langle\alpha_s G^2\rangle$ depends on the spin of the operators (ϕ^μ, in our example), to which it is attached; in general, it will differ for particles like the ρ or the f^0. Second, there are presently calculations of over 50 masses and parameters: the agreement with experiment is impressive if we remember that the only parameters required are the quark masses (u, d, s, c, and b), Λ, $\langle\alpha_s G^2\rangle$ and $\langle\bar{q}q\rangle$. The last three may even be taken from other sources.

We conclude this section with a detailed sample calculation of a nonperturbative contribution, viz., that of $\langle \bar{s}s \rangle$ to $\Pi_{\phi}^{\mu\nu}(q)$. From (36.1),

$$\Pi_{\phi}^{\mu\nu}(q) = iC_{\phi}^{2}\int d^{4}x\, e^{iq\cdot x}\langle T\bar{s}(x)\gamma^{\mu}s(x)\bar{s}(0)\gamma^{\nu}s(0)\rangle_{\text{vac}}. \tag{36.5}$$

Therefore,

$$\Pi_{\phi}^{\mu\nu}(q) = -iC_{\phi}^{2}\int d^{D}\hat{k}\, \mathrm{Tr}\,\gamma^{\mu}S_{s}(k)\gamma^{\nu}S_{s}(k+q). \tag{36.6}$$

If we considered only the perturbative $S_s = S_P$ propagator, we would have obtained the perturbative piece,

$$\Pi_{P}^{\mu\nu}(q) = \frac{8C_{\phi}^{2}n_{c}}{6}\cdot\frac{1}{16\pi^{2}}(-g^{\mu\nu}q^{2}+q^{\mu}q^{\nu})$$
$$\times\left(N_{\epsilon} - \log q^{2} + \text{finite constant} + O\left(m_{s}^{2}\right)\right). \tag{36.7}$$

The nonperturbative part is obtained by using the full expression, $S_s = S_P + S_{NP}$. The leading term is the mixed term,

$$\Pi_{NP}^{\mu\nu} = -iC_{\phi}^{2}\int d^{D}\hat{k}\, \mathrm{Tr}\{\gamma^{\mu}S_{NP}(k)\gamma^{\nu}S_{P}(k+q)$$
$$+\gamma^{\mu}S_{P}(k)\gamma^{\nu}S_{NP}(k+q)\}, \tag{36.8}$$

with S_{NP} given by (to leading order) Equation (35.3) and $S_P(k) = i/(\not{k} - m_s)$. We find

$$\Pi_{NP}^{\mu\nu} = \frac{-2C_{\phi}^{2}m_{s}\langle\bar{s}s\rangle_{\text{vac}}}{q^{4}}(-g^{\mu\nu}q^{2}+q^{\mu}q^{\nu}),$$

as stated above in Equation (36.4).

37 The $U(1)$ Problem. The Gluon Anomaly

In Section 33, we discussed the triangle anomaly in connection with the decay $\pi^0 \to \gamma\gamma$. As remarked there, the anomaly is not restricted to photons. In particular, we have a gluon anomaly. Defining the current

$$A_{0}^{\mu} = \sum_{f=1}^{n}\bar{q}_{f}\gamma^{\mu}\gamma_{5}q_{f}, \tag{37.1}$$

we find that it has an anomaly

$$\partial_{\mu}A_{0}^{\mu} = i\sum_{f=1}^{n}\bar{q}_{f}\gamma_{5}q_{f} + \frac{ng^{2}}{16\pi^{2}}\tilde{G}G, \tag{37.2}$$

where the *dual* $\tilde{G}$ is defined by

$$\tilde{G}_{a}^{\mu\nu} \equiv \tfrac{1}{2}\epsilon^{\mu\nu\alpha\beta}G_{a\alpha\beta},$$

and

$$\tilde{G}G \equiv \sum_{a}\tilde{G}_{a}^{\mu\nu}G_{a\mu\nu}.$$

The current (37.1) is the so-called $U(1)$ current (pure flavor singlet), and is atypical in more respects than one. In particular, it is associated with the so-called $U(1)$ problem, to which we now turn.

Assume that we have n light quarks; we only consider these and will neglect (as irrelevant to the problem at hand) the existence of eventual heavy flavors. We may take $n = 2$ (u, d) and then speak of "the $SU(2)$ $U(1)$ problem" or $n = 3(u, d, s)$, which is the $SU(3)$ $U(1)$ problem. Take the $n^2 - 1$ matrices in flavor space, $\lambda_1, \ldots, \lambda_{n^2-1}$; for $SU(3)$, they coincide with the Gell–Mann matrices, for $SU(2)$, with the Pauli matrices. Any Hermitian $n \times n$ matrix may be written as a combination of the n^2 matrices $\lambda_a, \ldots, \lambda_{n^2-1}, \lambda_0 \equiv 1$. It will prove convenient to let indices $a, b, c, \ldots$ vary from 1 to $n^2 - 1$, and let indices $\alpha, \beta, \delta, \ldots$ also include the value 0. Because of the completeness property just alluded to, it is sufficient to consider the currents

$$A_\alpha^\mu = \sum \bar{q}_f \gamma^\mu \gamma_5 \lambda_{ff'}^\alpha q_{f'}\,;$$

of course, only A_0 has an anomaly. Then let $N_1(x), \ldots, N_k(x)$ denote local operators (simple or composite), and consider the quantity

$$\langle \text{vac}| T A_\alpha^\mu(x) \prod_j N_j(x_j) |\text{vac}\rangle. \tag{37.3}$$

When $\alpha \neq 0$, the Goldstone theorem implies that the masses of the pseudoscalars P_a with the quantum numbers of the A_a vanish in the chiral limit: introducing a common parameter ϵ for all quark masses by letting $m_f = \epsilon r_f$, where the r_f $(f = 1, \ldots, n)$ remain fixed in the chiral limit, we have

$$m_a^2 \equiv m_{P_a}^2 \approx \epsilon. \tag{37.4}$$

This was shown in Section 31 in Equations (31.4) and (31.5). Therefore, in this limit, (37.3) develops a pole at $q^2 = 0$, for $\alpha = a$. To be precise, what this means is that in the chiral limit, i.e., for zero quark masses,

$$\lim_{q \to 0} \int d^4x \, e^{iq\cdot x} \partial_\mu \langle \text{vac}| T A_\alpha^\mu(x) \prod_j N_j(x_j) |\text{vac}\rangle$$
$$\approx (\text{constant}) q_\mu \frac{1}{q^2}\,. \tag{37.5}$$

If we neglect anomalies, the derivation of (37.4) can be repeated for the case $\alpha = 0$, and thus we would find that the $U(1)$ particle would also have vanishing mass in the chiral limit [Glashow (1968)]. In fact, this statement was made more precise by Weinberg (1975) who showed that $m_0 \leqslant \sqrt{n} \times$ (average m_a). Now, this is a catastrophe, since for the $SU(2)$ case, $m_\eta \gg \sqrt{2}\, m_\pi$, and for $SU(3), m_{\eta'}$ also violates the bound. In addition, Brandt and Preparata (1970) proved that under these conditions, the decay $\eta \to 3\pi$ is forbidden, also in contradiction with experiment. We therefore *assume* that (37.3) remains *regular* as $\epsilon \to 0$ for $\alpha = 0$. If we could *prove* that this is so, we would have solved the $U(1)$ problem. This will be discussed further later on; for the moment, we shall assume that there are no $U(1)$ bosons, without questioning whether or not this may be proved from QCD. It is quite clear that if there was no anomaly, this assumption would be inconsistent, so perhaps it is a good strategy to see what we can obtain from the interplay of the absence of P_0 Goldstone bosons and the existence of an anomaly for the A_0 current. We will proceed to do this, following the excellent review of Crewther (1979*b*).

The current A_0, as defined in (37.1), is gauge invariant but not $U(1)$ invariant in the chiral limit due to the anomaly of Equation (37.2). We may construct another current which is $U(1)$ invariant, as shown by Adler (1969) for the Abelian case and by Bardeen (1974) in general. We define

$$\hat{A}_0^\mu = A_0^\mu - 2nK^\mu, \tag{37.6}$$

where we have introduced the purely gluonic current

$$K^\mu = \frac{2g^2}{32\pi^2}\epsilon^{\mu\nu\rho\sigma}\sum B_{a\nu}\left\{\partial_\rho B_{a\sigma} + \frac{1}{3}f_{abc}B_{b\rho}B_{c\sigma}\right\}. \tag{37.7}$$

That this is the correct answer is easily checked by noting that

$$\partial_\mu K^\mu = \frac{g^2}{32\pi^2}\tilde{G}G, \tag{37.8}$$

so that, from (37.2), we obtain, in the chiral limit,

$$\partial_\mu \hat{A}_0^\mu = 0. \tag{37.9}$$

It should be remarked that K is not unique, even requiring (37.8): indeed, it is gauge dependent. In principle, Equation (37.6) is defined for the bare quantities; but we may always renormalize in such a way that it remains valid for the dressed ones. The reason, of course, is that the anomaly does not become renormalized.

The generator of $U(1)$ rotations must be the current which is conserved, viz., $\hat{A}_0$. We therefore define the *chiralities* χ:

$$\delta(x^0 - y^0)\left[\hat{A}_0^0(x), N_j(y)\right] = -\chi_j\delta(x-y)N_j(y), \tag{37.10a}$$

or, in integrated form,

$$\left[\hat{Q}_0, N_j\right] = -\chi_j N_j, \tag{37.10b}$$

and we have defined the $U(1)$ chiral charge

$$\hat{Q}_0 = \int d\vec{x}\,\hat{A}_0^0(x). \tag{37.11}$$

Since $\hat{A}$ verifies (37.9), $\hat{Q}_0$ is time independent and, hence, we will expect not only that (37.10) makes sense, but that the numbers χ_j will not become renormalized. To prove this more formally, consider the VEV

$$\langle\text{vac}|T\hat{A}_0^\mu(x)\prod_j N_j(x_j)|\text{vac}\rangle,$$

and apply ∂_μ to it. We obtain the Ward identity,

$$\partial_\mu \langle \text{vac}| T\hat{A}_0^\mu(x) \prod_j N_j(x_j)|\text{vac}\rangle$$
$$= -\left\{ \sum_l \chi_l \delta(x - x_l) \right\} \langle \text{vac}| T \prod_j N_j(x_j)|\text{vac}\rangle; \qquad (37.12)$$

we have used (37.9) and (37.10a). Since $\hat{A}$ is (partially) conserved, we know that it is not renormalized, and so the χ must share this property. In the following section we will see that (37.12) plus the absence of $U(1)$ bosons lead to peculiar properties of the QCD vacuum.

38 The θ Parameter; The QCD Vacuum; The Effect of Massless Quarks; Solution to the $U(1)$ Problem

So far, we have been working with the QCD Lagrangian (omitting gauge fixing and ghost terms)

$$\mathscr{L} = \sum_q \bar{q}(i\not{D} - m)q - \frac{1}{4} GG; \qquad (38.1)$$

now we ask what would be the modifications introduced by adding a term

$$\mathscr{L}_{1\theta} = -\frac{\theta g^2}{32\pi^2} \tilde{G}G, \qquad (38.2a)$$

obtaining

$$\mathscr{L}_\theta = \mathscr{L} + \mathscr{L}_{1\theta} \qquad (38.2b)$$

In fact, $\mathscr{L}_{1\theta}$ is the only extra term we can add to $\mathscr{L}$ which is allowed by gauge invariance and renormalizability. Moreover, as shown in the previous section, it is a four-divergence, and thus leaves the equations of motion unchanged. Of course, we can dispose of it by setting $\theta = 0$; but although there are indications that θ is very small indeed, there are also reasons why it may be nonzero. At any rate, it is of interest to find the implications of choosing the more general form (38.2).

First, because we are adding a new interaction, we expect the physical vacuum to depend on θ, so we write $|\theta\rangle$ for it. Our next task is to explore the θ dependence of the Green's functions.

To do this, consider the *topological charge operator*[99]

$$Q_K = \frac{g^2}{32\pi^2} \int d^4x\, \tilde{G}G; \qquad (38.3)$$

we may use (37.8) and Gauss's theorem to write it as a surface integral,

$$Q_K = \int d\sigma_\mu K^\mu.$$

[99] More about θ-vacua and the topics of this section will be found in Sections 43–45, where the reasons for some seemingly peculiar names will become apparent.

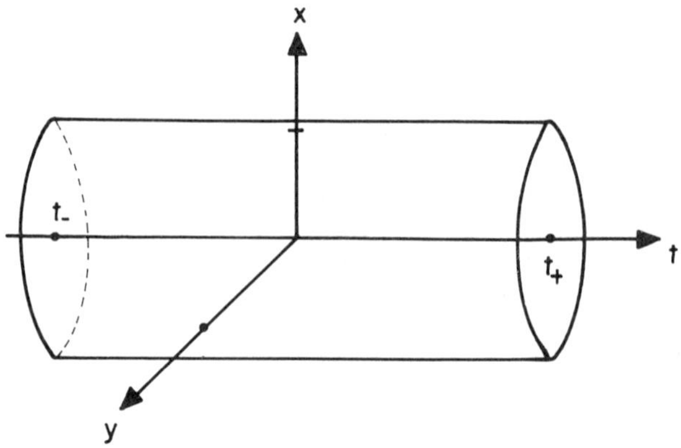

Figure 29. Region of integration for the topological charge.

We will choose as a surface that of a cylinder oriented along the time axis, with bases at $t_+ \to +\infty$ and $t_- \to -\infty$ (Figure 29). We allow the sides to approach infinity, and thus

$$Q_K = \int d\vec{x}\, K^0(t_+ \to +\infty, \vec{x}) - \int d\vec{x}\, K^0(t_- \to -\infty, \vec{x}) \equiv K_+ - K_- . \quad (38.4)$$

The operators $K_\pm$ are self-adjoint, and related to one another by time reversal, so their spectra coincide. We label their eigenstates $|n_\pm\rangle \equiv |n, t_\pm \to \pm\infty\rangle$, with

$$K_\pm |n_\pm\rangle = n|n_\pm\rangle. \quad (38.5)$$

Because of the hermiticity of the $K_\pm$, we may expand the physical vacuum as

$$|\theta\rangle = \sum c_n(\theta)|n_+\rangle = \sum c_n(\theta)|n_-\rangle; \quad (38.6)$$

the c_n are the same. Indeed, the vacuum is invariant by time translations and, hence, we may take it at $t = 0$. Thus, applying time reversal $U(T)$, we find (38.6) with equal c_n. We then have to determine the c_n. To do so, apply the operation $i\partial/\partial\theta$ to a Green's function; recalling the formalism of Section 2,

$$\begin{aligned} i\frac{\partial}{\partial\theta}&\langle\theta|T\prod N_j(x_j)|\theta\rangle \\ &= i\frac{\partial}{\partial\theta}\langle 0|T\prod N_j^0(x_j)e^{i\int d^4x\{\mathscr{L}^0_{\text{int}}(x)+\mathscr{L}^0_{1\theta}(x)\}}|0\rangle \\ &= \frac{g^2}{32\pi^2}\int d^4x\,\langle 0|T\tilde{G}^0(x)G^0(x)\prod N_j^0(x_j)e^{i\int d^4x\{\mathscr{L}^0_{\text{int}}(x)+\mathscr{L}^0_{1\theta}(x)\}}|0\rangle \\ &= \frac{g^2}{32\pi^2}\int d^4x\,\langle\theta|T\tilde{G}(x)G(x)\prod N_j(x_j)|\theta\rangle. \end{aligned} \quad (38.7)$$

In other words, the operation $i\partial/\partial\theta$ is equivalent to the insertion of Q_K. Using (38.3) and (38.4) and because $+\infty$ is later and $-\infty$ earlier than any time, (38.7) becomes

$$i\frac{\partial}{\partial\theta}\langle\theta|T\prod N_j(x_j)|\theta\rangle$$
$$=\langle\theta|K_+T\prod N_j(x_j)|\theta\rangle-\langle\theta|T\prod N_j(x_j)K_-|\theta\rangle:$$

expanding as in (38.6), we find[100]

$$i\frac{\partial}{\partial\theta}\sum_{n,m}c_n^*(\theta)c_m(\theta)=\sum_{n,m}(n-m)c_n^*(\theta)c_m(\theta)$$

with solution

$$c_n(\theta)=Ce^{in\theta}. \tag{38.8}$$

The constant C is arbitrary and may be taken to be unity.

A first consequence of (38.8) is that different θ-vacua are orthogonal:

$$\langle\theta|\theta'\rangle=\delta(\theta-\theta'), \tag{38.9}$$

so each value of θ yields a different world (up to periodicity).

Until now we have not taken into account the existence of fermions. We will now describe how the analysis is modified if we introduce n fermions of vanishing mass. We begin by rewriting our familiar Ward identity (37.12):

$$\partial_\mu\langle\theta|T\hat{A}_0^\mu(x)\prod_j N_j(x_j)|\theta\rangle$$
$$=-\left\{\sum_l\chi_l\delta(x-x_l)\right\}\langle\theta|T\prod_j N_j(x_j)|\theta\rangle,$$

which we now integrate with d^4x:

$$\int d^4x\,\partial_\mu\langle\theta|T\hat{A}_0^\mu(x)\prod_j N_j(x_j)|\theta\rangle$$
$$=-\left(\sum\chi_l\right)\langle\theta|T\prod_j N_j(x_j)|\theta\rangle.$$

Using (37.6) and (37.8), we find

$$\int d^4x\,\partial_\mu\langle\theta|T\sum_f\bar{q}_f(x)\gamma^\mu\gamma_5 q_f(x)\prod_j N_j(x_j)|\theta\rangle$$
$$=2n\frac{g^2}{32\pi^2}\int d^4x\,\langle\theta|T\tilde{G}(x)G(x)\prod_j N_j(x_j)|\theta\rangle$$
$$-\left(\sum\chi_l\right)\langle\theta|T\prod_j N_j(x_j)|\theta\rangle. \tag{38.10}$$

[100] A more rigorous derivation may be found in Crewther (1979*a*); later, in Section 45, we will present an alternate discussion.

Two remarks are in order. Clearly,

$$\int d^4x\,\partial_\mu\langle\theta|T\sum_f \bar{q}_f(x)\gamma^\mu\gamma_5 q_f(x)\prod N_j(x_j)|\theta\rangle$$

$$= -\lim_{q\to 0} iq_\mu\int d^4x\,e^{iq\cdot x}\langle\theta|T\sum_f \bar{q}_f(x)\gamma^\mu\gamma_5 q_f(x)\prod N_j(x_j)|\theta\rangle.$$

Now, if there are no $U(1)$-bosons, the VEV here has no pole as $q^2 = 0$, so it vanishes. Then, as we saw earlier, insertion of Q_K is equivalent to $i\partial/\partial\theta$. Thus, (38.10) becomes

$$2ni\frac{\partial}{\partial\theta}\langle\theta|T\Pi N_j(x_j)|\theta\rangle = \left(\sum\chi_l\right)\langle\theta|T\Pi N_j(x_j)|\theta\rangle. \tag{38.11}$$

Now, for massless quarks, the vacuum is invariant under chiral rotations:

$$|\theta\rangle = U_\phi|\theta\rangle, \qquad U_\phi = e^{-i\phi\hat{Q}_0}; \tag{38.12}$$

using (37.10*b*), on the other hand, we have

$$i\frac{\partial}{\partial\phi}U_\phi^{-1}\prod N_j U_\phi = \left(\sum\chi_l\right)U_\phi^{-1}\prod N_j U_\phi; \tag{38.13}$$

so the right-hand side of (38.11) may be rewritten as

$$i\frac{\partial}{\partial\phi}\langle\theta|T\prod_j N_j(x_j)|\theta\rangle.$$

We find that the operation

$$2ni\frac{\partial}{\partial\theta} - i\frac{\partial}{\partial\phi}$$

annihilates all Green's functions. This means that a change in θ may be compensated for by a change in ϕ. Therefore, the theory is equivalent to one with $\theta = 0$, because it is certainly chiral invariant. Thus, in the special case where the quarks are massless,[101] the θ parameter may be taken to zero, and the old QCD Lagrangian $\mathscr{L}$ of (38.1) is actually the most general one.

One may argue that quark masses are of weak origin generated in the manner discussed by Higgs and Weinberg so, for pure QCD, quarks should be assumed to be massless. However, we are interested in the real world, and thus the effects of perturbing QCD by weak interactions (at least to first order) cannot be eschewed.[102] Another possible escape is to note that the $\mathscr{L}_{1\theta}$ term violates time reversal and parity invariance. Thus we can take θ to zero by imposing T and P invariance. Again, this view cannot be

[101] A more detailed analysis shows that it is enough that *one* quark be massless. This result was first obtained by Peccei and Quinn (1977).

[102] Another possibility is to use suitable Higgs systems that imply $\theta = 0$ [Peccei and Quinn (1977)]; this can be shown to lead to the existence of a new pseudoscalar boson [the "axion," cf., Wilczek (1978); Weinberg (1978*b*)]. There is not enough evidence to decide whether or not it exists.

maintained. Weak interactions violate T and P, and some of this may seep into strong interactions. If this is the origin of θ, however, there are reasonable arguments [Ellis and Gaillard (1979)] that the effect is small, provided θ_{QCD} is originally zero.

Perhaps it is more profitable to discuss experimental bounds on θ. As will be shown later (Section 45), the effects of $\mathscr{L}_{1\theta}$ on processes like deep inelastic scattering are quite negligible; the only place where one can obtain a substantial effect is in T and P violating effects. The best such quantity is the neutron dipole moment, d_n. The calculation was performed by Crewther *et al.* (1980), which refined a previous estimate of Baluni (1979). One finds,

$$d_n \approx 4 \times 10^{-16} |\theta| \quad (\text{in e} - \text{cm}).$$

Experimentally,

$$d_n^{\text{exp}} \leqslant 1.6 \times 10^{-24},$$

so we obtain $|\theta| \lesssim 10^{-8}$, a very small value indeed.

Let us return to the vacuum problem. We have discussed the effect of massless quarks; now we need to study the influence of chiral symmetry breaking by "small" mass terms. That is to say, what happens after introducing the perturbation

$$\sum m_f \bar{q}_f q_f ,$$

at least to first order in ϵ (recall that $m_f = \epsilon r_f, r_f$ fixed). We will not enter into the details here; the interested reader is referred to the lectures of Crewther (1979*b*). We merely summarize the results. Consider the inequality

$$m_u^{-1} > \sum_{f=2}^{n} m_f^{-1}; \tag{38.14}$$

note that the results of Section 31 imply that it is probably satisfied in the real world. Then, (i) if (38.14) holds, the topological charge is quantized in integer units; that is to say, the difference ν between any two eigenvalues of K_+ and K_- is an integer, and (ii) if (38.14) does *not* hold, then there are at least fractional values of ν. In fact, for some particular values of the masses, ν must take irrational values.

We end this section with two comments. First, we have obtained constraints on the spectrum of $K_\pm$, or the expression of the vacuum in terms of the eigenvectors $|n_\pm\rangle$; but we have not proved that the spectrum is nontrivial. Indeed, one could imagine the possibility that all the n coincide, and thus the contents of these sections would be much ado about nothing. Luckily (or unluckily, according to the point of view), the existence of instantons implies that at least there exists a denumerable infinity, $\ldots$, $-1, 0, 1, 2, \ldots$ of different values of n. This will be shown later (Section 45).

Second, we have assumed that no zero-mass $U(1)$ bosons exist. The mass of a pseudoscalar meson may be evaluated as in Section 31. If we repeat the calculation for the singlet current, A_0^μ, we find that, because of the anomaly, Equation (31.5) is modified by a term

$$n_f^2\left(\frac{g^2}{32\pi^2}\right)^2\int d^4x\,\langle TG(x)\tilde{G}(x)G(0)\tilde{G}(0)\rangle_{\text{vac}}. \tag{38.15}$$

This would vanish for the perturbative vacuum; but the existence of instanton solutions (Sections 43–45) shows that, at least in the semiclassical approximation, (38.15) remains nonzero in the chiral limit ['t Hooft (1976)]. One may question the validity of this approximation. Alternatively, the same result is obtained in the large n_c(= number of colors) limit [Witten (1979*a*)]. Thus, though we do not have a totally rigorous proof, it appears extremely likely that QCD solves the $U(1)$ problem.

CHAPTER V

Functional Methods, Nonperturbative Solutions

Alice laughed. "There is no use trying", she said. "One cannot believe impossible things".

"I daresay you haven't had much practice", said the Queen . . . , "Why, sometimes I have believed as many as six impossible things before breakfast!"

LEWIS CARROLL, 1896

39 Path Integral Formulation of Field Theory

Until now, we have been concerned mainly with the perturbative aspects of QCD. For these, the use of either a canonical or path integral formulation of field theory is largely a matter of taste. Nonperturbative aspects of QCD, however, can be formulated with greater clarity in a functional language. In this section, we briefly review the Feynman path integral formalism, in particular as applied to field theory. Of course, this is no substitute for a detailed treatment, for which the interested reader may consult the lectures of Fadeyev (1976) and Lee (1976) or the textbooks of Itzykson and Zuber (1980) and Fadeyev and Slavnov (1980).

Let us start with the case of the nonrelativistic quantum mechanics, [Feynman and Hibbs (1965)], in one dimension. We have a Hamiltonian, $\hat{H}$, a function of $\hat{P}$, $\hat{Q}$ (we place carets over operators temporarily); we assume that it has been written in "normal form" with all $\hat{P}$s to the left of all $\hat{Q}$s. The classical Hamiltonian, H, may be obtained as

$$\langle p|\hat{H}|q\rangle = \frac{e^{-ipq}}{\sqrt{2\pi}} H(p,q), \tag{39.1}$$

where $\hat{P}|p\rangle = p|p\rangle$, $\hat{Q}|q\rangle = q|q\rangle$, $\langle p|q\rangle = e^{-ipq}/\sqrt{2\pi}$. We then evaluate the matrix elements of the evolution operator,

$$\langle q''|e^{-i(t''-t')\hat{H}}|q'\rangle. \tag{39.2}$$

To do so, we write

$$e^{-it\hat{H}} = \lim_{N\to\infty}\left(1 - \frac{it}{N}\hat{H}\right)^N, \qquad t = t'' - t',$$

and insert sums over complete sets of states:

$$\langle q''|e^{-it\hat{H}}|q'\rangle = \lim_{N\to\infty}\int\prod\frac{dp_n}{2\pi}\prod\frac{dq_n}{2\pi}\langle q''|p_N\rangle\langle p_N|1 - \frac{it}{N}\hat{H}|q_N\rangle$$

$$\times\langle q_N|p_{N-1}\rangle\langle p_{N-1}|1 - \frac{it}{N}\hat{H}|q_{N-1}\rangle\cdots\langle p_1|1 - \frac{it}{N}\hat{H}|q'\rangle.$$

Now, using (39.1), we find

$$\langle p_n|1 - \frac{it}{N}\hat{H}|q_n\rangle = \frac{\exp\{-ip_nq_n - (it/N)H(p_n,q_n)\}}{\sqrt{2\pi}} + O\left(\frac{1}{N^2}\right),$$

so

$$\langle q''|e^{-it\hat{H}}|q'\rangle = \lim_N\int\prod\frac{dp_n}{2\pi}\prod dq_n$$

$$\times\exp i\left\{p_N(q_N'' - q_N) + \cdots + p_1(q_1 - q')\right.$$

$$\left.- \frac{t}{N}(H(p_N, q_N)\ldots H(p_1,q'))\right\}. \qquad (39.3)$$

Feynman's trick consists of defining two functions, $p(t), q(t)$, with $p(t_n) = p_n$, $q(t_n) = q_n$, so we may replace the integrals

$$\prod_n\frac{dp_n}{2\pi}\to\prod_t\frac{dp(t)}{2\pi}, \qquad \prod_n\frac{dq_n}{2\pi}\to\prod_t\frac{dq(t)}{2\pi}, \qquad (39.4)$$

i.e., we now integrate over all *functions*, and the term in brackets in (39.3) becomes

$$\int_{t'}^{t''}dt\,\{p(t)\dot{q}(t) - H(p(t),q(t))\}, \qquad \dot{f}\equiv\frac{df}{dt}.$$

The entire (39.3) is then

$$\langle q''|e^{-it\hat{H}}|q\rangle = \int\prod_t\frac{dq(t)\,dp(t)}{2\pi}\exp i\int_{t',q'}^{t'',q''}dt\,(p\dot{q} - H). \qquad (39.5)$$

Of course, this expression is formal,[103] and it only makes sense as a limit of (39.3), but in this it is not so very different from the usual Riemann definition of an ordinary integral. The important thing about (39.5) is that only classical c-number functions enter: we have traded the complexities of operator calculus for those of functional integrations.

Equation (39.5) may be simplified. If $H = p^2/2m + V(q)$, then the integral over dp is Gaussian and can be explicitly evaluated. Shifting the

[103] See, however, Wiener (1923) for a rigorization of functional integrals similar to (39.5).

integration variables by $p \to p - m\dot{q}$,

$$\int \prod_t \frac{dp(t)}{2\pi} \exp i \int dt \left(p\dot{q} - \frac{p^2}{2m} \right)$$

$$= \int \prod_t \frac{dp(t)}{2\pi} \exp\left(-i \int dt \frac{p^2(t)}{2m} \right) \exp\left(i \int dt \frac{m\dot{q}^2(t)}{2} \right);$$

therefore,

$$\langle q''|e^{-i(t''-t')\hat{H}}|q'\rangle = N \int \prod_t dq(t) \exp i \int_{q',t'}^{q'',t''} dt\, L[q(t), \dot{q}(t)]. \qquad (39.6)$$

Here we have identified $m\dot{q}^2 - V$ with the (integrated) Lagrangian, and defined the normalization factor, independent of the dynamics,

$$N = \int \prod_t \frac{dp(t)}{2\pi} \exp\left\{ -i \int dt \frac{p^2(t)}{2m} \right\}.$$

The generalization of (39.6) to several degrees of freedom is obvious; let us write $q(t,k)$ instead of $q_k(t)$, $k = 1, \ldots, n$, to facilitate the transition to the field-theoretic case; and let us also introduce the Lagrangian (density), writing $L = \sum_k \mathscr{L}$. We get

$$\langle q''|e^{-i(t''-t')\hat{H}}|q'\rangle$$

$$= N \int \prod_{t,k} dq(t,k) \exp\left\{ i \int_{q',t'}^{q'',t''} dt \sum_k \mathscr{L}[q(t,k); \dot{q}(t,k)] \right\}. \qquad (39.7)$$

The generalization to the field-theoretic case is then straightforward. Considering, for the sake of simplicity, a theory with a single field, ϕ, the role of k is now played by $\vec{x}$. If we choose the state $|\phi(t,\vec{x})\rangle$ such that

$$\hat{\phi}(x)|\phi(x)\rangle = \phi(x)|\phi(x)\rangle,$$

then

$$\langle \phi(t'',\vec{x})|e^{-i(t-t')\hat{H}}|\phi(t',\vec{x}')\rangle = N \int \prod_x d\phi(x) \exp\left\{ i \int_{t'}^{t''} d^4x\, \mathscr{L}(\phi, \partial\phi) \right\}. \qquad (39.8)$$

Of course, just as in the ordinary quantum mechanical case, the functional integral has to be interpreted via a limiting procedure. Consider a large volume of space, V, and divide the four-dimensional volume ($t'' - t'$, V) into a finite number, n, of cells. Let x_j, $j = 1, \ldots, n$ be points inside each cell, and let δ be the four-dimensional volume of each cell. Then the right-hand side of (39.8) is defined as

$$\lim_{\substack{V\to\infty \\ n\to\infty \\ \delta\to 0}} \int d\phi(x_1) \ldots d\phi(x_n)\, e^{i\delta \sum_j \mathscr{L}[\phi(x_j), \partial\phi(x_j)]} \qquad (39.9)$$

(later we will see that the normalizing factor N can be disposed of when considering transition amplitudes). When evaluating S matrix elements or

Green's functions, we require VEVs $\langle T\phi(x) \ldots \phi(z)\rangle_0$. For this, we consider the vacuum-to-vacuum amplitude,

$$\langle 0|\hat{S}|0\rangle = \lim_{\substack{t' \to -\infty \\ t'' \to +\infty}} \langle 0|e^{-i(t''-t')\hat{H}}|0\rangle,$$

and obtain the Green's functions through the introduction of sources. According to (39.7)

$$\langle 0|\hat{S}|0\rangle = N\int \prod_x d\phi(x) \exp i\mathscr{A}, \qquad \mathscr{A} = \int d^4x \, \mathscr{L}; \tag{39.10}$$

$\mathscr{A}$ is the action. We add a source term to $\mathscr{L}$,

$$\mathscr{L}_\eta = \mathscr{L} + \eta(x)\phi(x), \qquad \mathscr{A}_\eta = \int d^4x \, \mathscr{L}_\eta ,$$

and define the *generating functional*

$$Z[\eta] = N\int \prod_x d\phi(x) \exp i\mathscr{A}_\eta \, . \tag{39.11}$$

We will see that from this it follows that

$$\left.\frac{\delta^n \log Z[\eta]}{\delta\eta(x_1) \ldots \delta\eta(x_n)}\right|_{\eta=0} = \frac{i^n \langle T\hat{\phi}(x_1) \ldots \hat{\phi}(x_n)\rangle_0}{\langle \hat{S}\rangle_0}, \tag{39.12}$$

where the right-hand side is the *connected Green's function*, which we have until now denoted simply by

$$\langle T\hat{\phi}(x_1) \ldots \hat{\phi}(x_n)\rangle_0 ,$$

absorbing the phase $\langle \hat{S}\rangle_0$ in the definition of the physical $\hat{S}$. We will prove (39.12) in the free-field case (for interactions, see below). The Lagrangian is thus

$$\mathscr{L} = \tfrac{1}{2}\partial_\mu\phi\,\partial^\mu\phi - \tfrac{1}{2}m^2\phi^2 = -\tfrac{1}{2}\phi\{\partial^2 + m^2\}\phi + \text{four-divergence}.$$

The trick lies in reducing the integral to a Gaussian. For this, define ϕ' so that

$$\phi'(x) = (\partial^2 + m^2)^{1/2}\phi(x),$$

which is valid provided

$$\begin{aligned} \phi'(x) &= \int d^4x \, K^{-1/2}(x-y)\phi(y), \\ K(z) &= \frac{-1}{(2\pi)^4}\int d^4k \frac{e^{ik\cdot z}}{(k^2 - m^2 + i0)} = i\Delta(z). \end{aligned} \tag{39.13}$$

The $+i0$ prescription guarantees that we will obtain time-ordered products. Then,

$$\begin{aligned} Z[\eta] = N\int \prod_x d\phi'(x) \det(\partial\phi/\partial\phi') \\ \times \exp i\int d^4x \left\{ \frac{-1}{2}\phi'(x)\phi'(x) + \int d^4y \, \eta(x)K^{1/2}(x-y)\phi'(y)\right\}; \end{aligned}$$

$\det(\partial\phi/\partial\phi')$ is the (infinite dimensional) Jacobian of the change of variables. The final step is a shift of the integration variable

$$\phi'(x) = \phi''(x) + \int d^4y\, K^{1/2}(x-y)\eta(y)$$

so that

$$Z[\eta] = \left\{ N \int \prod_x d\phi''(x) \det(\partial\phi/\partial\phi'') e^{-i\int d^4x\, \phi''^2/2} \right\} \times e^{(i^2/2)\int d^4x\, d^4y\, \eta(x)\Delta(x-y)\eta(y)}, \tag{39.14}$$

where $\Delta(x-y)$ is the propagator

$$\Delta(x) = \frac{i}{(2\pi)^4} \int d^4k \frac{e^{-ik\cdot x}}{k^2 - m^2 + i0} = \langle T\phi(x)\phi(0)\rangle_0 .$$

The term in braces in the right-hand side of (39.14) is independent of η: hence, it will cancel when taking the logarithmic derivative. So we may write

$$Z[\eta] = \bar{N} \exp\left\{ \frac{i^2}{2} \int d^4x\, d^4y\, \eta(x)\Delta(x-y)\eta(y) \right\}, \tag{39.15}$$

from which (39.12) may be checked directly.

The introduction of vector fields presents no problem; likewise, operator insertions are dealt with by the introduction of extra sources (an example will be found in Section 42); only fermion fields require some elaboration. These necessitate the introduction, at the classical level, of *anticommuting c-numbers*,[104] defined by the relations

$$\psi(x)\psi(y) = -\psi(y)\psi(x), \qquad [\psi(x)]^2 = 0.$$

A functional of (classical) fermion fields will be of the general form

$$F[\psi] = K_0 + \int dx_1\, K_1(x_1)\psi(x_1) + \cdots + \int dx_1 \ldots dx_n\, K_n(x_1, \ldots, x_n)\psi(x_1) \ldots \psi(x_n) + \cdots,$$

where K_1 is an anticommuting function, and the K_n, $n \geqslant 2$, may be taken as fully antisymmetric in their arguments. The extension of the definition

$$\frac{\delta F[\psi]}{\delta\psi(x)} = \lim_{\epsilon\to 0} \frac{F[\psi + \epsilon\delta_x] - F[\psi]}{\epsilon},$$

where ϵ is an anticommuting number,

$$\epsilon\psi = -\psi\epsilon, \qquad \epsilon^2 = 0,$$

[104]The corresponding structure is known as a Grassmann algebra in the standard mathematical literature. More details may be found in the treatise of Berezin (1966).

yields

$$\left.\frac{\delta^n F[\psi]}{\delta\psi(x_n)\ldots\delta\psi(x_1)}\right|_{\psi=0} = n!\,K_n(x_1,\ldots,x_n).$$

Note the reversed order of the x; this is so because

$$\frac{\delta^2}{\delta\psi_1\delta\psi_2} = -\frac{\delta^2}{\delta\psi_2\delta\psi_1}.$$

The integration over anticommuting functions also presents peculiarities, because in order to be consistent, we have to define

$$\int d\psi(x) = 0, \qquad \int d\psi(x)\psi(y) = \delta(x-y).$$

Finally, if we want to generate *one particle irreducible* (1PI) Green's functions, i.e., Green's functions that remain connected when cutting one internal line, we do so by functional differentiation with respect not to η, but to the new field $\bar{\phi}$, a functional $\Gamma[\bar{\phi}]$:

$$\Gamma[\bar{\phi}] = \frac{1}{i}\log Z[\eta] - \int d^4x\,\eta(x)\bar{\phi}(x), \tag{39.16a}$$

$$\bar{\phi}(x) \equiv \frac{-i\delta\log Z[\eta]}{\delta\eta(x)}. \tag{39.16b}$$

Note that $\bar{\phi}$ is the VEV of $\hat{\phi}$.

The proof that Γ generates 1PI Green's functions is apparent from an identity that we now prove. Differentiating Γ twice,

$$\frac{\delta^2\Gamma}{\delta\bar{\phi}(x)\,\delta\bar{\phi}(y)} = -\frac{\delta\eta(x)}{\delta\bar{\phi}(y)} = \left[-\frac{\delta\bar{\phi}(y)}{\delta\eta(x)}\right]^{-1} = -i\Delta^{-1}(x-y),$$

so that, in particular, $\Delta\{\delta^2\Gamma/\delta\bar{\phi}(x)\delta\bar{\phi}(y)\}\Delta = i\Delta$: up to an i, the propagator is obtained by dressing the 1PI Green's function with propagators. More generally,

$$\frac{\delta}{\delta\bar{\phi}} = \left[\frac{\delta\eta}{\delta\bar{\phi}}\right]\frac{\delta}{\delta\eta} = -i\Delta^{-1}(x-y)\frac{\delta}{\delta\eta}, \tag{39.17}$$

which is the required equation.

40 The WKB Approximation in the Path Integral Formalism: Tunnelling

In ordinary quantum mechanics, the WKB approximation consists of expanding in powers of Planck's constant, $\hbar$. To zero order, we have the classical trajectory; higher orders yield the quantum corrections to this trajectory. The path integral formulation lends itself particularly well to the extension of the method to the field-theoretic case. To accomplish this, we reintroduce $\hbar$ into (39.11).

$$Z[\eta] = \int \prod_x d\phi(x) \exp \frac{i}{\hbar} \mathscr{A}_\eta[\phi], \tag{40.1}$$

write

$$\phi(x) = \phi_{\text{cl}}(x) + \hbar^{1/2}\tilde{\phi}(x) + \cdots, \qquad \pi(x) = \partial_0\phi_{\text{cl}}(x) + \hbar^{1/2}\tilde{\pi}(x) + \cdots, \tag{40.2}$$

and match the powers of $\hbar$. The term ϕ_{cl} is the solution of the *classical* equation of motion,

$$\partial^2\phi_{\text{cl}} + m^2\phi_{\text{cl}} = \left.\frac{\partial \mathscr{L}_{\text{int}}}{\partial\phi}\right|_{\phi=\phi_{\text{cl}}}, \tag{40.3a}$$

or, equivalently,

$$\phi_{\text{cl}}(x) = \phi_0(x) + i\int d^4y\, \Delta(x-y) \left.\frac{\partial \mathscr{L}_{\text{int}}}{\partial\phi}\right|_{\phi=\phi_{\text{cl}}}, \tag{40.3b}$$

where ϕ_0 is a free classical field, $(\partial^2 + m^2)\phi_0 = 0$. Because ϕ_{cl} satisfies the equation of motion, $\mathscr{A}[\phi_{\text{cl}}]$ is stationary: we are expanding (40.1) around the stationary phase. The zero-order approximation yields the tree approximation; higher orders correspond to an approximation in the number of loops. The usefulness of the method lies in the fact that, to each order, the functional integral is of Gaussian type and can therefore be evaluated. Let us show this for the first correction. To order $\hbar$,

$$\mathscr{A} = \mathscr{A}[\phi_{\text{cl}}] - \frac{1}{2}\int d^4x \left\{ \tilde{\phi}(x)(\partial^2 + m^2)\tilde{\phi}(x) - \left.\frac{\partial^2\mathscr{L}_{\text{int}}(\phi)}{\partial\phi^2}\right|_{\phi=\phi_{\text{cl}}} \tilde{\phi}(x)\tilde{\phi}(x) \right\}.$$

Next we perform the change of variables

$$\tilde{\phi}(x) \to \phi'(x) = \left\{ \partial^2 + m^2 - \frac{\partial^2\mathscr{L}_{\text{int}}}{\partial\phi^2} \right\}^{1/2} \tilde{\phi}(x),$$

and thus find

$$Z = (\text{constant})\exp\left\{ -\frac{1}{2}\operatorname{Tr}\log\left[1 - (\partial^2 + m^2)^{-1} \left.\frac{\partial^2\mathscr{L}_{\text{int}}}{\partial\phi^2}\right|_{\phi=\phi_{\text{cl}}} \right] \right\} Z_{\text{tree}}, \tag{40.4a}$$

where, using (40.3) and the relation $i(\partial^2 + m)\Delta(x) = \delta(x)$, we can write

$$Z_{\text{tree}} = N \exp \frac{i}{\hbar} \left\{ \int d^4x\, \mathscr{L}_{\text{int}}(\phi_{\text{cl}}) - \frac{i}{2}\int d^4x\, d^4y \left.\frac{\partial\mathscr{L}_{\text{int}}}{\partial\phi(x)}\right|_{\phi=\phi_{\text{cl}}} \times \Delta(x-y) \left.\frac{\partial\mathscr{L}_{\text{int}}}{\partial\phi(y)}\right|_{\phi=\phi_{\text{cl}}} \right\}. \tag{40.4b}$$

The constant in (40.4a) contains the term

$$\int \prod d\phi'(x)\, e^{-(i/2)\int dx\, \phi'^2(x)} \det(\partial^2 + m^2)^{1/2},$$

and we have used the relation

$$\det(A^{-1/2}) = \exp\left\{ \frac{-1}{2} \operatorname{Tr} \log A \right\}.$$

It is known that there are quantum mechanical situations for which no classical trajectory exists. This occurs whenever there is tunnelling through a potential barrier. However, one can still adopt the WKB method to cope with this situation. We will exemplify this with the typical case of a particle in one dimension, subject to a potential $V(x)$. The wave function is, in the WKB approximation [see, e.g., Landau and Lifschitz (1958)]

$$\psi(x) = Ce^{i\mathscr{A}(x)}, \tag{40.5}$$

where $\mathscr{A}$ is the action calculated along the classical trajectory

$$\tfrac{1}{2} m\ddot{x} + V(x) = E. \tag{40.6}$$

Take a potential with two minima, corresponding to $V = 0$ and located at $x = x_0, x_1$ (Figure 30a). If $E > \max V$, the motion from x_0 to x_1 is possible, and (40.5) yields the "diffusion" amplitude. However, if $E < \max V$, the correct WKB matching gives a result in which the transition amplitude

$$\langle x_1 | x_0 \rangle = Ce^{i\mathscr{A}(x_2, x_0)} \tag{40.7}$$

is to be replaced by the *tunnelling amplitude*

$$\langle x_1 | x_0 \rangle = Ce^{-\mathscr{A}(x_1, x_0)}, \tag{40.8}$$

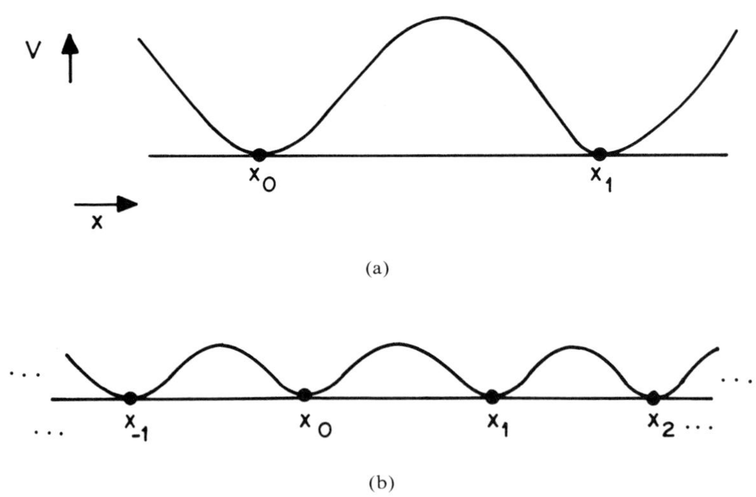

Figure 30. Potentials with multiple minima. (a), two minima; (b), periodic case.

where $\underline{\mathscr{A}}$ is not calculated along the solution of (39.2), but for

$$-\tfrac{1}{2}m\ddot{x} + V(x) = E. \tag{40.9}$$

We see that to obtain a tunnelling amplitude we can use the same formula as that for a transition, making only the formal replacement of t by it both in the expression for the action

$$\mathscr{A} = \int_{t(x_0)}^{t(x_1)} dt\, L \rightarrow i\underline{\mathscr{A}}$$

and in the equations of motion, (40.6)–(40.9).

Equations (40.5) and (40.8) do not give the normalization which may, however, be readily obtained by dividing by $\langle x_0 | x_0 \rangle$. Therefore, we infer that in quantum field theory, the leading tunnelling amplitude will be

$$\langle \Psi_1, t = +\infty | \Psi_0, t = -\infty \rangle \approx C \exp\left\{ -\int d^4\underline{x}\, \underline{\mathscr{L}}(\underline{\phi}_{\text{cl}}) \right\}, \tag{40.10}$$

where $\underline{\phi}_{\text{cl}}$ is the field solution to the Euclidean equations of motion, i.e., with x^0 replaced by ix_4, x_4 real.

According to the discussion at the beginning of this section, we may consider this to be the leading order of the exact expression,

$$\langle \Psi_1, t = +\infty | \Psi_0, t = -\infty \rangle = N \exp\left\{ -\int d^4\underline{x}\, \underline{\mathscr{L}}(\underline{\phi}) \right\}, \tag{40.11}$$

when expanding the field $\underline{\phi}$ in powers of $\hbar$ around $\underline{\phi}_{\text{cl}}$.

An important property of the states of a system in a situation when tunnelling is possible is that the stationary states (in particular the ground state, to be identified with the vacuum in field theory) are not those in which the system is localized in one minimum of the potential, but is shared by all minima. We will see an example of this in QCD for a periodic case like that of Figure 30*b*.

41 Functional Formalism for QCD: Gauge Invariance

The formalism developed in the preceding sections may be applied directly to QCD, provided we first tackle the question of gauge invariance. A possibility is to choose a physical gauge

$$u \cdot B_a(x) = 0, \qquad u^2 \leqslant 0, \tag{41.1}$$

so we have to integrate over all B compatible with (41.1), i.e., we define, with N as an arbitrary normalization factor,

$$Z = N \int (dq)(d\bar{q})(dB) \prod_{a,x} \delta(u \cdot B_a(x)) \exp i \int d^4x\, \mathscr{L}_u, \tag{41.2}$$

where we have introduced the notation, to be used systematically, (dq)

$\equiv \Pi_{x,f,i,\alpha} dq^i_{f\alpha}(x)$, $(dB) \equiv \Pi_{x,\mu,a} dB^\mu_a(x)$, etc. and $\mathscr{L}_u$ is the QCD Lagrangian (without the gauge-fixing term). This is all we need if we want to work with a physical gauge. However, we will also want to extend the formalism to other gauges, particularly covariant ones. We may write the gauge condition as

$$K_a[B(x)] = 0, \tag{41.3}$$

where K is the gauge-fixing functional. For example, the Lorentz gauges are

$$K_a[B(x)] = \partial_\mu B^\mu_a(x) - \phi_a(x), \tag{41.4}$$

where ϕ is a prescribed function (in particular, we could take $\phi = 0$).

Let $T(\theta)$ be a gauge transformation with parameters $\theta(x)$, and let B_T be the transform of B under T:

$$B^\mu_{Ta}(x) = B^\mu_a(x) + g\sum f_{abc}\theta_b(x)B^\mu_c(x) - \partial^\mu\theta_a(x)$$

(cf. Section 3). The quantity

$$\Delta_K^{-1}[B] = \int \prod_{x,a} d\theta_a(x) \prod_{x,a} \delta(K_a[B_T(x)]) \tag{41.5}$$

is independent of the gauge

$$\Delta_K^{-1}[B_{T'}] = \Delta_K^{-1}[B].$$

The proof only requires the fact that the integration element $\prod_{x,a} d\theta_a(x)$ is a gauge-invariant measure. This is obvious for infinitesimal θ (all we really need) because then

$$T(\theta)T(\theta') = T(\theta + \theta').$$

Let us temporarily neglect the quarks, which play no role in gauge shifts, We may rewrite (41.2) as

$$Z = N\int (dB)(d\theta) \prod \delta(u \cdot B_a(x)) \prod \delta(K_b[B_T])\Delta_K[B_T]e^{i\mathscr{A}_{\mathrm{YM}}}, \tag{41.6}$$

where $\mathscr{A}_{\mathrm{YM}}$ is the pure Yang–Mills action,

$$\mathscr{A}_{\mathrm{YM}} = -\frac{1}{4}\int d^4x \sum G_{a\mu\nu}(x)G^{\mu\nu}_a(x).$$

Suppose we change variables in (40.6) via a gauge transformation,

$$B(x) \to B_{T_0}(x),$$

choosing $T_0 = T^{-1}$. We find

$$Z = N\int (dB)(d\theta)\Delta_K[B] \prod \delta(u \cdot B_{T_0 a}(y)) \prod \delta(K[B(y)])e^{i\mathscr{A}_{\mathrm{YM}}}.$$

Let B_u be the gluon fields that verify (41.1). We may find the B_{T_0} by a gauge transformation, $U(\theta_u)$. Then,

$$\delta(u \cdot B_{T_0}) = \delta(u \cdot B_{uU}),$$

and thus

$$\int (d\theta) \prod \delta(u \cdot B_{T_0 a}(y)) = \int (d\theta) \prod \delta(-u \cdot \partial^\mu \theta_{ua}(y)),$$

which is independent of B and can therefore be absorbed into the normalization factor N. We have obtained

$$Z = N' \int (dB) \Delta_K[B] \prod \delta(K[B]) e^{i\mathscr{A}_{\mathrm{YM}}}. \tag{41.7}$$

We have to eliminate the δ function and evaluate Δ_K. For the former, choose, for example, a Lorentz gauge (41.4); integrating (41.7) over $d\phi$ with the weight

$$\exp\left\{ \frac{-i\lambda}{2} \int d^4x \left[\phi_a(x)\right]^2 \right\},$$

we obtain, on the left-hand side, Z times a factor independent of B, namely,

$$\int (d\phi) \exp\left\{ \frac{-i\lambda}{2} \int d^4x \left[\phi_a(x)\right]^2 \right\},$$

which can again be lumped into N', and on the right-hand side, the integral over $d\phi$ may be carried trivially with the help of the δ-function:

$$Z = N'' \int (dB) \Delta_K[B] e^{i(\mathscr{A}_{\mathrm{YM}} + \mathscr{A}_{\mathrm{GF}})}, \tag{41.8}$$

where the gauge-fixing action is

$$\mathscr{A}_{\mathrm{GF}} = \frac{-\lambda}{2} \int d^4x \left[\partial_\mu B_a^\mu(x)\right]^2.$$

Let us then turn to Δ_K. Because of Equation (41.7), we only require B's that verify (41.3). Thus, for infinitesimal θ, $K[B_T] = K[B] + (\delta K/\delta B)\delta B \sim (\delta K/\delta B)\delta B$, $\delta B = B_T - B$, so that

$$\Delta_K^{-1}[B] = \int (d\theta) \Pi \delta\left(\frac{\delta(\partial B_a)}{\delta B_b^\mu} \left(\partial^\mu \theta_b - g \sum f_{bcd} B_a^\mu \theta_c\right)\right).$$

This may be cast in a more convenient form by introducing the Fadeyev–Popov ghost fields ω, $\bar{\omega}$ as anticommuting c-number functions, for then, with $\bar{N}$ a number independent of B and ω,

$$\Delta_K[B] = \bar{N} \int (d\omega)(d\bar{\omega}) \times \exp\left\{ -i \int d^4x\, d^4y\, \bar{\omega}_a(y) \frac{\delta(\partial B_a)}{\delta B_b^\mu} \left[\partial^\mu \omega_b(x) - g \sum f_{bcd} B_d^\mu \omega_c(x)\right]\right\}. \tag{41.9}$$

The proof is based on the formula

$$\int \prod_i d\bar{c}_i \prod_j dc_j\, e^{\Sigma \bar{c}_k A_{kk'} c_{k'}} = (\text{constant}) \det A,$$

which is valid[105] for anticommuting c-numbers c_j, and on the fact that due to the relation

$$\int dx_1 \ldots dx_k \prod_{i=1}^{k} \delta(f_i(x_1, \ldots, x_k)) = \frac{1}{\det(\partial f_i/\partial x_j)},$$

Δ_K is simply the determinant of the (infinite) matrix

$$\frac{\partial}{\partial \theta}\left\{ \frac{\delta(\partial B_a)}{\delta B_b^\mu}\left(\partial^\mu \theta_b - g \sum f_{bcd} B_d^\mu \theta_c\right)\right\}.$$

There only remains one additional step to complete the analysis. The functional derivative entering (41.9) is (cf. Appendix H)

$$\frac{\delta(\partial B_a(x))}{\delta B_b^\mu(y)} = \delta_{ab}\, \partial_\mu \delta(x-y),$$

so we may transfer the ∂_μ to the left-hand side and integrate d^4y. We finally obtain

$$Z = \bar{N} \int (dB)(d\omega)(d\bar{\omega})\, e^{i(\mathscr{A}_{\text{YM}} + \mathscr{A}_{\text{GF}} + \mathscr{A}_{\text{FP}})}, \tag{41.10a}$$

where the Fadeyev–Popov ghost action is

$$\mathscr{A}_{\text{FP}} = \int d^4x \sum (\partial_\mu \bar{\omega}_a(x))\left[\delta_{ab}\,\partial^\mu - g f_{abc} B_c^\mu(x)\right]\omega_b(x), \tag{41.10b}$$

in agreement with the result we obtained in Section 5 using unitarity.

To generate Green's functions, we have to introduce anticommuting sources $\bar{\eta}_a, \eta_a; \bar{\xi}_{if}, \xi_{if}$ for the ghost $\omega_a, \bar{\omega}_a$ and quark $q_f^i, \bar{q}_f^i$ fields, and commuting sources λ_a^μ for the gluons B_a^μ. Thus, our starting point will be the functional

$$Z[\eta, \bar{\eta}; \xi, \bar{\xi}; \lambda] = \int (dq)(d\bar{q})(d\omega)(d\bar{\omega})(dB) \exp i \int d^4x \left\{\mathscr{L}_{\text{QCD}}^{\xi} + \mathscr{L}_{\text{source}}\right\}, \tag{41.11a}$$

where $\mathscr{L}_{\text{QCD}}^{\xi}$ is given in Equation (5.11) and

$$\mathscr{L}_{\text{source}} = \sum \left\{\bar{\eta}_a \omega_a + \bar{\omega}_a \eta_a + \bar{\xi}_{if} q_f^i + \bar{q}_f^i \xi_{if} + \lambda_{a\mu} B_a^\mu\right\}. \tag{41.11b}$$

[105] PROOF.

$$\int \prod_{i=1}^{N_0} dc_i \prod_{j=1}^{N_0} d\bar{c}_j e^{\sum \bar{c}_k A_{kk'} c_{k'}} = \int \prod_{i=1}^{N_0} dc_i \prod_{j=1}^{N_0} d\bar{c}_j \sum_{N=0}^{\infty} \left\{\sum \bar{c}_k c_{k'} A_{kk'}\right\}^N \frac{1}{N!}.$$

Due to the rules for integrating "fermion" variables, only the term with $N = N_0$ will not vanish; and there we obtain

$$\frac{(-1)^{N_0}}{N_0!} \sum \operatorname{sign}(k_1, \ldots, k_{N_0}) \operatorname{sign}(k_1', \ldots, k_{N_0}') A_{k_1 k_1'} \ldots A_{k_{N_0} k_{N_0}'},$$

where the sum is extended over all permutations $k_1, \ldots, k_{N_0}; k_1', \ldots, k_{N_0}'$ of $1, 2, \ldots, N_0$. This is nothing but $(-1)^{N_0} \det A / N_0!$. The extra $-i$ of the exponent in (41.9) only contributes an overall factor; this means the phase of the FP term is arbitrary. We have chosen it to agree with that of conventional scalar fields.

The functional formalism allows us to introduce a very elegant method, the so-called *background field method*,[106] which presents the advantage that the effective action (Section 39) is gauge invariant. It amounts to taking the condition $K[B] = \sum(\partial_\mu B_a^\mu + g\sum f_{adc} b_d^\mu B_{c\mu})^2$ to be gauge fixing, where b is a classical "background" field, shifting B to $B + b$, and computing functional derivatives with respect to b. The details and references may be found in the paper of Abbott (1981).

42 Feynman Rules

In Section 39 we stated that the expansion of the Green's functions generated by Equation (41.11) in powers of g reproduces the usual Feynman rules which were previously obtained with Wick's theorem and the decomposition of field operators in creation-annihilation operators. Alternatively, we could have derived the Feynman rules from (41.11). We will exemplify this with three typical cases: the gluon propagator, the ghost-gluon vertex, and the nonsinglet composite operators of deep inelastic scattering.

With respect to the first, we consider

$$\langle T\hat{B}_a^\mu(x)\hat{B}_b^\nu(y)\rangle_0|_{g=0} = (-i)^2 \frac{\delta^2 \log Z}{\partial\lambda_{a\mu}(x)\,\delta\lambda_{b\nu}(y)}\Bigg|_{\substack{\text{source}=0\\ g=0}}. \tag{42.1}$$

We have returned to the use of carets to denote operators. Imitating the discussion of Section 39, we write, up to a four-divergence, and with $\lambda = a^{-1}$ for the gauge parameter,

$$\begin{aligned}
&\frac{-1}{4}\sum\left[\partial^\rho B_a^\sigma(x) - \partial^\sigma B_a^\rho(x)\right]\left[\partial_\rho B_{a\sigma}(x) - \partial_\sigma B_{a\rho}(x)\right] - \frac{a^{-1}}{2}\sum\left[\partial_\tau B_a^\tau(x)\right]^2\\
&\quad= \frac{1}{2}\sum B_a^\sigma(x)\left[\partial^2 B_{a\sigma}(x) - (1-a^{-1})\,\partial_\sigma\partial^\rho B_{a\rho}(x)\right] + \partial_\mu f^\mu,\\
&\quad= \frac{1}{2}\sum_{a,b} B_{a\sigma}(x)(K^{-1})_{ab}^{\sigma\rho} B_{b\rho}(x) + \partial_\mu f^\mu,
\end{aligned}$$

where

$$(K^{-1})_{ab}^{\sigma\rho} = \delta_{ab}\left\{g^{\sigma\rho}\frac{\partial^2}{\partial x^2} - (1-a^{-1})\frac{\partial}{\partial x_\sigma}\frac{\partial}{\partial x_\rho}\right\}. \tag{42.2}$$

Thus, setting η, $\bar{\eta}$, ξ, $\bar{\xi}$, and g in (41.11) to zero,

$$\begin{aligned}
Z = &\int (dq)(d\bar{q})(d\omega)(d\bar{\omega})(dB)\\
&\times \exp i\int d^4x\left\{i\bar{q}(x)\not{\partial} q(x) + \frac{1}{2}\sum B_{a\sigma}(x)(K^{-1})_{ab}^{\sigma\rho}B_{b\rho}(x)\right.\\
&\qquad\left. + \sum \lambda_{a\mu}(x)B_a^\mu(x)\right\}.
\end{aligned} \tag{42.3}$$

[106] This method was first introduced by B. De Witt, and extended (in particular, to gauge theories) by G. 't Hooft.

The integrals over q, $\bar{q}$, ω, and $\bar{\omega}$ yield a constant that will drop when the logarithmic derivative is taken. If we change variables,

$$B \to B' = K^{-1/2}B,$$

(42.3) becomes

$$Z = (\text{constant}) \int (dB')J(K)$$
$$\times \exp i \int d^4x \sum \left\{ \frac{1}{2} B'_{a\mu}(x) B'^{\mu}_{a}(x) + \lambda_{a\mu}(x)(K^{1/2}B')^{\mu}_{a}(x) \right\},$$

where $J(K)$ is the Jacobian. Finally, we displace the integration variable

$$B' \to B'' = B' + K^{1/2}\lambda,$$

so that

$$Z = (\text{constant}) \int (dB'')J(K)$$
$$\times \exp i \int d^4x \sum \left\{ \frac{1}{2} B''_{a\mu}(x) B''^{\mu}_{a}(x) - \frac{1}{2} \lambda_{a\mu}(x)(K\lambda)^{\mu}_{a}(x) \right\}. \quad (42.4a)$$

It is convenient to write K in integral form

$$(Kf)^{\mu}_{a}(x) = -i \sum \int d^4y \, D^{\mu\nu}_{ab}(x-y) f_{b\nu}(y); \quad (42.4b)$$

we then obtain

$$\left. \frac{\delta^2 \log Z}{\delta\lambda_{a\mu}(x)\delta\lambda_{b\nu}(y)} \right|_{\substack{\text{sources}=0 \\ g=0}} = -D^{\mu\nu}_{ab}(x-y)$$

The form of D is obtained from its definition. It is such that

$$\widetilde{(K^{-1}f')}^{\mu}_{a}(x) = \sum \delta_{ab} \left\{ g^{\mu\nu}\partial^2 - (1-a^{-1})\partial^{\mu}\partial^{\nu} \right\} f'_{b\nu}(x),$$

so, with the help of the Fourier transform which we denote by a tilde,

$$(K^{-1}f')^{\mu}_{a}(k) = \sum \delta_{ab} \left\{ -g^{\mu\nu}k^2 + (1-a^{-1})k^{\mu}k^{\nu} \right\} \tilde{f}'_{b\nu}(k);$$

letting $\widetilde{Kf'} = f$, this immediately yields

$$(\widetilde{Kf})^{\mu}_{a}(k) = \sum \delta_{ab} \frac{-g^{\mu\nu} + (1-a)k^{\mu}k^{\nu}/k^2}{k^2} \tilde{f}_{b\nu}(k),$$

and therefore we see that

$$\langle T\hat{B}^{\mu}_{a}(x)\hat{B}^{\nu}_{b}(y)\rangle_0 \big|_{g=0} = D^{\mu\nu}_{ab}(x-y)$$
$$= \delta_{ab} \frac{i}{(2\pi)^4} \int d^4k \, e^{-ik\cdot(x-y)} \frac{-g^{\mu\nu} + (1-a)k^{\mu}k^{\nu}/k^2}{k^2},$$
$$a = \lambda^{-1}, \quad (42.5)$$

as was expected. The proof that the poles in (42.5) have to be circumvented with the $+i0$ prescription requires either a consideration of asymptotic states or any other similar boundary condition; it may be found in the lectures of Fadeyev (1976).

For the vertex, we require

$$\langle T\hat{\bar{\omega}}_a(x_1)\hat{\omega}_b(x_2)\hat{B}_c^{\mu}(x_3)\rangle_0\Big|_{\text{order } g} = \frac{i\delta^3 \log Z}{\delta\eta_a(x_1)\delta\bar{\eta}_b(x_2)\delta\lambda_{c\mu}(x_3)}\Bigg|_{\substack{\text{sources}=0\\ \text{first order } g}}. \tag{42.6}$$

Let us now denote by k the Klein–Gordon operator, $kf(x) = \partial^2 f(x)$. We perform the changes of variables $B \to B' = K^{-1/2}B$, $\omega \to \omega' = k^{-1/2}\omega$, $\bar{\omega} \to \bar{\omega}' = k^{-1/2}\bar{\omega}$, and integrate the quarks, which play no role here. Then Z becomes

$$Z = (\text{constant})\int (d\omega')(d\bar{\omega}')(dB')J(K)J(k)$$
$$\times \exp i\int d^4x \sum \Big\{ g\big[\partial_\mu (k^{1/2}\bar{\omega}')_a(x)\big] f_{abc}(K^{1/2}B')_c^{\mu}(x)(k^{1/2}\omega')_b(x)$$
$$+\tfrac{1}{2}B'^2 - \bar{\omega}'\omega + \bar{\eta}_a(x)(k^{1/2}\omega')_a(x)$$
$$+ (k^{1/2}\bar{\omega}')_a(x)\eta_a(x) + \lambda_a^{\mu}(x)(K^{1/2}B')_{a\mu}(x) + \cdots \Big\}$$

where the dots represent terms that will vanish for $g^2 \approx 0$, sources $= 0$. Next, we translate:

$$B' \to B'' = B' - K^{1/2}\lambda, \qquad \omega' \to \omega'' = \omega' + k^{1/2}\eta, \qquad \bar{\omega}' \to \bar{\omega}'' = \bar{\omega}' + k^{1/2}\bar{\eta}.$$

The only term that will yield a contribution is the one containing a product of all three sources. This is

$$g\sum (\partial_\mu (k\bar{\eta})_a(x)) f_{abc}(K\lambda)_c^{\mu}(x)(k\eta)_b(x),$$

and therefore

$$\langle T\hat{\bar{\omega}}_a(x_1)\hat{\omega}_b(x_2)\hat{B}_c^{\mu}(x_3)\rangle_0\Big|_{\text{order } g}$$
$$= \int \frac{d^4p_1}{(2\pi)^4} e^{-ix_1\cdot p_1}\frac{i}{p_1^2}\int \frac{d^4p_2}{(2\pi)^4} e^{-ix_2\cdot p_2}\frac{i}{p_2^2}\int \frac{d^4p_3}{(2\pi)^4} e^{-ix_3\cdot p_3}$$
$$\times i\frac{-g^{\mu\nu} + (1-\lambda^{-1})p_3^{\mu}p_3^{\nu}/p_3^2}{p_3^2}(2\pi)^4\delta(p_1+p_2+p_3)\, gf_{cba}p_{1\nu},$$

again, as expected.

Finally, we consider the vertex

$$\langle T\hat{\bar{q}}_1(x_1)\hat{N}_{NS}^{\mu_1\ldots\mu_n}(x_2)\hat{q}_2(x_3)\rangle_0 \tag{42.7}$$

to order zero in g and where (cf. Section 19)

$$\hat{N}_{NS}^{\mu_1\ldots\mu_n}(x) = \tfrac{1}{2}i^{n-1}\mathscr{S}\ :\hat{\bar{q}}_2(x)\gamma^{\mu_1}\hat{D}^{\mu_2}\ldots\hat{D}^{\mu_n}\hat{q}_1(x):-\text{traces}. \tag{42.8}$$

To calculate (42.7), we introduce into (41.11) a new source term,

$$j_{\mu_1 \dots \mu_n} N_{NS}^{\mu_1 \dots \mu_n}(x),$$

so that

$$\left\langle T\hat{\bar{q}}_1(x_1)\hat{N}_{NS}^{\mu_1\dots\mu_n}(x_2)\hat{q}_2(x_3)\right\rangle_0 = \left.\frac{i\delta^3 \log Z}{\delta\xi_1(x_1)\delta\bar{\xi}_2(x_3)\delta j_{\mu_1\dots\mu_n}(x_2)}\right|_{\substack{g=0\\ \text{sources}=0}}. \tag{42.9}$$

To zero order in g, the gluons or ghosts play no role and can be eliminated through integration. Similarly, the covariant derivatives of N may be replaced by ordinary derivatives. The quark fields are treated in the same way we treated the gluon or ghost fields. Using the definitions

$$q_f' = S^{-1/2}q_f, \qquad \bar{q}_f' = \bar{q}_f\bar{S}^{-1/2}, \qquad f = 1,2,$$

where

$$S^{-1}q_f(x) = \not\partial q_f(x), \qquad \bar{q}_f(x)\bar{S}^{-1} = \bar{q}_f(x)\overleftarrow{\not\partial},$$

we find, to zero order in g,

$$\begin{aligned} Z = (\text{constant})\int (dq)\,(d\bar{q})\,J(S)J(\bar{S})& \\ \times \exp i\int d^4x\,\Big\{\bar{q}_1'q_1' + \bar{q}_2'q_2 + \bar{\xi}_1 S^{1/2}q_1' + \bar{\xi}_2 S^{1/2}q_2'& \\ + (\bar{q}_1'\bar{S}^{1/2})\xi + (\bar{q}_2'\bar{S}^{1/2})\xi_2 + (\bar{S}^{1/2}N_{NS}'^{\mu_1\dots\mu_n}S^{1/2})j_{\mu_1\dots\mu_n}\Big\}.& \end{aligned} \tag{42.10}$$

Then we shift:

$$q_f'' = q_f' + S^{1/2}\xi_f, \qquad \bar{q}_f'' = \bar{q}_f' + \bar{\xi}_f\bar{S}^{1/2}.$$

The only term that contains all three sources ξ_1, ξ_2, and j is

$$\begin{aligned} &\bar{S}N_{NS}^{\mu_1\dots\mu_n}Sj_{\mu_1\dots\mu_n} \\ &\equiv \tfrac{1}{2}i^{n-1}\Big\{\mathscr{S}(\bar{\xi}_2\bar{S}^{-1})(x)\gamma^{\mu_1}\overleftrightarrow{\partial}^{\mu_2}\dots\overleftrightarrow{\partial}^{\mu_n}(S^{-1}\xi_1)(x) - \text{traces}\Big\}j_{\mu_1\dots\mu_n}(x), \end{aligned}$$

so that, using the explicit expression for S, we find

$$\begin{aligned} &\left\langle T\hat{\bar{q}}_1(x_1)\hat{N}_{NS}^{\mu_1\dots\mu_n}(x_2)\hat{q}_2(x_3)\right\rangle_0 \\ &= \int\frac{d^4p_2}{(2\pi)^4}e^{-ip_2\cdot x_2}\int\frac{d^4p_1}{(2\pi)^4}e^{-ip_1\cdot x_1}\frac{i}{\not p_1}\,\frac{1}{2}\{\mathscr{S}\gamma^{\mu_1}p_3^{\mu_2}\dots p_3^{\mu_n} - \text{traces}\} \\ &\quad\times\int\frac{d^4p_3}{(2\pi)^4}e^{ip_3\cdot x_3}\frac{i}{\not p_3}(2\pi)^4\delta(p_1+p_2-p_3). \end{aligned} \tag{42.11}$$

We can simplify the formula by introducing a vector Δ^{μ} with $\Delta^2 = 0$, and contracting (42.11) with it:

$$\Delta_{\mu_1} \ldots \Delta_{\mu_n} \langle T\hat{\bar{q}}_1(x_1) \hat{N}_{NS}^{\mu_1 \ldots \mu_n}(x_2) \hat{q}_2(x_3) \rangle_0$$

$$= \int \frac{d^4 p_2}{(2\pi)^4} e^{-ip_2 \cdot x_2} \int \frac{d^4 p_1}{(2\pi)^4} e^{-ip_1 \cdot x_1} \frac{i}{\not{p}_1} \not{\Delta} (\Delta \cdot p_3)^{n-1} \int \frac{d^4 p_3}{(2\pi)^4} e^{ip_3 \cdot x_3} \frac{i}{\not{p}_3}$$

$$\times (2\pi)^4 \delta(p_1 + p_2 - p_3): \tag{42.12}$$

the symmetrization is automatic and the traces yield zero (they contain terms $g^{\mu\mu'} \Delta_{\mu} \Delta_{\mu'}$). The vertex may, of course, be recovered by differentiation, $(\partial / \partial \Delta_{\mu_1}) \ldots (\partial / \partial \Delta_{\mu_n})$. Equation (42.12) generates the Feynman rule given in Appendix E and used in Section 20.

43 Euclidean QCD

Consider the energy momentum tensor of the pure Yang–Mills QCD, given in Equation (10.2), leaving quarks aside: they are irrelevant for the considerations of this and the next two sections. We can rewrite it as

$$\Theta^{\mu\nu} = -\frac{1}{2} g_{\alpha\beta} \sum_a G_a^{\mu\alpha} G_a^{\nu\beta} - \frac{1}{2} g_{\alpha\beta} \sum_a \tilde{G}_a^{\mu\alpha} \tilde{G}_a^{\nu\beta}. \tag{43.1}$$

It follows that Θ^{00} is positive for *real* gluon fields,

$$\Theta^{00} = \frac{1}{2} \sum_{k,\alpha} \left\{ \left(G_a^{0k} \right)^2 + \left(\tilde{G}_a^{0k} \right)^2 \right\}. \tag{43.2}$$

Therefore, $\Theta^{\mu\nu} = 0$ requires that $G \equiv 0$, and thus only the zero-field configurations can be identified with the vacuum. However, (43.2) no longer has a definite sign if we allow for complex $G^{\mu\nu}$. Particularly important is the case where a complex Minkowskian $G^{\mu\nu}$ corresponds to a real $\underline{G}_{\mu\nu}$ in Euclidean space; for, according to the discussion at the end of Section 40, this will indicate a tunnelling situation. This is the rationale for seeking solutions to the QCD equations in Euclidean space.[107]

Another point is that in Minkowski space,

$$\tilde{\tilde{G}}_a^{\mu\nu} = -G_a^{\mu\nu},$$

so only the trivial $G = 0$ may be *dual*,

$$\tilde{G} = \pm G. \tag{43.3}$$

(If the sign is $(+)$ we say G is *self-dual*, if $(-)$ *anti-dual*.) However, in

[107] This is usually referred to as Euclidean QCD or, more generally, Euclidean field theory. We will distinguish Euclidean quantities from the corresponding Minkowskian ones by underlining the first. Also, sums over repeated space-time indices will be written explicitly.

Euclidean space,

$$\tilde{\tilde{\underline{G}}} = \underline{G},$$

so nontrivial dual values of G may, and indeed do, exist. In addition, Euclidean dual G automatically satisfy the equations of motion. This comes about as follows: the equations of motion for the G read [recall Equation (3.6)]:

$$D_\mu G_a^{\mu\nu} \equiv \partial_\nu G_a^{\mu\nu} + g \sum f_{abc} B_{b\mu} G_c^{\mu\nu} = 0; \tag{43.4}$$

the condition

$$D_\mu \tilde{G}_a^{\mu\nu} = 0 \tag{43.5}$$

is nothing but the Bianchi identity, identically satisfied by any $G = D \times B$ whether or not B solves the equations of motion. However, if $\underline{G}$ is *dual*, (43.5) implies (43.4), as was to be shown [Polyakov (1977)].

The connection with the problem of the vacuum occurs because in the Euclidean case, (43.1) is replaced by

$$\underline{\Theta}_{\mu\nu} = -\frac{1}{2} \sum_{a,\lambda} \{ \underline{G}^a_{\mu\lambda} \underline{G}^a_{\nu\lambda} - \tilde{\underline{G}}^a_{\mu\lambda} \tilde{\underline{G}}^a_{\nu\lambda} \}, \tag{43.6}$$

so for dual fields, $\underline{\Theta}_{\mu\nu} = 0$: dual $\underline{G}$ may represent nontrival vacuum states.

Another property of dual fields has to do with a condition of minimum of the Euclidean action. We can write

$$\begin{aligned} \underline{\mathscr{A}} &= \frac{1}{4} \int d^4 \underline{x} \sum \underline{G}^a_{\mu\nu} \underline{G}^a_{\mu\nu} \\ &= \frac{1}{4} \sum \int d^4 \underline{x} \left\{ \frac{1}{2} \left(\underline{G}^a_{\mu\nu} \pm \tilde{\underline{G}}^a_{\mu\nu} \right)^2 \mp \underline{G}^a_{\mu\nu} \tilde{\underline{G}}^a_{\mu\nu} \right\} \geqslant \frac{1}{4} \left| \int d^4 \underline{x} \sum G\tilde{G} \right|. \end{aligned} \tag{43.7}$$

Thus the action is positive-definite and reaches its minimum for dual fields where one has equality:

$$\underline{\mathscr{A}} = \frac{1}{4} \left| \int d^4 \underline{x} \sum \underline{G}^a_{\mu\nu} \tilde{\underline{G}}^a_{\mu\nu} \right| = \frac{1}{4} \int d^4 \underline{x} \sum_{\mu,\nu,a} \left(\underline{G}^a_{\mu\nu} \right)^2. \tag{43.8}$$

Now, and *at least in situations where the semi-classical approximation WKB holds*, we know that the tunnelling amplitude is given by $\exp(-\underline{\mathscr{A}})$ so the leading tunnelling effect, if it exists, will be provided by dual configurations.

We have talked about "nontrivial vacuum states." It is not difficult to see that nonzero values of B exist for which $G = 0$. In fact, the general form of such B is what is called a *pure gauge*, and may be obtained from $B = 0$ by a gauge transformation. To see this, write a finite gauge transformation as

$$B_a^\mu(x) \to B_a'^\mu(x) = 2 \operatorname{Tr} t^a U^{-1}(x) t^b U(x) B_b^\mu(x) - \frac{2}{ig} \operatorname{Tr} t^a U^{-1}(x) \partial^\mu U(x); \tag{43.9}$$

[cf. Equation (3.1)]. Here $U(x)$ is any x-dependent matrix with $U^+(x)$

$= U^{-1}(x)$, $\det U(x) = 1$. Now, if $B = 0$,

$$B_a'^{\mu}(x) = -\frac{2}{ig}\operatorname{Tr} t^a U^{-1}(x)\partial^{\mu}U(x): \tag{43.10}$$

the gauge covariance of $G_a^{\mu\nu}$ ensures that $G'^{\mu\nu} = G^{\mu\nu} = 0$. Nontrivial solutions will be those for which $G \neq 0$.

44 Instantons

We now seek Euclidean field configurations that lead to a dual field strength tensor, G. To simplify the notation, we will assume summation over repeated or omitted color indices.

We are interested in fields with finite action. This means that we require, in particular,

$$\lim_{x\to\infty} |x|^2 \underline{G}_{\mu\nu}(x) = 0, \tag{44.1}$$

where we define the Euclidean length

$$|x| \equiv +\left\{\sum_{\mu=1}^{4}(x_{\mu})^2\right\}^{1/2}.$$

Let $U(x)$ be a gauge transformation, i.e., a 3×3 matrix with $\det U = 1$, $U^{-1} = U^{+}$. The condition (44.1) will be satisfied provided, at large x, B is the gauge transform of a null field, i.e., that it is asymptotically *pure gauge*. Thus,

$$\begin{aligned} &B_a^{\mu}\xrightarrow[|x|\to\infty]{}\frac{-2}{ig}\operatorname{Tr} t^a U^{-1}(x)\partial^{\mu}U(x),\\ &G_a^{\mu\nu}\xrightarrow[|x|\to\infty]{}0, \end{aligned} \tag{44.2}$$

and we try the ansatz

$$\underline{B}_{\mu}^{a} = \varphi(|x|^2)\,\hat{\underline{B}}_{\mu}^{a}, \qquad \hat{\underline{B}}_{\mu}^{a} = \frac{-2}{ig}\operatorname{Tr} t^a U^{-1}\partial_{\mu}U, \qquad \varphi\xrightarrow[|x|\to\infty]{}1. \tag{44.3}$$

It is instructive to check that the $\hat{\underline{G}}$ corresponding to $\hat{B}$ is zero: for this, we define the matrices

$$\underline{\mathscr{B}}_{\mu} \equiv t^a \underline{B}_{\mu}^{a}, \qquad \underline{\mathscr{G}}_{\mu\nu} \equiv t^a \underline{G}_{\mu\nu}^{a}. \tag{44.4a}$$

Clearly,

$$\underline{B}_{\mu}^{a} = 2\operatorname{Tr} t^a \underline{\mathscr{B}}_{\mu}, \qquad \underline{G}_{\mu\nu}^{a} = 2\operatorname{Tr} t^a \underline{\mathscr{G}}_{\mu\nu}, \tag{44.4b}$$

and

$$\underline{\mathscr{G}}_{\mu\nu} = \partial_{\mu}\underline{\mathscr{B}}_{\nu} - \partial_{\nu}\underline{\mathscr{B}}_{\mu} - ig[\underline{\mathscr{B}}_{\mu}, \underline{\mathscr{B}}_{\nu}]. \tag{44.4c}$$

Of course, Equations (44.4) also hold in the Minkowskian case. Now, if $\underline{B}$ is given by (44.3),

$$\mathscr{B}_\mu \underset{|x|\to\infty}{\simeq} -\frac{1}{ig} U^{-1}\partial_\mu U, \tag{44.5}$$

so that

$$\begin{aligned}\mathscr{G}_{\mu\nu} \underset{x\to\infty}{\simeq} & \frac{1}{-ig}\left\{\partial_\mu(U^{-1}\partial_\nu U) - \partial_\nu(U^{-1}\partial_\mu U)\right\} \\ & -ig\left(\frac{1}{-ig}\right)^2\left[U^{-1}\partial_\mu U, U^{-1}\partial_\nu U\right] \\ = & \frac{1}{-ig}\left\{-U^{-1}(\partial_\mu U)U^{-1}(\partial_\nu U) + U^{-1}(\partial_\nu U)U^{-1}(\partial_\mu U)\right\} \\ & + \frac{1}{-ig}\left[U^{-1}\partial_\mu U, U^{-1}\partial_\nu U\right] = 0.\end{aligned}$$

Note that the bilinear and quadrilinear terms cancel one another; the factor $1/g$ is essential because of the nonlinear form of $\underline{G}$. Its appearance heralds the nonperturbative character of the solutions.

If U is a group element that can be continuously connected to the identity then $\mathscr{G}$ vanishes not only asymptotically, but $\mathscr{G} \equiv 0$. So we will need U values that are not of the form $\exp it\theta(x)$. The only possibility is to couple space-time and color indices. This can be managed only because the dimension of space-time is four. Its group of invariance is $SO(4)$, whose Lie algebra (actually its *complex* Lie algebra) is isomorphic to the product of the Lie algebra of $SU(2)$ by itself. Thus, we may couple $SO(4)$ with an $SU(2)$ subgroup of color $SU(3)$. In view of this, we seek a matrix U of the form

$$U = \begin{pmatrix} u & 0 \\ 0 & 1 \end{pmatrix}$$

where u is a 2×2 matrix in $SU(2)$. Let $\sigma_4 = \left(\begin{smallmatrix} 1 & 0 \\ 0 & 1 \end{smallmatrix}\right)$, and let σ_i be the Pauli matrices. Any 2×2 matrix A can be written as $A = \sum a_\mu\sigma_\mu$. If we let $\hat{a}_i = -a_i$, $\hat{a}_4 = a_4$, then

$$\left(\sum a_\mu\sigma_\mu\right)\left(\sum \hat{a}_\mu\sigma_\mu\right) = \sum a_\mu\hat{a}_\mu,$$

and

$$\det A = \sum a_\mu\hat{a}_\mu;$$

we obtain that the most general u may be written as

$$u_f = \frac{1}{|f(x)|}\left\{\sigma_4 f_4(x) + i\vec{\sigma}\vec{f}(x)\right\}, \qquad f_\mu(x) = \text{real}. \tag{44.6}$$

Thus, the simplest solution is to take $f_\mu(x) = x_\mu$; hence,

$$u(x) = \frac{1}{|x|}(\sigma_4 x_4 + i\vec{\sigma}\vec{x}). \tag{44.7a}$$

The space-time and color indices are coupled in a nontrivial way, for one cannot write $u(x) = \exp(i/2)\vec{\sigma}\vec{\theta}(x)$. One then tries,[108] as stated,

$$\underline{\mathscr{B}}_\mu(x) = \varphi(|x|^2)\underline{\hat{\mathscr{B}}}_\mu(x), \quad \underline{\hat{\mathscr{B}}}_\mu(x) = \frac{1}{-ig}U^{-1}(x)\partial_\mu U(x), \quad U = \begin{pmatrix} u & 0 \\ 0 & 1 \end{pmatrix}. \tag{44.7b}$$

It is useful to remember that because $\underline{\hat{\mathscr{B}}}$ is pure gauge, its corresponding $\underline{\hat{\mathscr{G}}}$ vanishes. Then,

$$\begin{aligned}
\underline{\mathscr{G}}_{\mu\nu} &= \partial_\mu\underline{\mathscr{B}}_\nu - \partial_\nu\underline{\mathscr{B}}_\mu - ig[\underline{\mathscr{B}}_\mu, \underline{\mathscr{B}}_\nu] \\
&= (\partial_\mu\varphi)\underline{\hat{\mathscr{B}}}_\nu - (\partial_\nu\varphi)\underline{\hat{\mathscr{B}}}_\mu + \varphi(\partial_\mu\underline{\hat{\mathscr{B}}}_\nu - \partial_\nu\underline{\hat{\mathscr{B}}}_\mu) \\
&\quad - ig\varphi^2[\underline{\hat{\mathscr{B}}}_\mu, \underline{\hat{\mathscr{B}}}_\nu] \\
&= 2\varphi'\{x_\mu\underline{\hat{\mathscr{B}}}_\nu - x_\nu\underline{\hat{\mathscr{B}}}_\mu\} + (\varphi - \varphi^2)\{\partial_\mu\underline{\hat{\mathscr{B}}}_\nu - \partial_\nu\underline{\hat{\mathscr{B}}}_\mu\}; \qquad \varphi' = \frac{d\varphi(|x|^2)}{d|x|^2}.
\end{aligned}$$

This is most easily calculated by noting that $\hat{B}^a_\mu = -(2/g|x|^2)\sum\eta^a_{\rho\mu}x_\rho$, with η given below. Then,

$$\begin{aligned}
\underline{G}^a_{\mu\nu} &= \frac{4i^2}{|x|^2 g}\left(\varphi' - \frac{\varphi - \varphi^2}{|x|^2}\right)\sum_\rho(\eta^a_{\rho\nu}x_\rho x_\mu - \eta^a_{\rho\mu}x_\rho x_\nu) \\
&\quad + \frac{4i^2}{|x|^2 g}(\varphi - \varphi^2)\eta^a_{\mu\nu}.
\end{aligned}$$

The mixed tensor η is defined by

$$\eta^a_{\alpha\beta} = \begin{cases} \epsilon_{a\alpha\beta 4} + \delta_{\alpha 4}\delta_{a\beta} - \delta_{\beta 4}\delta_{a\alpha}, & a = 1,2,3 \\ 0, & a = 4, \ldots, 8. \end{cases} \tag{44.8}$$

We note that it is self-dual, $\eta_{\alpha\beta} = \tilde{\eta}_{\alpha\beta}$: therefore, the condition of self-duality for $\underline{G}$ is met if φ satisfies the equation

$$\varphi' - \frac{\varphi - \varphi^2}{|x|^2} = 0,$$

i.e.,

$$\underline{\mathscr{B}}_\mu(x) = \frac{|x|^2}{|x|^2 + \lambda^2}\cdot\frac{1}{-ig}U^{-1}(x)\partial_\mu U(x), \qquad \lambda \text{ arbitrary.} \tag{44.9}$$

This is the original *instanton* solution found by Belavin *et al.* (1975). We note that it is concentrated around $x \approx 0$, i.e., in space *and* time (hence, the name instanton). Solutions concentrated around $x \approx y$, any y, are obtained from (44.9) by displacing $x \to x - y$. This will be useful later. Equation (44.9) can be made more explicit by substituting U; we find that $\underline{B}$ is real:

$$\underline{B}^a_\mu(x) = \frac{1}{g}\cdot\frac{-2}{|x|^2 + \lambda^2}\sum_\rho \eta^a_{\rho\mu}x_\rho. \tag{44.10}$$

[108] More general ansätze have been described by Corrigan and Fairlie (1977) and Wilczek (1977).

The coupling of space-time and color is obvious from the form of η. The corresponding G is

$$\underline{G}^{a}_{\mu\nu}(x) = \frac{1}{g} \cdot \frac{-4\lambda^2 \eta^{a}_{\mu\nu}}{(|x|^2 + \lambda^2)^2} . \tag{44.11}$$

As was to be expected, both B and G become singular (and complex!) when continued to Minkowski space because x^2 is no longer positive and hence $x^2 + \lambda^2$ may vanish. A remarkable property of the instantons is that, whereas $B \approx 1/|x|$ for large x, a sufficient number of cancellations occur when forming G so that $G \approx 1/|x|^4$, well within the requirements of (44.1).

The solution (44.9) is all we will use here; but other solutions[109] have been found by De Alfaro, Fubini, and Furlan (1976, 1977) and by Cerveró, Jacobs, and Nohl (1977). It turns out that there is perfect symmetry between self-dual and anti-dual solutions: the anti-dual solutions corresponding to (44.10) use the tensor

$$\bar{\eta}^{a}_{\alpha\beta} = \eta^{a}_{\alpha\beta}, \qquad \alpha, \beta = 1,2,3, \qquad \bar{\eta}^{a}_{\alpha\beta} = -\eta^{a}_{\alpha\beta}, \qquad \alpha \text{ or } \beta = 4. \tag{44.12}$$

They are called *anti-instantons.*

Next let us compute the action corresponding to the instanton. Using $\sum \eta^{a}_{\mu\nu}\eta^{a}_{\mu\nu} = 12$ and the formulas of Appendix B,

$$\begin{aligned} \underline{\mathscr{A}} &= \frac{1}{4} \int d^4 \underline{x} \sum \underline{G}^{a}_{\mu\nu} \underline{G}^{a}_{\mu\nu} \\ &= \frac{48\lambda^2}{g^2} \int d^4 \underline{x} \frac{1}{(|x|^2 + \lambda^2)^4} = \frac{8\pi^2}{g^2} . \end{aligned} \tag{44.13}$$

In the following section, we will show that instantons produce tunnelling between states $|n_{\pm}\rangle$ and $|n_{\pm} + \nu\rangle$, where ν is an integer. In this sense, they provide the "existence proof" for the reality of the complicated vacuum structure discussed in Section 38. One may wonder about the necessity of the sophisticated discussion there, since we have found explicit solutions. The answer lies in the requirement of "finite action" under which instanton solutions were found. As discussed in Section 40, the observable tunnelling amplitude between two states $|a\rangle$ and $|b\rangle$ is

$$\langle a \,|\, b \rangle_{\text{phys}} = \langle a | e^{-\mathscr{A}} | b \rangle / \langle b | e^{-\mathscr{A}} | b \rangle, \tag{44.14}$$

so even configurations with infinite action may yield a finite tunnelling, provided the infinities in numerator and denominator of (44.14) cancel. The requirement of finite action may be appealing, but it is not compelling. In fact, we will see in Section 45 that instantons lead to *integer* values of ν, while we know from the work of Crewther (1977b), discussed in Section 38, that some patterns of quark masses lead to noninteger values of ν.[110] The importance of instantons lies in the fact that they provide explicit tunnel-

[109] With finite Minkowski action, but infinite Euclidean action.

[110] "Semi-instantons" with finite Euclidean action and half-integer topological charge seem to have been recently found by Forgács, Horvath, and Palla (1981).

ling effects, and they thus give indications of how to estimate these, but it is unlikely that they exhaust all possibilities. With this proviso in mind, we continue with the study of instantons and the requirement of finite action.

45 Connection with the Topological Quantum Number and the QCD Vacuum

Consider the quantity (cf., Equation [38.3])

$$Q_K = \frac{g^2}{32\pi^2} \int d^4 \underline{x} \sum \underline{\tilde{G}^a_{\mu\nu}} \underline{G^a_{\mu\nu}}. \tag{45.1}$$

The gluon fields that approach zero at infinity are, as we discussed, of the form

$$\underline{\mathscr{B}_\mu} \underset{x\to\infty}{\simeq} \frac{-1}{ig} T_B^{-1}(x)\partial_\mu T_B(x), \tag{45.2}$$

where T_B is a general matrix in $SU(3)$. Consider x varying on the boundary of a four-dimensional sphere, ∂S_4. The gauge field maps each point x into a $T_B(x)$ in the gauge group: so we have a mapping of ∂S_4 into $SU(3)$. We may say that two fields are *homotopic*, and $\underline{\mathscr{B}} \approx \underline{\mathscr{B}}'$ if they can be continuously deformed one into another. Clearly, this relation is an equivalence relation, and thus we may split all gauge fields into homotopy classes. The number of homotopy classes[111] is a *countable* infinity, so we may label fields by an integer n according to their homotopy class. Our next task is to show that n coincides with Q_K as given in (45.1). The quantity Q_K is called the *topological*, *winding*, or *Pontryagin* quantum number. The second name refers to the number of times the mapping wraps the sphere around the group.

To see this, we first note that (45.1) is invariant under continuous gauge transformations, as can be seen by direct computation. Next, we note that the integrand there is actually a four divergence. In fact, as shown in Section 38,

$$\frac{g^2}{32\pi^2} \sum \underline{G^a_{\mu\nu}} \underline{\tilde{G}^a_{\mu\nu}} = \sum \partial_\mu \underline{K_\mu}, \tag{45.3}$$

where K is the "chiral current,"

$$\underline{K_\mu} = \frac{g^2}{16\pi^2} \sum \epsilon_{\mu\nu\rho\sigma} \left\{ \left(\partial_\rho \underline{B^a_\sigma}\right) \underline{B^a_\nu} + \frac{1}{3} g f_{abc} \underline{B^a_\rho} \underline{B^b_\sigma} \underline{B^c_\nu} \right\}. \tag{45.4}$$

Because of Gauss's theorem,

$$\begin{aligned} Q_K &= \frac{g^2}{32\pi^2} \int d^4x \sum \underline{G^a_{\mu\nu}} \underline{\tilde{G}^a_{\mu\nu}}, \\ &= \int_{\partial S_4} \sum d\sigma_\mu \underline{K_\mu}, \end{aligned}$$

[111]This holds for *any* gauge group that is simple and contains an $SU(2)$ subgroup.

where $d\sigma_\mu$ is the surface element in ∂S_4. Using (45.4), we find

$$Q_K = \frac{g^3}{48\pi^2} \sum \epsilon_{\mu\nu\rho\sigma} f_{abc} \int_{\partial S_4} d\sigma_\mu \, \underline{B}_\rho^a \underline{B}_\sigma^b \underline{B}_\nu^c .$$

The calculation simplifies if we assume $\underline{B}_a = 0$ except for $a = 1,2,3$; this is possible because the homotopy relation is dependent only on an $SU(2)$ subgroup. Now, in this case, we let

$$\underline{\mathscr{B}}_\mu = \frac{1}{2} \sum \sigma_k \underline{B}_\mu^k ,$$

and (45.2) holds with T in $SU(2)$. We thus have

$$Q_K = \frac{1}{12\pi^2} \sum \epsilon_{\mu\nu\rho\sigma} \int_{\partial S_4} d\sigma_\mu \operatorname{Tr}\{(T^{-1}\partial_\rho T)(T^{-1}\partial_\sigma T)(T^{-1}\partial_\nu T)\}. \quad (45.5)$$

Suppose we parametrize the elements of $SU(2)$ by three Euler angles ξ_i; the invariant measure over the group is

$$d\mu = \operatorname{Tr}\left\{ T^{-1}\frac{\partial T}{\partial \xi_1} T^{-1}\frac{\partial T}{\partial \xi_2} T^{-1}\frac{\partial T}{\partial \xi_3} \right\} d\xi_1 \, d\xi_2 \, d\xi_3 ,$$

$$\int_{SU(2)} d\mu = 12\pi^2.$$

We see that (45.5) gives precisely the number of times the surface of the sphere is wrapped around $SU(2)$. Thus, our instanton/anti-instanton solution has $Q_K = \pm 1$, as is clear from (44.13) and using the self-dual/anti-dual property. It is also not difficult to construct solutions for any ν. Suppose ν is positive. Consider the dilute gas of ν instanons,

$$\underline{B}_\mu^{a(\nu)}(x) = \sum_{k=1}^{\nu} \underline{B}_\mu^a(x - y_k), \quad (45.6a)$$

let $\underline{B}$ equal the value given by (44.10), and let $|y_j - y_k| \to \infty$. Clearly, the overlap between two different terms in (45.6) when building $G^{(\nu)}$ or $G^{(\nu)}G^{(\nu)}$ tends toward zero as $|y_j - y_k| \to \infty$; hence, in this limit,

$$\frac{g^2}{32\pi^2} \int d^4\underline{x} \; \underline{G}^{(\nu)} \underline{\tilde{G}}^{(\nu)} = \nu. \quad (45.6b)$$

We have succeeded in finding a representative in each homotopy class. What is more interesting, the multi-instanton field configurations are dual; hence, the corresponding energy-momentum tensor vanishes, $\underline{\Theta}^{(\nu)} = 0$. This means that in QCD (at least, in the Euclidean version) there is not a single vacuum, but an infinity of vacuum configurations, $|\nu\rangle$, $\nu = \ldots, -1, 0, 1, 2, \ldots$, that are topologically inequivalent. The situation is like that of Figure 30b.

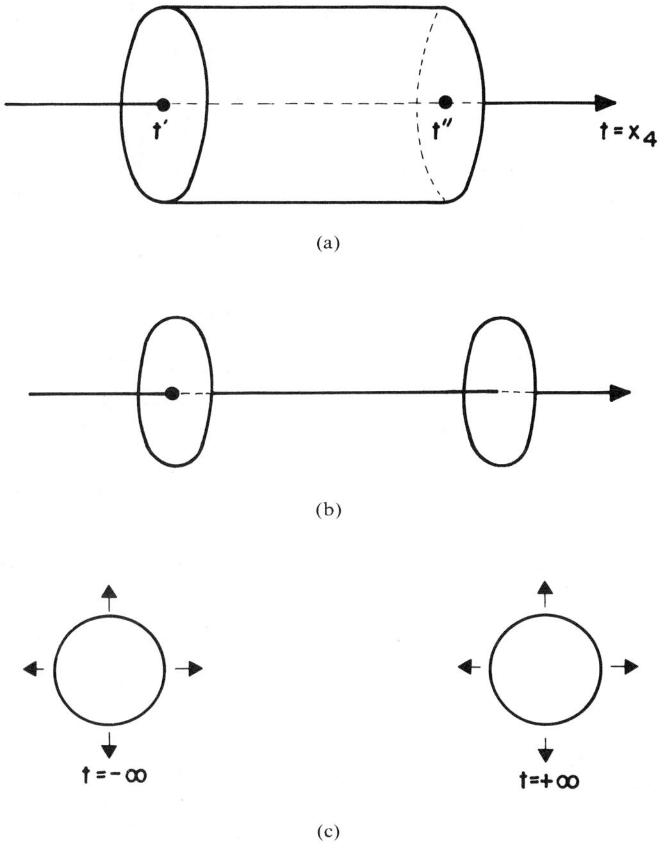

Figure 31. Regions of integration for the instantons.

To explore this in greater detail, let us use a different hypersurface; to be precise, we take a cylinder along the time axis, as in Figure 31*a*. First, we choose a Coulomb-like gauge so that $\underline{B}_4 =_{x\to\infty} 0$. Thus, only the integrals along the bases of the cylinder remain,

$$\nu = \left\{ \int_{t''} - \int_{t'} \right\} dx_1\, dx_2\, dx_3\, \underline{K}_4^{(\nu)}\,.$$

Since the field vanishes at infinity, we may identify the points at spatial infinity of the bases of the cylinder (Figure 31*b, c*), so we obtain integrals over large three-dimensional spheres, one at $t = -\infty$, the other at $t = +\infty$.

Then choose the gauge so that

$$\int_{t'\to-\infty} dx_1\, dx_2\, dx_3\, \underline{K}_4^{(\nu)} = n(-\infty) = \text{integer.}$$

The proof that these properties may be achieved by a gauge choice, continuously connected to the identity, may be found in the lectures of Sciuto (1979). In view of (45.6b), we see that

$$\int_{t''} dx_1\, dx_2\, dx_3\, \underline{K}_4^{(\nu)} = n(t''), \qquad n(+\infty) - n(-\infty) = \nu. \tag{45.7}$$

A multi-instanton $\underline{B}^{(\nu)}$ connects vacua separated by ν units of the topological quantum number between $-\infty$ and $+\infty$. So, in the quantum case, and according to the discussion of Section 40, we expect that these vacua will be connected by the tunnel effect, the leading amplitude for tunnelling being

$$\langle n(+\infty) \,|\, n(-\infty)\rangle = (\text{constant})\exp(-\underline{\mathscr{A}}^{(\nu)}).$$

As we discussed earlier, the minimum of the action is reached for self-dual/anti-dual solutions, i.e., for the instanton or anti-instanton (if $|n(+\infty) - n(-\infty)| = 1$). Thus, to leading order,

$$\langle n(+\infty) \,|\, n(-\infty)\rangle \approx (\text{constant})\exp\left\{ \frac{-8\pi^2|\nu|}{g^2} \right\}. \tag{45.8}$$

The higher corrections can be calculated ['t Hooft (1976)] by expanding not around $\underline{B}_{cl} = 0$, but around $\underline{B}_{cl} = \underline{B}_{cl}^{(\nu)} = \underline{B}^{(\nu)}$. They are important in that they yield the constant in (45.8). Indeed,

$$\exp\left\{ \frac{-8\pi^2|\nu|}{g^2}\left(1 + a\frac{g^2}{16\pi^2}\right)\right\} = e^{-a/2}\exp\left\{ \frac{-8\pi^2|\nu|}{g^2} \right\},$$

but they do not qualitatively alter the result. What occurs is that in order to verify the calculation (45.8), one has to consider situations where g is small, and then the $\exp(-2\pi/\alpha_g)$ term overwhelms any constant.

Let us now turn to the vacuum. The generating functional was defined in Sections 39 to 41. Neglecting gauge-fixing and ghost terms, it was (converted to the Euclidean case)

$${}_+\langle 0 \,|\, 0\rangle_- = \underline{Z} = \int (d\underline{B})\exp\left\{ -\int d^4x\, \underline{\mathscr{L}}(\underline{B})\right\}. \tag{45.9a}$$

However, we now have to decide which homotopy classes to integrate. We may recall that in (45.9a), the left-hand side was $\langle 0, t = +\infty \,|\, 0, t = -\infty\rangle$; so it appears that we should reinterpret (45.9a) as

$$\langle n(+\infty) \,|\, m(-\infty)\rangle = \int (d\underline{B}_{n-m})\exp\left\{ -\int d^4x\, \underline{\mathscr{L}}(\underline{B})\right\}. \tag{45.9b}$$

In perturbation theory, only the vacuum $|n = 0\rangle$ is considered; but because of tunnelling, it is clear that all the $|n\rangle$ are connected [Callan, Dashen, and Gross (1976); Jackiw, Nohl, and Rebbi (1977); 't Hooft (1976)], so none of them is a true vacuum state, in that it is not stationary. Stationary states are formed like the Bloch states in solids, by considering the superpositions

$$\sum_n e^{in\theta}|n\rangle \equiv |\theta\rangle.$$

This is certainly invariant under changes of topological charge, for if we let Γ_k be the operator that changes n by k units,

$$\Gamma_k|\theta\rangle = \sum_n e^{in\theta}|n+k\rangle = \sum_m e^{i(m-k)\theta}|m\rangle = e^{-ik\theta}|\theta\rangle,$$

i.e., the vacuum undergoes only a change of phase. The generating functional is now, in terms of the θ-vacua,

$$\langle\theta(+\infty)\,|\,\theta'(-\infty)\rangle = N\delta(\theta-\theta')\sum_\nu e^{-i\nu\theta}\int(d\underline{B}^{(\nu)})e^{-\int d^4\underline{x}\,\underline{\mathscr{L}}(\underline{B}^{(\nu)})}. \quad (45.10)$$

We may drop the $\delta(\theta-\theta')$, which only expresses the fact that worlds corresponding to different values of θ are unconnected. Moreover, we can extend the integral over $\underline{B}$ to all field configurations, introducing a factor of

$$\delta\left[\nu-(g^2/32\pi^2)\int d^4x\sum\underline{G}\,\underline{\tilde{G}}\right];$$

but then the sum over ν may be carried over trivially and we obtain

$$Z = N\int(d\underline{B})e^{-\int d^4\underline{x}\,\underline{\mathscr{L}}_\theta}, \quad (45.11a)$$

where

$$\underline{\mathscr{L}}_\theta = -\frac{1}{4}\sum\underline{G}\,\underline{G} + \frac{ig^2\theta}{32\pi^2}\sum\underline{G}\,\underline{\tilde{G}}. \quad (45.11b)$$

Now we can finally return to Minlowski space and conclude that the existence of instantons suggests that the true Lagrangian is actually

$$\mathscr{L}_\theta = -\frac{1}{4}\sum_a G_a^{\mu\nu}G_{a\mu\nu} - \frac{\theta g^2}{32\pi^2}\sum_a G_a^{\mu\nu}\tilde{G}_{a\mu\nu}, \quad (45.12)$$

thereby justifying the necessity of the introduction, in the general case, of the $\mathscr{L}_{1\theta}$ term (recall the discussion at the beginning of Section 38).

One may wonder to what extent the phenomena we have discussed will modify the results found previous to Section 37. First, the bounds obtained for the value of θ (Section 38) force it to be so small that $\mathscr{L}_{1\theta}$ by itself should have practically no effect. In addition, instanton and related effects are *long-distance* effects; field configurations that vanish sufficiently rapidly as $x\to\infty$ have $Q_K=0$. Since we have always discussed short-distance effects (for $\pi^0\to 2\gamma$, deep inelastic scattering, etc.), we would think that the perturbative regime should continue to be relevant. This may be seen very clearly if one considers the tunnelling effect due to an instanton:

$$\langle 0|\pm 1\rangle \approx (\text{constant})\exp\left(-\frac{2\pi}{\alpha_g}\right).$$

After renormalization, α_g should be replaced by the running coupling constant so that, up to logarithmic corrections,

$$\langle 0| \pm 1\rangle \approx \left(\frac{\Lambda^2}{Q^2}\right)^{(33-2n_f)/G} \tag{45.13}$$

This shows that at large Q^2, tunnelling is negligible and we may work with $|0\rangle$ as if it were the true vacuum: the error induced by expression (45.13) is much smaller than, for example, twist-four or -six effects. Indeed, estimates by Baulieu *et al.* (1979) show that the instanton corrections for e^+e^- annihilations or deep inelastic scattering are utterly negligible for $Q^2 \gtrsim 1$ GeV^2. Thus, when instantons effects are important, the calculation does not apply; when it does apply, the effect is unobservable. In this, the instantons resemble that mythical animal, the basilisk, whose sight was supposed to cause the death of the beholder.

46 Other Topics

i Lattice QCD

In principle, the path integral formalism appears to fulfill the theorist's dream: quantum field theory reduced to quadratures. It would appear sufficient to discretize space-time in a lattice with some spacing, δ, and size N^4, and integrate the generating functional. In practice, the situation is more difficult. Only Gaussian integrations, or integrations over fermion fields, can be carried over explicitly, and thus we are forced to rely on numerical methods. This may explain why, after the work of Wilson (1975) and until recently, almost no new results have been obtained.

However, the situation has changed dramatically in the past two years. Not only has evidence of confinement been gathered, but after the breakthrough represented by the introduction of fermions, a number of basic quantities (including $\langle \alpha G^2\rangle$, $\langle qq\rangle$, and, particularly m_ρ, m_p, f_π) have been computed within reasonable accuracy (~30%). Unfortunately, we are not able to go into details and will thus have to neglect all of these very exciting results. The interested reader should consult the original work.[112]

ii $1/N$ Expansions

QCD is formulated with three colors, but the theory becomes simpler if we take the number of colors N to infinity ['t Hooft (1974*a*, *b*)]. With some luck this limit will respect the basic properties of the theory and the corrections, $O(1/N)$ will be small. The main problem is that one does not

[112] Lautrup and Nauenberg (1981); Creutz (1981); Hamber and Parisi (1981); Marinari, Parisi, and Rebbi (1981) from where further references may be obtained.

know how to calculate even the zero-order term! However, this does not mean that the approximation is useless; it allows us to make contact with the so-called topological expansions in hadron physics [Rossi and Veneziano (1977); Chew and Rosenzweig, (1978)], and it sheds some light on the $U(1)$ limit and, perhaps, may be relevant to confinement. One also may use $1/N$ expansions to obtain qualitative estimates of various effects. For example, the mass of a nucleon (say, the proton) is expected to be $N\Lambda$; so target mass effects become

$$O(N^2\Lambda^2),$$

while higher twist (twist-four) effects are

$$O(N\Lambda^2).$$

Hence, to leading order in $1/N$, the latter may be neglected against the former. Likewise, the mass degeneracy of the ρ, ω or the absence of $\pi\pi$ bound states may be qualitatively understood. We refer the reader to Witten (1979*b*, 1980) for a review and references.

iii Bag Models

It is probably safe to say that, at present, most physicists consider bag models as an approximation to QCD and as useful in particular circumstances.

The idea of the bag approximation is as follows: if quarks (and gluons) are confined at an average distance δ, it may be possible to simulate the effect of the confining forces by setting the boundary condition that $q(x)$, $B(x)$ vanish identically outside a sphere of radius δ. The rest of QCD could then be treated as a perturbation.

The bag model has proven useful in a number of phenomenological applications: not only in calculating static quantities (masses of various hadrons, magnetic moments), but also in obtaining the absolute normalization of structure functions $f(x, Q_0^2)$ [Jaffe and Ross, (1980)]. The interested reader is referred to the review of Hasenfratz and Kuti (1978) for details and references on the so-called MIT bag.

A related approach is that of strings, where one confines quarks and gluons not inside a bag, but along a string, which ties with the success of string dynamics [Gervais and Neveu, (1972)]. The report of Nambu (1978) may be consulted as well.

iv Infrared Properties of QCD

Whereas the ultraviolet limit of QCD appears to be well understood, perhaps as well as (if not better than) the same limit in QED, very little is known about the infrared properties. There is nothing in QCD like the theorem of Thirring (1950) or the Bloch-Nordsiek (1937) analysis, which essentially allow us to treat the long-distance effects in QCD classically: results like the Lee-Nauenberg (1964) theorem are only very partially applicable.[113] We will content ourselves with giving as references the papers of Muller (1978), and Zachariasen (1980), where some aspects of the IR problem are treated.

v Functional Methods

A long time ago, Schwinger and Dyson found a set of functional equations which express the equations of a quantum field theory in closed form. [The form of the equations may be found in the texts of Bogoliubov and Shirkov (1959) or Bjorken and Drell (1965)]. While, of course, the equations cannot be solved exactly, one may try to find self-consistent nonperturbative solutions consistent with confinement. Even after truncation of the equations, this is a formidable task, though perhaps not hopeless [Mandelstam (1979)].

vi Liberated Quarks and Gluons

It appears to be very difficult to prove confinement, perhaps because it is untrue. It should always be borne in mind that persistent candidates for liberated quarks exist. [LaRue, Phillips, and Fairbank, (1981)]. How would one modify QCD to cope with this and retain its successes? The earliest alternative, that of Pati and Salam, does not agree with current experiment, mainly due to the integer change of quarks. Perhaps the most appealing scheme is that of De Rújula, Giles, and Jaffe (1978).

vii High Temperature QCD

In this book we have only considered QCD at zero temperature—that is, we did not force a large number of quarks (and gluons) to be inside a small volume at high energy. In addition to the intrinsic interest in studying finite-temperature QCD, there are cosmological situations (like very heavy stars or the big bang) where this is relevant. What is more, similar situations may possibly be obtained in the laboratory via heavy ion collisions. We

[113] In fact, one may consider bag or string models as methods for trying to circumvent the infrared problem of QCD, intimately linked to confinement.

refer the interested reader to the review of Gross, Pisarski, and Yaffe (1981).

viii Potential Models

An important subject not touched upon at all here is that of QCD-inspired constituent models for hadrons, although the first successes of the quark model were in this field. There are two reasons for this omission. First, although QCD is essential to obtain some of the features of such models, it is difficult to justify with any rigor—in the present state of the art—many of the necessary assumptions. Second, a book [Flamm and Schoberl (1981)] dealing with precisely this topic has recently appeared.

ix QCD Corrections to Electroweak Processes

Besides what may be called "pure" hadronic physics, QCD also provides a means of estimating strong interaction corrections to electroweak processes. In a sense, one can interpret also in this way the QCD corrections to the purely partonic picture of e^+e^- annihilations on deep inelastic scattering; but we are referring now to corrections to processes like non-leptonic or semi-leptonic decays of heavy quarks, including (partial) explanations of the $\Delta I = 1/2$ rule, or to the bare GIM mechanism or proton decay. We direct the interested reader to the reviews of Gaillard (1978) or Altarelli (1982) for details and references.

Appendix A: γ-Algebra in Dimension *D*

The γ matrices are taken to be of dimension 4. We have D γ^μ matrices,

$$\gamma^0, \gamma^1, \ldots, \gamma^{D-1},$$

and the matrix[114] γ_5. They verify the anticommutation relations,

$$\{\gamma^\mu, \gamma^\nu\} = 2g^{\mu\nu}, \qquad \gamma_5^2 = 1$$

with

$$g^{\mu\nu} = 0, \qquad \mu \neq \nu, \qquad g^{00} = 1, \qquad g^{ii} = -1 \qquad \text{for} \quad i = 1, \ldots, D-1.$$

$$g_{\mu\nu} = g^{\mu\nu}.$$

[114]More about γ_5 may be found in Sections 7 and 33.

A few useful relations are, with $S_{\mu\nu\alpha\beta} = g_{\mu\nu}g_{\alpha\beta} + g_{\mu\beta}g_{\nu\alpha} - g_{\mu\alpha}g_{\nu\beta}$,

$$A_\mu = g_{\mu\nu}A^\nu, \qquad \not{A} = \gamma_\mu A^\mu,$$

$$\operatorname{Tr}\gamma^\mu\gamma^\nu = 4g^{\mu\nu}, \quad \operatorname{Tr}\gamma_5\gamma^\mu\gamma^\nu = 0, \quad \operatorname{Tr}\gamma^\mu \overset{\text{(odd)}}{\ldots} \gamma^\tau = 0, \quad \operatorname{Tr}\gamma_5\gamma^\mu \overset{\text{(odd)}}{\ldots} \gamma^\tau = 0,$$

$$\operatorname{Tr}\gamma^\mu\gamma^\nu\gamma^\alpha\gamma^\beta = 4S^{\mu\nu\alpha\beta} = 4\{g^{\mu\nu}g^{\alpha\beta} + g^{\mu\beta}g^{\nu\alpha} - g^{\mu\alpha}g^{\nu\beta}\};$$

$$\not{a}\not{a} = a^2; \qquad \not{a}\not{b}\not{a} = -a^2\not{b} + 2(a\cdot b)\not{a},$$

$$\gamma_\mu\gamma^\mu = D, \qquad \gamma_\mu\gamma^\alpha\gamma^\mu = (2-D)\gamma^\alpha; \qquad \gamma_\mu\gamma_5\gamma^\mu = -D\gamma_5;$$

$$\gamma^\mu\gamma^\alpha\gamma^\beta\gamma_\mu = 4g^{\alpha\beta} + (D-4)\gamma^\alpha\gamma^\beta,$$

$$\gamma^\mu\gamma^\alpha\gamma^\beta\gamma^\delta\gamma_\mu = -2\gamma^\delta\gamma^\beta\gamma^\alpha + (4-D)\gamma^\alpha\gamma^\beta\gamma^\delta.$$

For $D = 4$, $\gamma_5 = i\gamma^0\gamma^1\gamma^2\gamma^3$, and then, defining the totally antisymmetric tensor $\epsilon^{\mu\nu\rho\sigma}$ by

$$\epsilon^{0123} = -1, \qquad \epsilon_{0123} = +1,$$

and cyclically,

$$\gamma^\mu\gamma^\alpha\gamma^\nu = S^{\mu\alpha\nu\beta}\gamma_\beta - i\epsilon^{\mu\alpha\nu\beta}\gamma_\beta\gamma_5; \qquad \gamma_5\gamma^\mu\gamma^\nu = \gamma_5 g^{\mu\nu} + \frac{1}{2i}\epsilon^{\mu\nu\alpha\beta}\gamma_\alpha\gamma_\beta.$$

$$\operatorname{Tr}\gamma_5\gamma^\mu\gamma^\nu\gamma^\lambda\gamma^\sigma = 4i\epsilon^{\mu\nu\lambda\sigma};$$

$$g_{\alpha\beta}\epsilon^{\alpha\mu\rho\sigma}\epsilon^{\beta\nu\tau\lambda} = -g^{\mu\nu}(g^{\rho\tau}g^{\sigma\lambda} - g^{\rho\lambda}g^{\sigma\tau}) - g^{\mu\lambda}(g^{\rho\nu}g^{\sigma\tau} - g^{\rho\tau}g^{\sigma\nu})$$
$$+ g^{\mu\tau}(g^{\rho\nu}g^{\sigma\lambda} - g^{\rho\lambda}g^{\sigma\nu});$$

$$\epsilon^{\mu\nu\alpha\beta}\epsilon_{\alpha\beta}{}^{\rho\sigma} = 2(g^{\nu\rho}g^{\mu\sigma} - g^{\mu\rho}g^{\nu\sigma}).$$

Moreover, $\{\gamma^\mu, \gamma_5\} = 0$. In the Pauli or Weyl realizations, $\gamma_2\gamma_\mu\gamma_2 = -\gamma_\mu^*$ and, also, $\gamma_0\gamma_\mu^+\gamma_0 = \gamma_\mu$, $\gamma_0(i\gamma_5)^+\gamma_0 = i\gamma_5$. Finally, if w_1, w_2 are spinors and $\Gamma_t, \ldots, \Gamma_n$ are any of the matrices $\gamma_\mu, i\gamma_5$,

$$(\overline{w}_1\Gamma_1 \ldots \Gamma_n w_2)^* = \overline{w}_2\Gamma_n \ldots \Gamma_1 w_1.$$

Appendix B: Some Useful Integrals

In D dimensions,

$$\int \frac{d^Dk}{(2\pi)^D}\cdot\frac{(k^2)^r}{(k^2-R^2)^m} = i\frac{(-1)^{r-m}}{(16\pi^2)^{D/4}}\cdot\frac{\Gamma(r+D/2)\Gamma(m-r-D/2)}{\Gamma(D/2)\Gamma(m)(R^2)^{m-r-D/2}};$$

$$\int d^Dk\,\frac{1}{k^2+i0} = 0; \qquad \int d^D\underline{k}\,\delta(1-|\underline{k}|) = \frac{2\pi^{D/2}}{\Gamma(D/2)}.$$

Symmetric integration:

$$\int d^D k\, k^\mu k^\nu f(k^2) = \frac{g^{\mu\nu}}{D} \int d^D k\, k^2 f(k^2);$$

$$\int d^D k\, k^\mu k^\nu k^\lambda k^\sigma f(k^2) = \frac{g^{\mu\nu}g^{\lambda\sigma} + g^{\mu\lambda}g^{\nu\sigma} + g^{\mu\sigma}g^{\nu\lambda}}{D^2 + 2D} \int d^D k\, k^4 f(k^2);$$

$$\int d^D k\, k^{\mu_1} \dots k^{\mu_{2n+1}} f(k^2) \equiv 0.$$

As $\epsilon \to 0$,

$$\Gamma(1+\epsilon) = 1 - \gamma_E \epsilon + \sum_{n=2}^{\infty} \frac{(-\epsilon)^n}{n!} \zeta(n), \qquad (R^2)^{\epsilon/2} = 1 + \frac{\epsilon}{2}\log R^2 + O(\epsilon^2);$$

Γ is Euler's function, ζ Riemann's function and $\gamma_E \simeq 0.5772$ is the Euler–Mascheroni constant.

Feynman parameters:

$$\frac{1}{A^\alpha B^\beta} = \frac{\Gamma(\alpha+\beta)}{\Gamma(\alpha)\Gamma(\beta)} \int_0^1 dx \frac{x^{\alpha-1}(1-x)^{\beta-1}}{\{xA + (1-x)B\}^{\alpha+\beta}},$$

$$\frac{1}{A^\alpha B^\beta C^\gamma} = \frac{\Gamma(\alpha+\beta+\gamma)}{\Gamma(\alpha)\Gamma(\beta)\Gamma(\gamma)} \int_0^1 dx \cdot x \int_0^1 dy \frac{u_1^{\alpha-1}u_2^{\beta-1}u_3^{\gamma-1}}{\{u_1 A + u_2 B + u_3 C\}^{\alpha+\beta+\gamma}},$$

$$u_1 = xy, \quad u_2 = x(1-y), \quad u_3 = 1-x.$$

$$\frac{1}{A^\alpha B^\beta C^\gamma D^\delta} = \frac{\Gamma(\alpha+\beta+\gamma+\delta)}{\Gamma(\alpha)\Gamma(\beta)\Gamma(\gamma)\Gamma(\delta)} \int_0^1 dx \cdot x^2 \int_0^1 dy \cdot y$$

$$\times \int_0^1 dz \frac{u_1^{\alpha-1}u_2^{\beta-1}u_3^{\gamma-1}u_4^{\delta-1}}{(u_1 A + u_2 B + u_3 C + u_4 D)^{\alpha+\beta+\gamma+\delta}},$$

$$u_1 = 1-x, \quad u_2 = xyz, \quad u_3 = x(1-y), \quad u_4 = xy(1-z), \quad \text{etc.}$$

In general,

$$\frac{1}{A_1 \dots A_n} = (n-1)! \int_0^1 dx_1 \dots \int_0^1 dx_n\, \delta\left(\sum_1^n x_i - 1\right) \frac{1}{(x_1 A_1 + \cdots + x_n A_n)^n}.$$

More formulas may be found in Narison (1982).

Some numerical integrals:

$$\int_0^1 dx \log(1+x) = 2\log 2 - 1$$

$$\int_0^1 dx \frac{\log(1+x)}{2} = \frac{\pi^2}{12}.$$

Many useful integrals can be derived from Euler's formula:

$$\int_0^1 dx\, x^{\alpha}(1-x)^{\beta} = \frac{\Gamma(1+\alpha)\Gamma(1+\beta)}{\Gamma(2+\alpha+\beta)}.$$

For example, by differentiation, we obtain

$$\int_0^1 dx\, x^{\alpha}\log x = \frac{-1}{(\alpha+1)^2};$$

$$\int_0^1 dx\, x^{\alpha}(1-x)^{\beta}\log x = \left[S_1(\alpha) - S_1(1+\alpha+\beta)\right]\frac{\Gamma(1+\alpha)\Gamma(1+\beta)}{\Gamma(2+\alpha+\beta)},$$

$$\int_0^1 dx\, \frac{x^{\alpha}-1}{1-x} = -S_1(\alpha),$$

$$\int_0^1 dx\, x^{\alpha}\log x\log(1-x) = \frac{S_1(1+\alpha)}{(1+\alpha)^2} + \frac{S_2(1+\alpha)}{1+\alpha} - \frac{\pi^2}{6}\cdot\frac{1}{1+\alpha};$$

$$\int_0^1 dx\, x^{\alpha}\frac{\log^2 x}{1-x} = 2\zeta(3) - 2S_3(\alpha),$$

$$\int_0^1 dx\, \frac{x^{\alpha}}{1-x}\log x\log(1-x) = \frac{\pi^2}{6}S_1(\alpha) - S_1(\alpha)S_2(\alpha) - S_3(\alpha) + \zeta(3),$$

$$\int_0^1 dx\, x^{\alpha}(1-x)^{\beta}\log x\log(1-x)$$

$$= \frac{\Gamma(1+\alpha)\Gamma(1+\beta)}{\Gamma(2+\alpha+\beta)}\Big\{S_2(1+\alpha+\beta) - \frac{\pi^2}{6} + \left[S_1(\alpha) - S_1(\alpha+\beta+1)\right]$$

$$\times\left[S_1(\beta) - S_1(\alpha+\beta+1)\right]\Big\},$$

$$\int_0^1 dx\, x^{\alpha}(1-x)^{\beta}\log^2 x$$

$$= \frac{\Gamma(1+\alpha)\Gamma(1+\beta)}{\Gamma(2+\alpha+\beta)}\Big\{\left[S_1(\alpha) - S_1(\alpha+\beta+1)\right]^2$$

$$+ S_2(\alpha+\beta+1) - S_2(\alpha)\Big\};$$

etc.

Here, $S_l(\alpha) = \zeta(l) - \sum_{k=1}^{\infty}[1/(k+\alpha)^l]$, $l > 1$; $S_l(\alpha) = \sum_{j=1}^{\alpha} 1/j^l$ for $\alpha =$ positive integer, any l. Note that $S_2(\infty) = \pi^2/6$, $S_l(\infty) = \zeta(l)$ where ζ is Riemann's function. For $l = 1$, the above formula for S_l may be replaced by $S_1(\alpha) = \alpha\sum_{k=1}^{\infty}[1/k(k+\alpha)] = \sum_{j=1}^{\alpha} 1/j$, $\alpha =$ integer > 0. Also, $S_1(\alpha)$

$= \psi(\alpha + 1) + \gamma_E$, with $\psi(z) = d \log \Gamma(z)/dz$. For the special functions Γ, ψ, ζ, see Abramowicz and Stegun (1965).

Appendix C: Group-Theoretic Quantities

For $SU(3)$, $t^a = \lambda^a/2$, with

$$\lambda^j = \begin{pmatrix} \sigma^j & 0 \\ 0 & 0 \end{pmatrix}, \quad j = 1,2,3; \quad \lambda^4 = \begin{bmatrix} 0 & 0 & 1 \\ 0 & 0 & 0 \\ 1 & 0 & 0 \end{bmatrix}; \quad \lambda^5 = \begin{bmatrix} 0 & 0 & -i \\ 0 & 0 & 0 \\ i & 0 & 0 \end{bmatrix};$$

$$\lambda^6 = \begin{bmatrix} 0 & 0 & 0 \\ 0 & 0 & 1 \\ 0 & 1 & 0 \end{bmatrix}; \quad \lambda^7 = \begin{bmatrix} 0 & 0 & 0 \\ 0 & 0 & -i \\ 0 & i & 0 \end{bmatrix}; \quad \lambda^8 = \frac{1}{\sqrt{3}} \begin{bmatrix} 1 & & 0 \\ & 1 & \\ 0 & & -2 \end{bmatrix};$$

$$\sigma^1 = \begin{pmatrix} 0 & 1 \\ 1 & 0 \end{pmatrix}, \quad \sigma^2 = \begin{pmatrix} 0 & -i \\ i & 0 \end{pmatrix}, \quad \sigma^3 = \begin{pmatrix} 1 & 0 \\ 0 & -1 \end{pmatrix}.$$

We can introduce the matrices C^a with elements $C^a_{bc} = -if_{abc} \equiv -if^{abc}$. The commutation relations of the t and C are

$$[t^a, t^b] = i \sum f^{abc} t^c, \qquad [C^a, C^b] = i \sum f^{abc} C^c,$$

and anti-commutation relations are

$$\{t^a, t^b\} = \sum d^{abc} t^c + \frac{1}{3} \delta^{ab}.$$

The f are totally anti-symmetric, the $d_{abc} \equiv d^{abc}$ are totally symmetric, and the only nonzero elements (up to permutations) are as follows:

$$1 = f_{123} = 2f_{147} = 2f_{246} = 2f_{257} = 2f_{345}$$

$$= -2f_{156} = -2f_{367} = \frac{2}{\sqrt{3}} f_{458} = \frac{2}{\sqrt{3}} f_{678};$$

$$\frac{1}{\sqrt{3}} = d_{118} = d_{228} = d_{338} = -d_{888}, \qquad \frac{-1}{2\sqrt{3}} = d_{448} = d_{558} = d_{668} = d_{778},$$

$$\frac{1}{2} = d_{146} = d_{157} = d_{247} = d_{256} = d_{344} = d_{355} = -d_{366} = -d_{377}.$$

For an arbitrary $SU(N)$ group, we define the invariants C_A, C_F, T_F by

$$\delta_{ab} C_A = \operatorname{Tr} C^a C^b = \sum_{cc'} f^{acc'} f^{bcc'},$$

$$\delta_{ik} C_F = \Big(\sum_a t^a t^a \Big)_{ik} = \sum_{a,l} t^a_{il} t^a_{lk},$$

$$\delta_{ab} T_F = \operatorname{Tr} t^a t^b = \sum_{k,i} t^a_{ik} t^b_{ki}.$$

One has

$$C_A = N, \qquad C_F = \frac{N^2 - 1}{2N}, \qquad T_F = \frac{1}{2}.$$

A useful relation is

$$\mathrm{Tr}\, t^a t^b t^c = \frac{i}{4} f^{abc} + \frac{1}{4} d^{abc},$$

and other useful invariants are

$$\sum_{abc} d_{abc}^2 = \frac{40}{3}, \qquad \sum_{abc} f_{abc}^2 = 24, \qquad \sum_{rka} \epsilon_{irk} t_{jr}^a t_{kl}^a = -\frac{2}{3} \epsilon_{ijl}.$$

Appendix D: Feynman Rules for QCD

The Feynman rules are as follows:

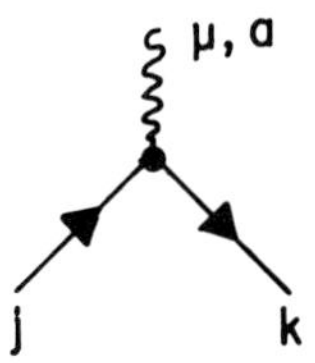

$$ig\gamma_\mu t_{kj}^a$$

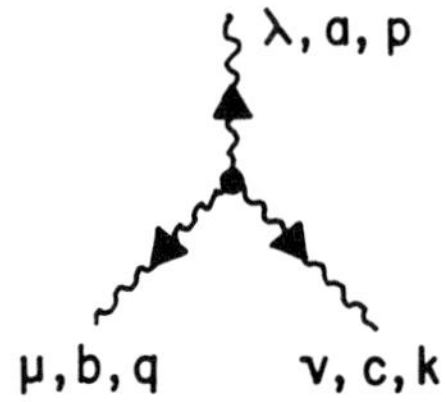

$$-gf^{abc}\left[(p-q)_\nu g_{\lambda\mu} + (q-k)_\lambda g_{\mu\nu} + (k-p)_\mu g_{\nu\lambda}\right]$$

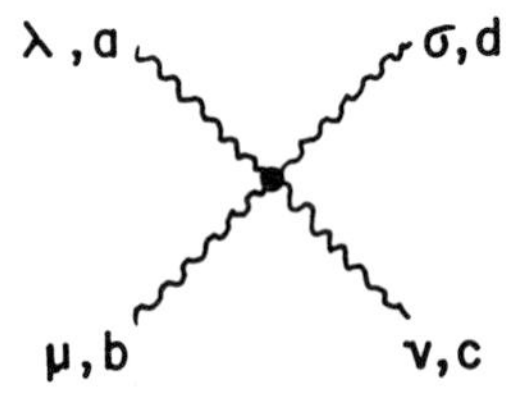

$$\begin{aligned} -ig^2 \sum_e \{ & f^{abe} f^{cde} (g_{\lambda\nu} g_{\mu\sigma} - g_{\lambda\sigma} g_{\mu\nu}) \\ & + f^{ace} f^{bde} (g_{\lambda\mu} g_{\nu\sigma} - g_{\lambda\sigma} g_{\mu\nu}) \\ & + f^{ade} f^{cbe} (g_{\lambda\nu} g_{\mu\sigma} - g_{\lambda\mu} g_{\sigma\nu}) \} \end{aligned}$$

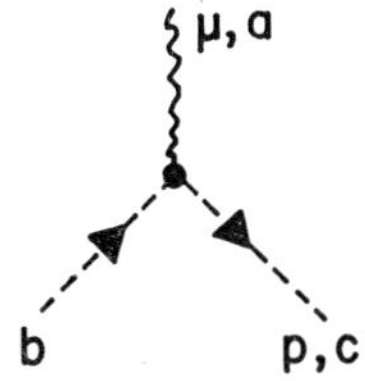

$$-gf^{acb}p_{\mu}$$

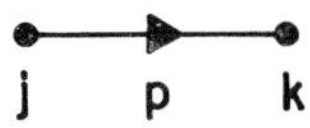

$$\frac{i}{\not{p} - m_j + i0}\,\delta_{jk}$$

$$i\,\frac{-g^{\mu\nu} + \xi k^{\mu}k^{\nu}/(k^2 + i0)}{k^2 + i0}\,\delta_{ab} \quad \text{(Lorentz gauges)}$$

$$i\,\frac{-g^{\mu\nu} + (n^{\mu}k^{\nu} + n^{\nu}k^{\mu})/n\cdot k - n^2(k^{\mu}k^{\nu}/n\cdot k)}{k^2 + i0}\,\delta_{ab}$$

(axial gauge)

$$\frac{i}{k^2 + i0}\,\delta_{ab}$$

An overall $(2\pi)^4\delta(P_i - P_f)$ is included for global energy-momentum conservation, and (-1) for closed fermion or ghost loops. Statistical factors are as follows:

$\frac{1}{2!}$ for  ;

$\frac{1}{3!}$ for  , etc.

Insert $\nu_0^{4-D}\int d^Dk/(2\pi)^D \equiv \int d^D\hat{k}$ for each loop integration. Diagrams with disconnected bubbles are excluded. A diagram is to be read against the directions of the arrows of its oriented lines. To obtain the S-matrix elements, amputate the amplitude and add, for external particles,

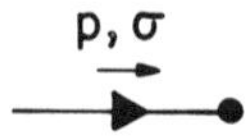

$$(2\pi)^{-3/2}u(p,\sigma)$$

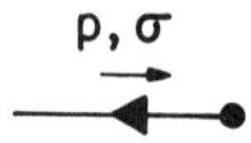

$$(2\pi)^{-3/2}\bar{v}(p,\sigma)$$

$$(2\pi)^{-3/2}\epsilon^{\mu}(k,\lambda)$$

$$(2\pi)^{-3/2}\epsilon^{\mu}(k,\lambda)^*$$

$$(2\pi)^{-3/2}\bar{u}(p,\sigma)$$

$$(2\pi)^{-3/2}v(p,\sigma)$$

$$(2\pi)^{-3/2}\epsilon^{\mu}(k,\lambda)^*$$

$$(2\pi)^{-3/2}\epsilon^{\mu}(k,\lambda)$$

The spinors and polarization vectors are normalized to

$$\sum_{\sigma} u(p,\sigma)\bar{u}(p,\sigma) = \not{p} + m,$$

$$\sum_{\lambda} \epsilon^{\mu}(k,\lambda)^* \epsilon^{\nu}(k,\lambda) = -g^{\mu\nu} \text{(Feynman gauge)}.$$

These rules differ from the ones in Bjorken and Drell (1965) by the normalization of the spinors,

$$\sum_{\sigma} u_{BD}\bar{u}_{BD} = \frac{\not{p}+m}{2m},$$

and the factors $(2\pi)^{-3/2}$ due to our normalization of $\mathscr{T}$ which differs from $\mathscr{T}_{BD}$ by precisely these factors.

Appendix E: Feynman Rules for Composite Operators

Let $\gamma_+ = 1$, $\gamma_- = \gamma_5$, and Δ an arbitrary four-vector with $\Delta^2 = 0$.

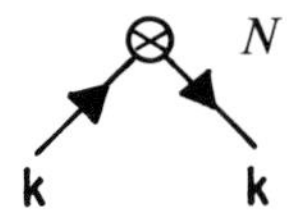

$$N = \bar{q}(0)\gamma^{\mu_1} \ldots \partial^{\mu_n}\gamma_\pm q(0)$$

$$\not{\Delta}(\Delta \cdot k)^{n-1}\gamma_\pm$$

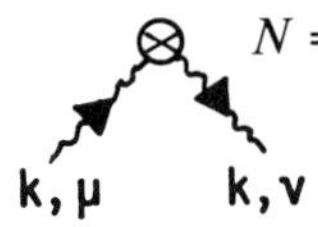

$$N = G^{\mu\mu_1}\partial^{\mu_2} \ldots \partial^{\mu}G$$

$$g_{\mu\nu}(\Delta \cdot k)^n + k^2\Delta_\mu\Delta_\nu(\Delta \cdot k)^{n-2} - (k_\mu\Delta_\nu + \Delta_\mu k_\nu)(\Delta \cdot k)^{n-1}$$

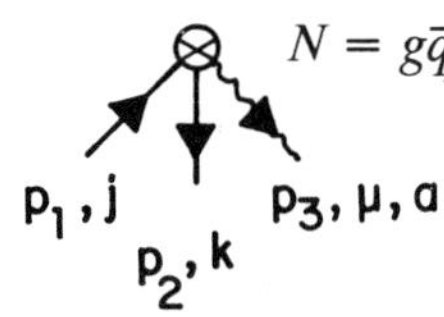

$$N = g\bar{q}_j(0)\gamma^{\mu_1} \ldots B_a^{\mu} t_{jk}^a \ldots \gamma^{\mu_n}\gamma_\pm q_k(0)$$

$$g t_{ij}^a \Delta^\mu \not{\Delta} \sum_{j=0}^{n-2} (\Delta \cdot p_1)^j (\Delta \cdot p_2)^{n-j-2}\gamma_\pm$$

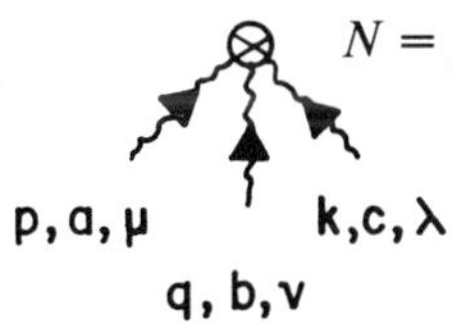

$$N = gG^{\mu\nu_1}\partial^{\mu_2} \ldots B^{\mu_i} \ldots G$$

$$\begin{aligned} \frac{ig}{3!} f_{abc} \Big\{ & \Delta_\nu\big[\Delta_\lambda k_\mu(\Delta \cdot p) + p_\lambda \Delta_p(\Delta \cdot k) \\ & - g_{\mu\lambda}(\Delta \cdot p)(\Delta \cdot k) - \Delta_\mu\Delta_\lambda(p \cdot k)\big] \\ & + \sum_{j=1}^{n-2} (-1)^j (\Delta \cdot p)^{j-1} (\Delta \cdot k)^{n-j-2} \\ & + \big[(g_{\mu\lambda}\Delta_\nu - \Delta_\mu g_{\nu\lambda})(\Delta \cdot k) \\ & + \Delta_\lambda(\Delta_\mu k_\nu - \Delta_\nu k_\mu)\big](\Delta \cdot k)^{n-2} \Big\} \end{aligned}$$

$$+ \text{ permutations.}$$

Cf. also Floratos, Ross, and Sachrajda (1977, 1979).

Appendix F: Some Singular Functions

We define x-space causal functions by

$$\Delta(x;m^2)=\int\frac{d^4k}{(2\pi)^4}e^{-ik\cdot x}\frac{i}{k^2-m^2+i0},$$

$$D_\xi^{\mu\nu}(x)=i\int\frac{d^4k}{(2\pi)^4}e^{-ik\cdot x}\frac{-g^{\mu\nu}+\xi k^\mu k^\nu/(k^2+i0)}{k^2+i0},$$

$$S(x;m)=i\int\frac{d^4k}{(2\pi)^4}e^{-ik\cdot x}\frac{\not{k}+m}{k^2-m^2+i0}.$$

We will at times omit the variable m from Δ, S. In terms of time ordered VEVs,

$$\langle T\phi(x)\phi(0)\rangle_0=\Delta(x;m);\qquad \langle Tq^j(x)\bar{q}^k(0)\rangle_0=\delta^{jk}S(x,m),$$

$$\langle TB_a^\mu(x)B_b^\nu(0)\rangle_0=\delta_{ab}D_\xi^{\mu\nu}(x).$$

The character of Green's functions of the propagators is exhibited clearly by the equations $(\partial_x^2+m^2)i\Delta(x-y)=\delta(x-y)$, etc. Furthermore,

$$S(x,m)=(i\not{\partial}+m)\Delta(x,m).$$

On the light cone,

$$\Delta(x,m^2)\underset{x^2\to 0}{\simeq}\frac{-1}{4\pi^2}\cdot\frac{1}{x^2-i0}+\frac{im^2\theta(x^2)}{16\pi}+\frac{m^2}{8\pi^2}\log\frac{m|x^2|^{1/2}}{2}+\cdots$$

$$S(x)\underset{x^2\to 0}{\simeq}\frac{2ix_\mu\gamma^\mu}{(2\pi)^2(x^2-i0)^2}+\cdots,\ \text{etc.}$$

Additional relations may be found in Bjorken and Drell (1965).[115] Fourier transforms of distributions are given in Gel'fand and Shilov (1962), pp. 277ff, 361ff. The ones used in the text are

$$\int d^4x\,e^{-ik\cdot x}\frac{1}{x^2\pm i0}=-4\pi^2\frac{i}{k^2\mp i0},$$

$$\int d^4x\,e^{-ik\cdot x}\frac{1}{(x^2\pm i0)^2}=-\pi^2 i\log(k^2\mp i0)+\text{constant}.$$

[115] Our causal functions differ by i from the Bjorken–Drell (1965) ones: $S=iS_{\text{BD}}$, $D=iD_{\text{BD}},\ldots$.

Equal-time and light-cone commutation relations for fermions:

$$\{q^i_\alpha(x), q^k_\beta(x)\} = 0; \qquad \delta(x^0 - y^0)\{q^i_\alpha(x), q^k_\beta(y)^+\} = \delta_{\alpha\beta}\delta_{ik}\delta(x-y),$$

$$\{q_\alpha(x), \bar{q}_\beta(0)\} \underset{x^2\to 0}{\simeq} (\partial\!\!\!/ - im)_{\alpha\beta}\left\{\frac{1}{2\pi}\epsilon(x^0)\delta(x^2) - \frac{m}{4\pi\sqrt{x^2}}\theta(x^2)\epsilon(x^0) + \cdots\right\}$$

Appendix G: Kinematics, Cross Sections, Decay Rates

The states of a particle with helicity λ and momentum p are normalized according to[116]

$$\langle p', \lambda' | p, \lambda\rangle = 2p^0\delta_{\lambda\lambda'}\delta(\vec{p} - \vec{p}'), \qquad P^\mu|p,\lambda\rangle = p^\mu|p,\lambda\rangle.$$

This corresponds to a density of particles per unit volume of

$$\rho(p) = \frac{2p^0}{(2\pi)^3}\,.$$

We define the scattering amplitude in terms of the S matrix:

$$S = 1 + i\mathcal{T}, \qquad \langle f|\mathcal{T}|i\rangle = \delta(P_f - P_i)F(i\to f).$$

For $|i\rangle$ a state of two particles with masses m_1, m_2, the cross section is then

$$d\sigma(i\to f) = \frac{2\pi^2}{\lambda^{1/2}(s, m_1^2, m_2^2)}\delta(P_f - P_i)|F(i\to f)|^2\frac{d\vec{p}_{f1}}{2p^0_{f1}}\cdots\frac{d\vec{p}_{fn}}{2p^0_{fn}},$$

where

$$\lambda(a,b,c) = a^2 + b^2 + c^2 - 2ab - 2ac - 2bc, \qquad s = P_i^2\,.$$

For the case $p_1 + p_2 \to p_2' + p_2'$, one obtains

$$\frac{d\sigma(i\to f)}{dt} = \frac{\pi^3}{\lambda(s, m_1^2, m_2^2)}|F(i\to f)|^2,$$

$$\left.\frac{d\sigma}{d\Omega}\right|_{\text{em}} = \frac{\pi^2}{4s}\cdot\frac{q'}{q}|F(i\to f)|^2,$$

$$\sigma(i - \text{all}) = \left[4\pi^2/\lambda^{1/2}(s, m_1^2, m_2^2)\right]\operatorname{Im} F(i\to i).$$

[116]The transformation properties for an arbitrary field are

$$U(a)\Phi(x)U^{-1}(a) = \Phi(x+a), \qquad U(a) = e^{iP\cdot a}.$$

Here

$$t=(p_2-p_2')^2, \qquad q=|\vec{p}_{1\,\mathrm{cm}}|=\frac{\lambda^{1/2}(s,m_1^2,m_2^2)}{2s^{1/2}},$$

$$q'=|\vec{p}_{1\,\mathrm{cm}}'|=\frac{\lambda^{1/2}(s,m_1'^2,m_2'^2)}{2s^{1/2}},$$

$$\Omega_{\mathrm{cm}}\approx(\theta_{\mathrm{cm}},\phi_{\mathrm{cm}}), \qquad d\Omega=d\cos\theta\,d\phi.$$

Likewise, the decay rate is[117]

$$d\Gamma(i\to f)=\frac{1}{4\pi m_i}\delta(P_i-P_f)|F(i\to f)|^2\frac{d\vec{p}_{f1}}{2p_{f1}^0}\cdots\frac{d\vec{p}_{fn}}{2p_{fn}^0}, \qquad P_i=\begin{pmatrix}m_i\\ \vec{0}\end{pmatrix}.$$

Our units are such that $\hbar=c=1$. Some useful formulas in this system are:

$1(\mathrm{MeV})^{-1}=1.973\times10^{-11}$ cm $=6.582\times10^{-22}$ sec.
$1(\mathrm{GeV})^{-4}=3.894\times10^{4}$ barn.
1 MeV $=1.783\times10^{-27}$ gr $=1.602\times10^{-6}$ erg;
1 cm $=5.068\times10^{10}\,(\mathrm{MeV})^{-1}$, 1 sec $=1.519\times10^{21}\,(\mathrm{MeV})^{-1}$
1 barn $=2.568\times10^{-3}\,(\mathrm{GeV})^{-2}$
1 gr $=5.610\times10^{26}$ MeV, 1 erg $=6.242\times10^{5}$ MeV.

Appendix H: Functional Derivatives

A functional is a application of the space of sufficiently smooth functions, $\{f(x)\}$, into the complex numbers:

$$F:f\to F[f].$$

Note that F need not be linear. We will treat functionals $F[f,g,\ldots]$ in the same way. We may consider a functional as a generalization of an ordinary function in the following sense: divide the space of the x values[118] in N cells, and let each x_j lie one in each cell. Then $F[f]$ is the limit for vanishing cell size of $F_N(f_1,\ldots,f_j,\ldots)$, $f_j\equiv f(x_j)$. The derivative $\partial F_N/\partial f_j$ is

$$\frac{\partial F_N(f_1,\ldots,f_j,\ldots)}{\partial f_j}=\lim_{\epsilon\to0}\frac{F_N(f_1,\ldots,f_j+\epsilon,\ldots)-F_N(f_1,\ldots,f_j,\ldots)}{\epsilon},$$

[117] All the formulas are valid for indistinguishable or distinguishable particles. When calculating integrated rates, however, we have to divide by the number of redundant permutations: For example, if we integrate over the momenta of j identical bosons or fermions, divide by $j!$.

[118] We take this space to be of finite size, L. Otherwise, an extra limit $L\to\infty$ has to be performed.

i.e., it may be obtained by shifting $f_i \to f_i + \epsilon\delta_{ij}$. So, in the limit, we define

$$\frac{\delta F[f]}{\delta f(y)} = \lim_{\epsilon\to 0} \frac{F[f+\epsilon\delta_y] - F[f]}{\epsilon},$$

where δ_y is the delta function at y: $\delta_y(x) = \delta(x-y)$. An important case is that of integral functionals:

$$F[f] = \int dx\, K_F(x) f(x);$$

then,

$$\frac{\delta F[f]}{\delta f(y)} = K_F(y).$$

Taylor series may be generalized to functional series. If the kernels $K_n(x_1, \ldots, x_n)$ are symmetric (anti-symmetric for fermionic f) and we consider the functional

$$F[f] = \sum_{n=0}^{\infty} \frac{1}{n!} \int dx_1 \ldots dx_n\, K_n(x_1, \ldots, x_n) f(x_1) \ldots f(x_n),$$

we may easily verify that

$$K_n(x_1, \ldots, x_n) = \frac{\delta^n F[f]}{\delta f(x_1) \ldots \delta f(x_n)}.$$

A concept related to that of the functional derivative is that of functional integration. We define

$$\int \prod_x df(x)\, F[f] \equiv \lim_{N\to\infty} \int df_1 \ldots df_N\, F_N(f_1, \ldots, f_N).$$

As in the case of functional differentiation, functional integration obeys rules analogous to those of ordinary integration. Both for differentiation and integration, some modifications are required to accommodate anti-commuting functions; they are described in Section 39.

Functional derivatives of expressions that do not involve integrals are found by reexpressing them as integrals. For example, the derivative entering Equation (41.9) is so evaluated:

$$\begin{aligned} \frac{\delta\, \partial B_a(x)}{\delta B_b^\rho(y)} &= \frac{\delta}{\delta B_b^\rho(y)} \frac{\partial}{\partial x^\mu} \sum_c \int d^4z\, \delta(z-x)\delta_{ac} B_c^\mu(z) \\ &= \delta_{ab} \frac{\partial}{\partial x^\rho} \delta(x-y). \end{aligned}$$

Appendix I: Gauge-Invariant Operator Product

It is intuitively obvious that in a gauge theory, an expression like those appearing in OPE,

$$\bar{q}(0)q(x) = \sum \frac{x^{\mu_1} \ldots x^{\mu_n}}{n!} \bar{q}(0)\partial_{\mu_1} \ldots \partial_{\mu_n} q(0),$$

should be replaced by another with derivatives substituted by covariant derivatives, $\partial^\mu \to D^\mu$. Here we sketch a formal proof of how this comes about. When the fields are interacting, the propagators are not free propagators. For example, for a fermion in the presence of the gluonic field, the propagator satisfies the equation, derived directly from the Lagrangian,

$$(i\not{D} - m)S_{\text{int}}(x, y) = i\delta(x - y).$$

Retaining only the more singular (lower twist) terms, the solution to this is

$$S_{\text{int}}(x, y) \approx \left\{ P \exp i \int_y^x dz_\mu \sum t^a B_a^\mu(z) \right\} S(x - y),$$

where S is the free propagator and P indicates ordering along the path from y to x. If we repeat the OPE taking this into account, we find that operator products $\bar{q}(x)q(y)$ are replaced by the gauge-invariant combination

$$\bar{q}(x) \left\{ P \exp i \int_y^x dz_\mu \sum t^a B_a^\mu(z) \right\} q(y),$$

whose expansion for $x \approx y$ is precisely that with covariant derivatives. The same is of course true for operators built from gluon fields. Additional details may be found in the paper of Wilson (1975) and the review of Efremov and Radyushkin (1980*b*).

References

Abad, J., and Humpert, B. (1978). *Phys. Lett.*, **B77**, 105.

Abarbanel, H. D., Goldberger, M. L., and Treiman, S. B. (1969). *Phys. Rev. Lett.* **22**, 500.

Abbott, L. F. (1981). *Nucl. Phys.* **B185**, 189.

Abbott, L. F., Atwood, W. B., and Barnett, R. M. (1980). *Phys. Rev.* **D22**, 582.

Abramowicz, M., and Stegun, I. E. (1965). *Handbook of Mathematical Functions*, Dover, New York.

Adler, S. L. (1966). *Phys. Rev.* **143**, 1144.

——— (1969). *Phys. Rev.* **177**, 2426.

——— (1971). In *Lectures in Elementary Particle and Quantum Field Theory* (Deser, Grisaru, and Pendleton, eds.), MIT Press.

Adler, S. L., and Bardeen, W. A. (1969). *Phys. Rev.* **182**, 1517.

Ali, A. (1982). *Phys. Lett.* **110B**, 67.

Alphonse X "The Wise" (1221–1284). Quoted in *The Harvest of the Quick Eye: A Selection of Scientific Quotations*, A. L. Mackay, The Institute of Physics, London, 1977.

Altarelli, G. (1982). *Phys. Reports* **81C**, 1.

Altarelli, G., and Parisi, G. (1977). *Nucl. Phys.* **B126**, 298.

Altarelli, G., Ellis, R. K., and Martinelli, G. (1978). *Nucl. Phys.* **B143**; 521 and Erratum **B146**, 544.

——— (1979). *Nucl. Phys.* **B157**, 461.

Amati, D. *et al.* (1980). *Nucl. Phys.* **B173**, 429.

Anaximander, (–546) Quoted in Aristotle's *Physics*. [From the German translation by W. Capelle, *Die Vorsokratiker*, Kröner-Verlag, Stuttgart, 1963.]

Anderson, H. L. *et al.* (1979). *Phys. Rev.* **D20**, 2645.

Applequist, T., and Carrazzone, J. (1975). *Phys. Rev.* **D11**, 2865.

Applequist, T., and Georgi, H. (1973). *Phys. Rev.* **D8**, 4000.

Applequist, T., and Politzer, H. D. (1975). *Phys. Rev. Lett.* **34**, 43.

Aubert, J. J. *et al.* (1981). *Phys. Lett.* **105B**, 315.

Baluni, V. (1979). *Phys. Rev.* **D19**, 2227.

Barbieri, R., Gatto, R., Kögerler, R., and Kunszt, Z. (1975). *Phys. Lett.* **57B**, 455.

Barbieri, R., Ellis, J., Gaillard, M. K., and Ross, G. G. (1976). *Nucl. Phys.* **B117**, 50.

Barbieri, R., Curci, G., d'Emilio, E., and Remiddi, E. (1979). *Nucl. Phys.* **B154**, 535.

Bardeen, W. A. (1974). *Nucl. Phys.* **B75**, 246.

Bardeen, W. A., and Buras, A. J. (1979). *Phys. Rev.* **D20**, 166.

Bardeen, W. A., Buras, A. J., Duke, D. W., and Muta, T. (1978). *Phys. Rev.* **D18**, 3998.

Barger, V. D., and Cline, D. B. (1969). *Phenomenological Theories of High Energy Scattering*, Benjamin, New York.

Barnett, R. M., Dine, M., and McLerran, L. (1980). *Phys. Rev.* **D22**, 594.

Bartels, J. (1979). In *Quantum Chromodynamics* (Alonso and Tarrach, eds.), Springer, Berlin.

Baulieu, L., Ellis, J., Gaillard, M. K., and Zakrewski, W. J. (1979). *Phys. Lett.* **81B**, 224.

Becchi, C., Rouet, A., and Stora, R. (1974). *Phys. Lett.* **52B**, 344.

——— (1975). *Commun. Math. Phys.* **42**, 127.

Becchi, C., Narison, S., de Rafael, E., and Ynduráin, F. J. (1981), *Z. Phys.* **C8**, 335.

Belavin, A., Polyakov, A., Schwartz, A., and Tyupkin, Y. (1975). *Phys. Lett.* **59B**, 85.

Bell, J. S., and Jackiw, R. (1969). *Nuovo Cimento* **60A**, 47.

Berezin, F. A. (1966). *Methods of the Second Quantization*, Academic Press, New York.

Bernabeu, J., Tarrach, R., and Ynduráin, F. J. (1979). *Phys. Lett.* **79B**, 464.

Bjorken, J. D. (1969). *Phys. Rev.* **179**, 1547.

Bjorken, J. D., and Drell, S. D. (1965). *Relativistic Quantum Fields*, McGraw Hill, London.

Bjorken, J. D., and Paschos, E. A. (1969). *Phys. Rev.* **185**, 1975.

Bloch, F., and Nordsiek, A. (1937). *Phys. Rev.* **52**, 54.

Bodek, A., *et al.* (1979). *Phys. Rev.* **D20**, 1471.

Bogoliubov, N. N., Logunov, A. A., and Todorov, I. T. (1975). *Axiomatic Field Theory*, Benjamin, New York.

Bogoliubov, N. N., and Shirkov, D. V. (1959). *Introduction to the Theory of Quantized Fields*, Interscience, London.

Bohr, N., and Rosenfeld, L. (1933). *Kong. Dansk Vid. Selsk. Matt.-Fys. Medd.* **12**, No. 8.

——— (1950). *Phys. Rev.* **78**, 794.

Bollini, C. G., and Gianbiagi, J. J., (1972). *Phys. Lett.* **40B**, 566.

Bollini, C. G., Gianbiagi, J. J., and González-Domínguez, A. (1964). *Nuovo Cimento*, **31**, 550.

Brandt, R., and Preparata, G. (1970). *Ann. Phys.* (N.Y.) **61**, 119.

——— (1971). *Nucl. Phys.* **B27**, 541 (1971).

Brodsky, S. J., and Farrar, G. (1973). *Phys. Rev. Lett.*, **31**, 1153.

Brodsky, S. J., Frishman, Y., Lepage, G. P., and Sachrajda, C. T. (1980). *Phys. Lett.* **91B**, 239.

Brodsky, S. J., and Lepage, G. P. (1980). *Phys. Rev.* **D22**, 2157.

Buras, A. J. (1980). *Rev. Mod. Phys.* **52**, 199.

——— (1981). In *Topical Questions in QCD*, *Phys. Scripta*, **23**, No. 5.

Buras, A. J., and Gaemers, K. J. F. (1978). *Nucl. Phys.* **B132**, 249.

Cabibbo, N., Parisi, G., Testa, M. (1970). *Nuovo Cimento Lett.* **4**, 35.

Callan, C. G. (1970). *Phys. Rev.* **D2**, 1541.

Callan, C. G., Coleman, S., and Jackiw, R. (1970). *Ann. Phys.* (N.Y.) **59**, 42.

Callan, C. G., Dashen, R., and Gross, D. J. (1976). *Phys. Lett.* **66B**, 375.

Callan, C. G., and Gross, D. J. (1969). *Phys. Rev. Letters*, **22**, 156.

Calvo, M. (1977). *Phys. Rev.* **D15**, 730.

Carroll, L. (1896). *Through the Looking Glass*. [Reprinted in *The Complete Works of Lewis Carroll*, Random House, New York.]

Caswell, W. E. (1974). *Phys. Rev. Lett.* **33**, 224.

Caswell, W. E., and Wilczek, F. (1974). *Phys. Lett.* **49B**, 291.

Cerveró, J., Jacobs, L., and Nohl, C. (1977). *Phys. Lett.* **69B**, 351.

Chetyrkin, K. G., Kataev, A. L., and Tkachov, F. V. (1979). *Phys. Lett.* **85B**, 277.

Chew, G. F., and Rozenzweig, C. (1978). *Phys. Reports* **C41**, No. 5.

Christ, N., and Lee, T. D. (1980). *Phys. Rev.* **D22**, 939.

Christ, N., Hasslacher, B., and Muller, A. (1972). *Phys. Rev.* **D6**, 3543.

Ciafalloni, M., and Curci, G. (1981). *Phys. Lett.* **102B**, 352.

Coleman, S. (1966). *J. Math. Phys.* **7**, 787.

Coleman, S., and Gross, D. J. (1973). *Phys. Rev. Lett.* **31**, 851.

Collins, J. C., Duncan, A., and Joglekar, S. D. (1977), *Phys. Rev.* **D16**, 438.

Conan Doyle, A. (1892). "Silver Blaze," Strand Magazine, London. [Reprinted in *The Memoirs of Sherlock Holmes*, Penguin Books, 1970.]

Coquereaux, R. (1980). *Ann. Phys.* (N.Y.) **125**, 401.

——— (1981). *Phys. Rev.* **D23**, 1365.

Cornwall, J. M., and Norton, R. E. (1969). *Phys. Rev.* **177**, 2584.

Corrigan, E., and Fairlie, D. B. (1977). *Phys. Lett.* **67B**, 69 (1977).

Creutz, M. (1981). In *Topical Questions in QCD*, *Phys. Scripta* **23**, No. 5.

Crewther, R. J. (1972). *Phys. Rev. Lett.* **28**, 1421.

——— (1979*a*). *Riv. Nuovo Cimento* **2**, No. 8.

——— (1979*b*). In *Field Theoretical Methods in Elementary Particle Physics*, Proc. Kaiserslautern School.

Crewther, R. J., Di Vecchia, P., Veneziano, G., and Witten, E. (1980). *Phys. Lett.* **88B**, 123; and Erratum, **91B**, 487.

Curci, G., Furmanski, W., and Petronzio, R. (1980). *Nucl. Phys.* **B175**, 27.

De Alfaro, V., Fubini, S., and Furlan, G. (1976). *Phys. Lett.* **65B**, 163.

——— (1977). *Phys. Lett.* **72B**, 203.

De Groot, J. G., *et al.* (1979). *Z. Phys.* **C1**, 143.

De Rújula, A., Ellis, J., Floratos, E. G., and Gaillard, M. K. (1978). *Nucl. Phys.* **B138**, 387.

De Rújula, A., Georgi, H., and Glashow, S. L. (1975). *Phys. Rev.* **D12**, 147.

De Rújula, A., Georgi, H., and Politzer, H. D. (1977*a*). *Ann. Phys.* (N.Y.) **103**, 315.

——— (1977*b*). *Phys. Rev.* **D15**, 2495.

De Rújula, A., and Glashow, S. L. (1975). *Phys. Rev. Lett.* **34**, 46.

De Rújula, A., Giles, R. C., and Jaffe, R. (1978). *Phys. Rev.* **17D**, 285.

De Witt, B. (1964). *Relativity, Groups and Topology*, p. 587 ff., Blakie & Son, London.

Dine, M., and Sapiristein, J. (1979). *Phys. Rev. Lett.* **43**, 668.

Di Giacomo, A., and Rossi, G. C. (1981). *Phys. Lett.* **100B**, 481.

Dixon, J. A., and Taylor, J. C. (1974). *Nucl. Phys.* **B78**, 552.

Dokshitzer, Yu. L., Dyakonov, D. I., and Troyan, S. I. (1980). *Phys. Reports* **C58**, 269.

Dominguez, C. A. (1978). *Phys. Rev. Lett.* **41**, 605.

Drell, S. D., and Yan, T. M. (1971). *Ann. Phys.* (N.Y.) **66**, 578.

Duncan, A. (1981). In *Topical Questions in QCD*, *Phys. Scripta*, **23**, No. 5.

Duncan, A., and Muller, A. (1980*a*). *Phys. Lett.* **93B**, 119.

——— (1980*b*). *Phys. Rev.* **D21**, 1636.

Eden, R. J., Landshoff, P. V., Olive, D. I., and Polkinghorne, J. C. (1966). *The Analytic S-Matrix*, Cambridge University Press.

Efremov, A. V., and Radyushin, A. V. (1980*a*). *Phys. Lett.* **94B**, 245.

——— (1980*b*). *Riv. Nuovo Cimento*, **3**, No. 2.

Ellis, J. (1976). In *Weak and Electromagnetic Interactions at High Energy*, North Holland, Amsterdam.

Ellis, J., and Gaillard, M. K. (1979). *Nucl. Phys.* **B150**, 141.

Ellis, J., and Sachrajda, C. T. (1980). In *Quarks and Leptions* (M. Lévy, *et al.*, eds.), p. 285; Plenum Press, London.

Ellis, R. K., *et al.* (1979). *Nucl. Phys.* **B152**, 285.

Epstein, H. Glaser, V., and Martin, A. (1969). *Commun. Math. Phys.* **13**, 257.

Fadeyev, L. D. (1976). In *Methods in Field Theory* (Balian and Zinn–Justin, eds.), North-Holland, Amsterdam.

Fadeyev, L. D., and Popov, Y. N. (1967). *Phys. Lett.* **25B**, 29.

Fadeyev, L. D., and Slavnov, A. A. (1980). *Gauge Fields*, Benjamin, New York.

Farhi, (1977), *Phys. Rev. Lett.* **39**, 1587.

Farrar, G., and Jackson, D. R. (1979). *Phys. Rev. Lett.* **43**, 246.

Ferrara, S. Gatto, R., and Grillo, A. F. (1972). *Phys. Rev.* **D5**, 5102.

Feynman, R. P. (1963). *Acta Phys. Polonica*, **24**, 697.

——— (1969). *Phys. Rev. Lett.* **23**, 1415.

——— (1972). *Photon Hadron Interactions*, Benjamin, New York.

Feynman, R. P., and Hibbs, A. R. (1965). *Quantum Mechanics and Path Integrals*, McGraw Hill, New York.

Feynman, R. P., and Field, R. D. (1977). *Phys. Rev.* **D15**, 2590.

Flamm, D., and Schoberl, F. (1981). *Introduction to the Quark Model of Elementary Particles*, Gordon and Breach, New York.

Floratos, E. G., Kounnas, C. and Lacaze, R. (1981). *Nucl. Phys.* **B192**, 417.

Floratos, E. G., Ross, D. A., and Sachrajda, C. T. (1977). *Nucl. Phys.* **B129**, 66; and Erratum (1978), **B139**, 545.

——— (1979). *Nucl. Phys.* **B152**, 493.

Forgács, P., Horváth, Z., and Palla, L. (1981). *Phys. Rev. Lett.* **46**, 392.

Fritzsch, H., and Gell-Mann, M. (1971). In *Broken Scale Invariance and the Light Cone*, (Dal Cin, Iverson, and Perlmutter, eds.), Gordon and Breach, London.

——— (1972). Proc. XVI Int. Conf. on High Energy Physics, Vol. 2, p. 135, Chicago.

Fritzsch, H., Gell-Mann, and Leutwyler, H. (1973). *Phys. Lett.* **B47**, 365.

Furmanski, W., and Petronzio, R. (1980). *Phys. Lett.* **97B**, 437.

Gaillard, M. K. (1978). Proc. SLAC Summer Institute on Particle Physics.

Gaillard, M. K., and Lee, B. W. (1974). *Phys. Rev.* **D10**, 897.

Gastmans, R. and Meuldermans, M. (1973). *Nucl. Phys.* **B63**, 277.

Gel'fand, I. M., and Shilov (Chilov) G. E. (1962). *Les Distributions*, Vol. I., Dunod, Paris.

Gell-Mann, M. (1961). Caltech preprint CTSL-20, unpublished.

——— (1962). *Phys. Rev.* **125**, 1067.

——— (1964*a*). *Phys. Lett.* **8**, 214.

——— (1964*b*). *Physics*, **1**, 63.

Gell-Mann, M., and Low, F. (1954). *Phys. Rev.* **95**, 1300.

Gell-Mann, M., Oakes, R. J., and Renner, B. (1968). *Phys. Rev.* **175**, 2195.

Georgi, H., and Politzer, H. D. (1974). *Phys. Rev.* **D9**, 416.

——— (1976). *Phys. Rev.* **D14**, 1829.

Gervais, J.-L., and Neveu, A. (1972). *Nucl. Phys.* **46B**, 381.

Glashow, S. L. (1968). In *Hadrons and Their Interactions*, p. 83, Academic Press, New York.

Glashow, S. L., Iliopoulos, J., and Maiani, L. (1970). *Phys. Rev.* **D2**, 1285.

Glashow, S. L., and Weinberg, S. (1968). *Phys. Rev. Lett.* **20**, 224.

Goldstone, J. (1961). *Nuovo Cimento* **19**, 154.

González-Arroyo, A., and López, C. (1980). *Nucl. Phys.* **B166**, 429.

González-Arroyo, A., López, C., and Ynduráin, F. J. (1979). *Nucl. Phys.* **B153**, 161.

——— (1980). *Nucl. Phys.* **B174**, 474.

Gordon, B. A., *et al.* (1979). *Phys. Rev.* **D20**, 2643.

Gottlieb, J. (1978). *Nucl. Phys.* **B139**, 125.

Greenberg, O. W. (1964). *Phys. Rev. Lett.* **13**, 598.

Gribov, V. N. and Lipatov, I. N. (1972). *Sov. J. Nucl. Phys.* **15**, 438.

Gross, D. J. (1974). *Phys. Rev. Lett.* **32**, 1071.

——— (1976). In *Methods in Field Theory* (Balian and Zinn-Justin, eds.), North-Holland, Amsterdam.

Gross, D. J., and Llewellyn Smith, C. H. (1969). *Nucl. Phys.* **B14**, 337.

Gross, D. J., Pisarski, R. D., and Yaffe, L. G. (1981). *Rev. Mod. Phys.* **53**, 43.

Gross, D. J., and Wilczek, F. (1973*a*). *Phys. Rev. Lett.* **30**, 1323.

——— (1973*b*), *Phys. Rev.* **D8**, 3635.

——— (1974). *Phys. Rev.* **D9**, 980.

Gürsey, F., and Radicati, L. A. (1964). *Phys. Rev. Lett.* **13**, 173.

Hamber, H., and Parisi, G. (1981). *Phys. Rev. Lett.* **47**, 1792.

Han, M., and Nambu, Y. (1965). *Phys. Rev.* **B139**, 1006.

Harada, K., and Muta, T. (1980). *Phys. Rev.* **D22**, 663.

Hasenfratz, P., and Kuti, J. (1978). *Phys. Reports* **C75**, 1.

Hinchliffe, I., and Llewellyn Smith C. H. (1977). *Nucl. Phys.* **B128**, 93.

Hubschmid, W., and Mallik, S. (1981). *Nucl. Phys.* **B193**, 368.

Humpert, B. and van Neerven, W. L. (1981). *Nucl. Phys.* **B184**, 225.

Ioffe, B. L. (1981). *Nucl. Phys.* **B188**, 317.

Itzykson, C. and Zuber, J. B. (1980). *Quantum Field Theory*, McGraw Hill, New York.

Izuka, J., Okada, K., and Shito, D. (1966). *Progr. Theor. Phys.* **35**, 1061.

Jackiw, R., Nohl, C., and Rebbi, C. (1977). *Phys. Rev.* **D15**, 1642.

Jackiw, R. and Rebbi, C. (1976). *Phys. Rev.* **D13**, 3398.

Jacob, M., and Landshoff, P. (1978). *Phys. Reports* **C48**, 285.

Jaffe, R., and Ross, G. G. (1980). *Phys. Lett.* **93B**, 313.

Jaffe, R., and Soldate, M. (1981). *Phys. Lett.* **105B**, 467.

Jones, D. T. R. (1974). *Nucl. Phys.* **B75**, 730.

Jost, R., and Luttinger, J. (1950). *Helvetica Physica Acta* **23**, 201.

Kingsley, R. L. (1973). *Nucl. Phys.* **B60**, 45.

Kinoshita, T. (1962). *J. Math. Phys.* **3**, 650.

Kluberg-Stern, H., and Zuber, J. B. (1975). *Phys. Rev.* **D12**, 467.

Kubar-André, J., and Paige, F. E. (1979). *Phys. Rev.* **D19**, 221.

Kummer, W. (1975). *Acta Physica Austriaca*, **41**, 315.

Landau, L. D., and Lifshitz, E. M. (1958). *Quantum Mechanics*, Pergamon Press, London.

LaRue, G. S., Phillips, J. D., and Fairbank, W. M. (1981). *Phys. Rev. Lett.* **46**, 967.

Lautrup, B., and Nauenberg, M. (1981). In *Topical Questions in QCD*, *Phys. Scripta*, **23**, No. 5.

Lee, B. W. (1976). In *Methods in Field Theory* (Balian and Zinn-Justin, eds.), North-Holland, Amsterdam.

Lee, B. W., and Zinn-Justin, J. (1972). *Phys. Rev.* **D5**, 3121.

Lee, T. D., and Nauenberg, M. (1964). *Phys. Rev.* **133**, B1549.

Leutwyler, H. (1974). *Nucl. Phys.* **B76**, 413.

Llewellyn Smith, C. H. (1972). *Phys. Reports* **C3**, 261.

López, C., and Ynduráin, F. J. (1981). *Nucl. Phys.* **B183**, 157.

Mackenzie, P. B., and Lepage, G. P. (1981). *Phys. Rev. Lett.* **47**, 1244.

Mandelstam, S. (1979). *Phys. Rev.* **D20**, 3223.

Marinari, E., Parisi, G., and Rebbi, C. (1981). *Phys. Rev. Lett.* **47**, 1795.

Marshak, R. E., Riazzuddin, and Ryan, C. P. (1969). *Theory of Weak Interactions in Particle Physics*, Wiley, New York.

Martin, F. (1979). *Phys. Rev.* **D19**, 1382.

Méndez, A. (1978). *Nucl. Phys.* **B145**, 199.

Muller, A. (1978). *Phys. Rev.* **D18**, 3705.

Nachtmann, O. (1973). *Nucl. Phys.* **B63**, 237.

Nambu, Y. (1960). *Phys. Rev. Lett.* **4**, 380.

——— (1978). Proc. Int. Conf. on High Energy Physics, Tokyo, p. 971.

Nambu, Y., and Jona-Lasinio, G. (1961*a*). *Phys. Rev.* **122**, 345.

——— (1961*b*). *Phys. Rev.* **124**, 246.

Nanopoulos, D. V. (1973). *Nuovo Cimento Lett.* **8**, 873.

Nanopoulos, D. V., and Ross, D. (1979). *Nucl. Phys.* **B157**, 273.

Narison, S. (1982). *Phys. Reports* **C82**, 263.

Narison, S., and de Rafael, E. (1981). *Phys. Lett.* **103B**, 57.

Ne'eman, Y. (1961). *Nucl. Phys.* **26**, 222.

Okubo, S. (1963). *Phys. Lett.* **5**, 165.

Pagels, H. (1975). *Phys. Reports*, **C16**, 219.

Pais, A. (1964). *Phys. Rev. Lett.* **13**, 175.

Parisi, G. (1979). *Phys. Letters* **84B**, 225.

Pascual, P., and de Rafael, E. (1982). *Z. Phys.* **C12**, 127.

Peccei, R. D., and Quinn, H. R. (1977). *Phys. Rev.* **16**, 1751.

Politzer, H. D. (1973). *Phys. Rev. Lett.* **30**, 1346.

Polyakov, A. M. (1977). *Nucl. Phys.* **120B**, 429.

Preparata, G. (1979). Proc. Int. Conf. on High-Energy Physics, Vol. 1, CERN, Geneva.

de Rafael, E. (1977). *Lectures on Quantum Electrodynamics*, Universidad Autónoma de Barcelona, UAB-FT-D1.

——— (1979). In *Quantum Chromodynamics* (Alonso and Tarrach, eds.), Springer, Berlin.

Reinders, L., Rubinstein, H., and Yazaki, S. (1981). *Nucl. Phys.* **B186**, 475.

Rossi, G., and Veneziano, G. (1977). *Nucl. Phys.* **B123**, 507.

Riordan, E. M. *et al.* (1978). *SLAC Report SLAC-PUB 1634.*

Sachrajda, C. T. (1979). In *Quantum Chromodynamics* (Alonso and Tarrach, eds.), Springer, Berlin.

Sciuto, S. (1979). *Riv. Nuovo Cimento*, **2**, No. 8.

Scadron, M. D. (1981). *Rep. Progr. Phys.*, **44**, 213.

Shifman, M. A., Vainshtein, A. I., and Zakharov, V. I. (1979*a*). *Nucl. Phys.* **B147**, 385.

——— (1979*b*). *Nucl. Phys.* **B147**, 448.

Siegel, W. (1979). *Phys. Lett.* **84B**, 193.

Slavnov, A. A. (1975). *Sov. J. Particles and Nuclei*, **5**, 303.

Speer, E. R. (1968). *J. Math. Phys.* **9**, 1404.

Steinberger, J. (1949). *Phys. Rev.* **76**, 1180.

Sterman, G., and Libby, S. (1978). *Phys. Rev.* **D18**, 3252 and 4737.

Sterman, G., and Weinberg, S. (1977). *Phys. Rev. Lett.*, **39**, 1436.

Stückelberg, E. C. G., and Peterman, A. (1953). *Helvetica Physica Acta*, **26**, 499.

Sutherland, D. G. (1967). *Nucl. Phys.* **B2**, 433.

Symanzik, K. (1970). *Commun. Mat. Phys.* **18**, 227.

——— (1973). *Commun. Mat. Phys.* **34**, 7.

Tarasov, O. V., Vladimirov, A. A., and Zharkov, A. (1980). *Phys. Lett.* **93B**, 429.

Tarrach, R. (1981). *Nucl. Phys.* **B183**, 384.

——— (1982). *Nucl. Phys.* **B196**, 45.

Taylor, J. C. (1971). *Nucl. Phys.* **B33**, 436.

Thirring, W. (1950). *Phil. Magazine*, **41**, 113.

Titchmarsh, E. C. (1939). *Theory of Functions*, Oxford University Press.

Tomboulis, E. (1973). *Phys. Rev.* **D8**, 2736.

't Hooft, G. (1971). *Nucl. Phys.* **B33**, 173.

——— (1973). *Nucl. Phys.* **B61**, 455.

——— (1974*a*). *Nucl. Phys.* **B72**, 461.

——— (1974*b*). *Nucl. Phys.* **B75**, 461.

——— (1976). *Phys. Rev. Lett.* **37**, 8.

't Hooft, G., and Veltman, M. (1972). *Nucl. Phys.* **B44**, 189.

Vainshtein, A. I., *et al.* (1978). *Yad. Fiz.* **27**, 514.

Veltman, M. (1967). *Proc. Roy. Soc. (London)* **A301**, 107.

Walsh, T. F., and Zerwas, P. (1973). *Phys. Lett.* **B44**, 195.

Weinberg, S. (1973*a*). *Phys. Rev. Lett.* **31**, 494.

——— (1973*b*). *Phys. Rev.* **D8**, 3497.

——— (1975). *Phys. Rev.* **D11**, 3583.

——— (1978*a*). In *A Festschrift for I. I. Rabi*, New York Academy of Sciences, New York.

——— (1978*b*). *Phys. Rev. Lett.* **40**, 223.

——— (1980). *Phys. Lett.* **91B**, 51.

Wess, J., and Zumino, B. (1971). *Phys. Lett.* **37B**, 95.

Wiener, N. (1923). *J. Math. and Phys.*, **2**, 131.

Wiik, B., and Wolf, G. (1979). *Electron-Positron Interactions*, Springer-Verlag, Berlin.

Wilczek, F. (1977). In *Quark Confinement and Field Theory* (Stump and Weingartner, eds.), J. Wiley, New York.

——— (1978). *Phys. Rev. Lett.* **40**, 279.

Wilson, R. (1969). *Phys. Rev.* **179**, 1499.

——— (1975). *Phys. Rev.* **D10**, 2445.

Wilson, R., and Zimmermann, W. (1972). *Commun. Math. Phys.* **24**, 87.

Witten, E. (1976). *Nucl. Phys.* **B104**, 445.

——— (1977). *Nucl. Phys.* **B120**, 189.

——— (1979*a*). *Nucl. Phys.* **B156**, 269.

——— (1979*b*). *Nucl. Phys.* **B160**, 57.

——— (1980). *Physics Today*, July, p. 38.

Yang, C. N., and Mills, R. L. (1954). *Phys. Rev.* **96**, 191.

Zachariasen, F. (1980). In *Hadronic Matter at Extreme Density* (Cabibbo and Sertorio, eds.), p. 313, Plenum Press.

Zee, A. (1973). *Phys. Rev.* **D8**, 4038.

Zee, A., Wilczek, F., and Treiman, S. B. (1974). *Phys. Rev.* **D10**, 2881.

Zepeda, A. (1978). *Phys. Rev. Lett.* **41**, 139.

Zimmermann, W. (1970). In *Lectures in Elementary Particles and Field Theory*, M.I.T. Press.

Zweig, G. (1964). CERN preprints Th. 401 and 412, unpublished.

Index

Texts and Monographs in Physics

J. Kessler: **Polarized Electrons** (1976).

W. Rindler: **Essential Relativity: Special, General, and Cosmological,** Revised Second Edition (1977).

K. Chadan and P.C. Sabatier: **Inverse Problems in Quantum Scattering Theory** (1977).

C. Truesdell and S. Bharatha: **The Concepts and Logic of Classical Thermodynamics as a Theory of Heat Engines: Rigourously Constructed upon the Foundation Laid by S. Carnot and F. Reech** (1977).

R.D. Richtmyer: **Principles of Advanced Mathematical Physics.** Volume I (1978). Volume II (1981).

R.M. Santilli: **Foundations of Theoretical Mechanics.** Volume I: The Inverse Problem in Newtonian Mechanics (1978). Volume II: Birkhoffian Generalization of Hamiltonian Mechanics (1983).

A. Böhm: **Quantum Mechanics** (1979).

H. Pilkuhn: **Relativistic Particle Physics** (1979).

M.D. Scadron: **Advanced Quantum Theory and Its Applications Through Feynman Diagrams** (1979).

O. Bratteli and D.W. Robinson: **Operator Algebras and Quantum Statistical Mechanics.** Volume I: C*- and W*-Algebras. Symmetry Groups. Decomposition of States (1979). Volume II: Equilibrium States. Models in Quantum Statistical Mechanics (1981).

J.M. Jauch and F. Rohrlich: **The Theory of Photons and Electrons: The Relativistic Quantum Field Theory of Charged Particles with Spin One-half,** Second Expanded Edition (1980).

P. Ring and P. Schuck: **The Nuclear Many-Body Problem** (1980).

R. Bass: **Nuclear Reactions with Heavy Ions** (1980).

R.G. Newton: **Scattering Theory of Waves and Particles,** Second Edition (1982).

G. Ludwig: **Foundations of Quantum Mechanics I** (1983).

G. Gallavotti: **The Elements of Mechanics** (1983).

F.J. Yndurain: **Quantum Chromodynamics: An Introduction to the Theory of Quarks and Gluons** (1983).